T0192718

**Chaotic Behaviour
of Deterministic Dissipative Systems**

Chaotic
Behaviour
of Deterministic
Dissipative
Systems

Miloš Marek
Igor Schreiber

Department of Chemical Engineering
Institute of Chemical Technology, Prague, Czech Republic

CAMBRIDGE
UNIVERSITY PRESS

CAMBRIDGE UNIVERSITY PRESS
Cambridge, New York, Melbourne, Madrid, Cape Town, Singapore, São Paulo

Cambridge University Press
The Edinburgh Building, Cambridge CB2 8RU, UK

Published in the United States of America by Cambridge University Press, New York

www.cambridge.org
Information on this title: www.cambridge.org/9780521321679

First published 1991
First paperback edition 1995

A catalogue record for this publication is available from the British Library

Library of Congress Cataloguing in Publication data
Marek, Miloš
    Chaotic behaviour of deterministic dissipative systems/Miloš Marek and Igor Schreiber.
    p. cm.
    Includes bibliographical references.
    ISBN 0 521 32167 0 (hb) – ISBN 0 521 43830 6 (pb)
    1. Chaotic behavior in systems. 2. Differentiable dynamical systems. 3. Nonlinear theories.
I. Schreiber, Igor. II. Title. III. Title: Deterministic dissipative systems.
QA845.M37 1991
003—dc 19 88-2958 CIP

ISBN 978-0-521-32167-9 hardback
ISBN 978-0-521-43830-8 paperback

Transferred to digital printing 2008

To our families in recognition of their patience

CONTENTS

PREFACE

Studies of nonlinear phenomena which occur in mathematical models and which are observed in experiments profit both from a general knowledge of the theory of dynamical systems and bifurcations, and from the experience accumulated in an interpretation of specific examples. The most interesting and important nonlinear phenomenon that has come to prominence recently is the chaotic behaviour of deterministic dissipative systems. The investigation of chaotic dynamics has undergone an explosive development over the past ten years but the results are still mostly scattered throughout the journal literature.

The number of interested students and research workers from diverse fields, ranging from mathematics and physics to engineering sciences and biology, increases continuously and many of them will find it useful to have an introductory text, that surveys both theoretical and experimental aspects of chaotic behaviour. We have attempted to provide this in the present book.

The introductory chapter discusses the significance of chaos as a model of many seemingly random processes in nature and a definition of the class of dissipative systems that we will study.

The second chapter considers basic notions of the theory of dynamical systems. The difference between linear and nonlinear systems is illustrated and asymptotic behaviour is discussed in more detail. Definitions of chaos and of strange attractors and a description of chaotic behaviour in the frame of ergodic theory are then surveyed.

The third chapter deals with qualitative changes of asymptotic behaviour as a chosen parameter is varied. These changes ('bifurcations') may lead to chaos in several well-defined routes. The role of bifurcation theory in understanding the onset of chaos is illustrated by a number of characteristic examples.

A review of the numerical methods used both in the treatment of mathematical models and in the interpretation of experimental data is provided in the fourth chapter. Methods for parametric dependences and for a characterization of chaotic behaviour are stressed.

The fifth chapter surveys some characteristic experimental observations of chaotic behaviour and includes data from mechanical systems, electronics, lasers, semiconductors, chemical and biological systems and hydrodynamics. It is stressed that most of these observations have many common features even though their physical nature is different.

The sixth chapter is based on our own experimental and numerical work and, using two detailed examples, it illustrates an interpretation of chaotic experimental data on the basis of one-dimensional mappings, and the role of numerical studies of bifurcations in the interpretation of complex periodic and chaotic behaviour in the system of two coupled cells.

In the final chapter we have tried to survey ways of approach to modelling of spatio-temporal chaotic behaviour in distributed systems. The results of analysis of cellular automata, coupled map lattices and partial differential equations are briefly reviewed and generally applicable methods are stressed.

The book contains two appendices. A brief survey of some normal forms of planar vector fields and the corresponding bifurcation diagrams can be found in the Appendix A. A program for continuation of stationary points and periodic orbits of dynamical systems and for the location and continuation of local bifurcation points is given in the Appendix B. Hence, an interested reader can with the help of these two appendices and the problems discussed in Chapters 5 and 6 obtain his own experience in the analysis of bifurcations and chaotic behaviour.

The individual chapters can be studied independently. References connected with the subjects discussed are given at the end of each chapter. In recent years the number of references has increased exponentially, and hence, we have had to limit ourselves to those references that, in our opinion, best illustrate the problems studied.

The study of chaotic behaviour often requires a profound knowledge of different branches of mathematics. In attempting to provide an objective explanation of various aspects of chaos which is also accessible to readers who do not possess such knowledge we have made a number of simplifications in the explanations and used graphical representation of the features studied. We would be grateful to readers for any comments that may improve the text.

The text reflects the results obtained in our research group over the last 10 years. We would like to express our sincere thanks to our colleagues, Alois Klíč, Milan Kubíček, Martin Holodniok, Miloš Dolník and many others (including a number of students) for the discussions and the friendly atmosphere which helped to write the book.

We are thankful to Dr. S. Capelin, Ing. B. Kyselová, Ing. J. Malátková and to other members of the staff of Academia Publishers, Prague and Cambridge University Press, Cambridge for their persistent and invaluable help with the transformation of the manuscript into book.

Last but not least we are indebted to Dana Dopitová and Edita Padušáková for their assistance in producing the manuscript.

M. Marek
I. Schreiber

1

Introduction

Observations of both natural and man-made systems evolving in time reveal an existence of various types of dynamics ranging from *steady time-independent* structures to very complicated *nonperiodic oscillations*. It is well known that the evolution of nonperiodic motions forms a basic problem in studies of hydrodynamical *turbulence*. However, both experimental and theoretical research in the last 20 years have clearly demonstrated that turbulent motion is in no case limited to fluids. It can exist in systems of different physical nature where oscillations occur, for example in mechanical vibrations, electronic circuits, chemical reactions, neurones, ecological systems, celestial mechanics, and so on.

An interpretation of experimental observations is closely coupled with the fast-developing theory of *nonlinear dynamical systems*. A typical mathematical model of an *evolution process* is in the form of a *differential equation*

$$\frac{dx}{dt} = v(\mathbf{x}, \alpha), \qquad \mathbf{x} \in \mathsf{R}^n \tag{1.1}$$

where the real variable t denotes *time* and α is a *parameter*. A *state* of the system (1.1) at a given time is determined by a point $\mathbf{x}$ in the *state space* R^n. Evolution of the variable $\mathbf{x}$ in time is given by a solution of Eq. (1.1). A *discrete time evolution process* can be described by a *difference equation*

$$\mathbf{x}_{k+1} = f(\mathbf{x}_k, \alpha), \qquad \mathbf{x} \in \mathsf{R}^n \tag{1.2}$$

where the time, denoted by k, is discrete. The solution of Eq. (1.2) is given by repeated *iterations* of the mapping f.

Actual states of the above systems are described by the vector variable $\mathbf{x}$ consisting of n independent components. However, the state variable is *spatially distributed* in the fluid flow as well as in a number of other systems. The state space then has an *infinite* dimension and the mathematical model is then formed by a system of *partial differential equations*. Evolution processes described by *integro-differential* equations and differential equations with a *time delay* also have infinite-dimensional state spaces.

If the evolution equations in the form of Eqs (1.1) and (1.2) are linear, then their solutions may be expressed in an explicit way and the evolution dynamics is relatively simple. It has become evident in the course of the last 20 years that even very simple but nonlinear equations can possess solutions, which from the statistical point of view look like random ones, although they are generated by a *deterministic system*. This behaviour is now called *deterministic chaos* and is generally believed to represent a valid model of low-dimensional turbulent behaviour in systems of various physical nature[1.56].

The onset of chaos can be studied by observing *asymptotic solutions* of (1.1) or (1.2) in dependence on the varying parameter **α**. An originally simple dynamic regime, represented, for example, by periodic oscillations, becomes more complex and finally leads to *chaotic behaviour*. In fact, a few degrees of freedom suffice to generate chaos and low-dimensional systems of the form of Eq. (1.1) and (1.2) are often used to study the onset of chaos. There exist several typical ways (routes) of *transition to chaos*[1.16]; the best known is the *period-doubling* route[1.18]. Further evolution of chaos may lead to an increased complexity of the chaotic motions as, for example, in hydrodynamical turbulence. On the other hand, *fully developed turbulence* in fluids[1.41] is a phenomenon too complex to be described by equations of the type (1.1) or (1.2) and its complete description is still an open problem. Thus the theory in its present state relates to the onset of turbulence and to the weak turbulence occurring in low-dimensional systems[1.62]. Nevertheless, the theory of dynamical systems in infinite-dimensional state spaces[1.63] as well as the theory of *cellular automata*[1.65] which are relevant to the problem of fully developed turbulence become more and more promising with a view of their applicability.

Here we shall study systems which *dissipate energy*; they are kept far from the thermodynamic equilibrium by an exchange of mass and/or energy with the environment. Mathematical models of *dissipative systems* possess the important property of the *contraction* of volumes in the state space.

Let A be a bounded set in the state space R^n and V be its volume. The set A is generally deformed under the time evolution according to Eqs (1.1) or (1.2). The time evolution of the volume V of set A is in the continuous time case given according to *Liouville's theorem*[1.43] as

$$\frac{dV(t)}{dt} = \int_{A(t)} \operatorname{div} v(x(t)) \, dx \; ; \tag{1.3}$$

if the time is discrete we have similarly

$$V_{k+1} = \int_{A_k} \left| \det \left(\frac{\partial f(x_k)}{\partial x} \right) \right| dx \; . \tag{1.4}$$

The global contraction of subsets of the state space will be guaranteed if div v (**x**) < 0 or $|\det \partial f/\partial \mathbf{x}| < 1$, respectively, for all **x**. Asymptotic motions will then occur on sets which have *zero volume*. If such asymptotic sets satisfy certain stability conditions, for example a stability against small bounded random perturbations, they are called *attractors*. Most of the points which are close to an attractor tend to it as time becomes large.

The structure of an attractor and of its dynamics can be simple, for example a point in the state space of continuous time systems corresponds to stationary behaviour and a closed curve to periodic oscillations. However, an overall contraction of volumes does not exclude complicated dynamics. The set A may be *expanding* in some directions in the state space even if its volume vanishes asymptotically. This may cause *folding* of different parts of A upon itself under time evolution and the asymptotic set may then have a very complicated geometric structure as well as complicated dynamics. In the case of a *chaotic attractor*, a locally *exponential expansion* of nearby points on the attractor is required.

In general dissipative systems need not contract all volumes in the state space; some sets may exist which expand their volumes with time, but such sets do not contain attractors. There is a special class of mathematical models characterized by the preservation of volumes under time evolution. Such systems are called *conservative* and are well known, for example from classical mechanics[1.2, 1.3]; they may also generate chaos but the phenomenon of attraction is lacking. Here we shall concentrate on *dissipative* systems.

One of the simplest discrete time systems which possesses a chaotic attractor is the following system of difference equations, studied first by M. Hénon[1.34]

$$x_{k+1} = f_1(x_k, y_k, a) = 1 - ax_k^2 + y_k \, ,$$

$$y_{k+1} = f_2(x_k, y_k, b) = bx_k \, , \tag{1.5}$$

where a, b are parameters.

The Jacobian of the mapping $f = (f_1, f_2)$ in (1.5) is $\det (\partial f/\partial \mathbf{x}) = -b$. Hence the system is dissipative if $|b| < 1$ and all areas of the state space R^2 are contracted uniformly. Numerical computations show that Eqs (1.5) have for many values of the parameters a and b solutions which tend asymptotically to an attractor with complicated dynamics. This attractor also has a very complex geometric structure, as illustrated in Fig. 1.1 (a) and (b). The structure does not disappear upon magnification and is repeated on arbitrarily small scales. Such sets are called *fractals* and are characterized by a *noninteger dimension*[1.46]. However, the most important feature of the Hénon attractor is the existence of a chaotic dynamics; subsequent iterations of (1.5) fill up the attractor *randomly*. This randomness appears to be equivalent to that of tossing a coin [1.19] and is caused mainly by the exponential divergence of nearby points on the attractor.

Another classical example of a chaotic attractor is the Lorenz attractor, arising in a continuous time system of three nonlinear ordinary differential equations (ODEs), studied first by E. N. Lorenz[1.44]. The severe truncation of the set of partial differential equations describing thermal convection in the atmosphere leads to the following set of ODEs

$$\frac{\mathrm{d}x}{\mathrm{d}t} = -\sigma x + \sigma y \,,$$

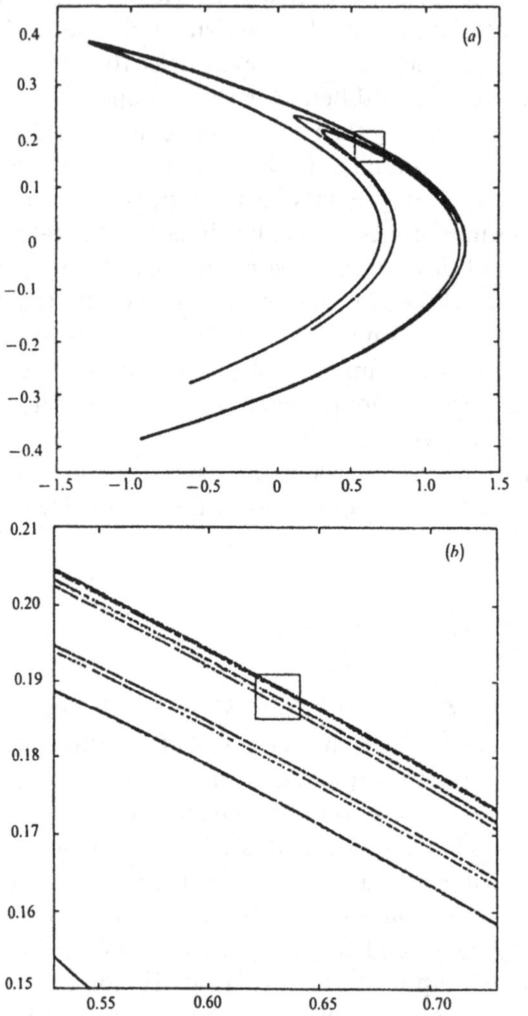

Fig. 1.1. *Several thousands of iterates of Eqs* (1.5) *for* $a = 1.4$, $b = 0.3$ plotted after ignoring several hundreds of initial iterations. (a) Entire attractor is covered by a single solution. (b) The magnified part of the attractor reveals a complex internal structure.

$$\frac{dy}{dt} = -xz + rx - y,\qquad\qquad (1.6)$$

$$\frac{dz}{dt} = xy - bz,$$

where σ, r, b are positive real-valued parameters. The state space of Eqs (1.6) is three-dimensional. The divergence of the vector field on the right hand sides of Eqs (1.6) is div $v(\mathbf{x}) = -\sigma - 1 - b$ and thus, as in the Hénon system, the flow of Eqs (1.6) uniformly contracts the volumes of the state space.

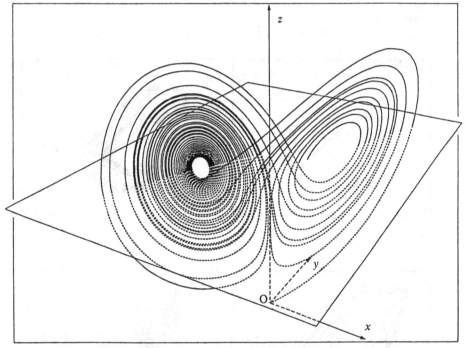

Fig. 1.2. The Lorenz attractor. The solution starting at $x = y = z = 0$ forms irregularly alternating loops to the left and to the right and quickly approaches the chaotic Lorenz attractor.

Despite its simplicity, the model yields chaotic oscillations at many parameter values, for example at $\sigma = 10$, $b = 8/3$, $r = 28$, see Fig. 1.2. Again this attractor has complex geometric structure as well as chaotic dynamics.

These two examples indicate the lowest possible dimension of the state space of a chaotic system. In the *discrete time* case given by Eq. (1.2), with f an *invertible* mapping, the dimension must be at least two, while the *continuous time* system (1.1) must have at least three-dimensional state space. If we consider a *noninvertible* mapping f in Eq. (1.2) then even a one-dimensional state space will admit the existence of chaos.

An example of an infinite-dimensional system may be provided by the Mackey–Glass mathematical model of haematologic disorders[1.45, 1.17] formed by a differential equation with a time delay

$$\frac{\mathrm{d}\mathbf{x}(t)}{\mathrm{d}t} = \lambda(\mathbf{x}(t - \tau)) - \gamma\,\mathbf{x}(t)\,, \tag{1.7}$$

where

$$\lambda(\mathbf{x}(t - \tau)) = \frac{\alpha\mathbf{x}(t - \tau)}{1 + (\mathbf{x}(t - \tau))^{\beta}}\,.$$

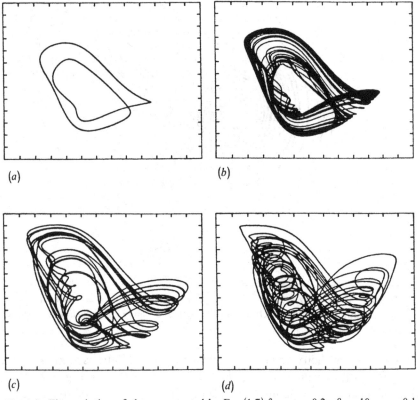

(a)   (b)

(c)   (d)

Fig. 1.3. The evolution of chaos generated by Eq. (1.7) for $\alpha = 0.2$, $\beta = 10$, $\gamma = 0.1$ and different values of the time delay τ. The figures represent a plot of $x(t)$ against $x(t - \tau)$: (a) periodic attractor, $\tau = 14$; (b) chaotic attractor, $\tau = 17$; (c) chaotic attractor, $\tau = 23$; (d) chaotic attractor, $\tau = 300$.

Here $x(t)$ is the concentration of blood cells at time t. Cells are lost from the circulation at a rate γ and the flux λ of cells into the circulation from the stem cell compartment depends on x at a delayed time $(t - \tau)$. Hence, the time derivative of x at time t is dependent both on the actual state $x(t)$ and on the delayed state $x(t - \tau)$. To generate a solution of Eq. (1.7) we have to know all states in the time interval $[t - \tau, t]$. The evolution of asymptotic regimes from periodic to chaotic attractors is shown in Fig. 1.3 $(a)-(d)$. The onset of chaos is here accompanied by a sequence of periodic oscillations with a successively doubled period.

The existence of complex solutions of a system of ODEs was already known to H. Poincaré at the end of the last century[1.53]. Similar systems were later studied by Birkhoff[1.5] and in the 1940s by Littlewood and Cartwright[1.11]. The theory of chaos was developed during the 1960s for conservative systems (Kolmogorov–Arnold–Moser theorem[1.3]) and for special types of dissipative systems (Smale[1.58]). First demonstrations of chaos in numerical solutions of the systems of the type (1.1) and (1.2) appeared at the same time[1.35, 1.44]. At the beginning of the 1970s Ruelle and Takens[1.56] expressed an idea about the possibility of the description of turbulence by chaotic solutions (strange attractors). The term 'chaos' was first used by Li and Yorke[1.42].

Then there was a large increase in the number of studies on chaos. A number of review articles, conference proceedings, reprint collections and several monographs on the subject of chaos have appeared in the last several years. A list of some of them is provided in the literature references to this chapter.

REFERENCES

1.1   Anishchenko V. S. *Dynamical Chaos in Physical Systems — Experimental Investigations of Self—Oscillating Circuits.* Leipzig, Teubner, 1988.

1.2   Anosov D. V., Arnold V. I., Novikov S. P. and Sinai Ya. G. (Eds.) *Dynamical Systems.* Vols. I — V. Berlin, Springer, 1987,1988,1989. (Russian original Moscow, VINITI, 1985.)

1.3   Arnold V. I. and Avez A. *Ergodic Problems of Classical Mechanics.* New York, Benjamin, 1968.

1.4   Bergé P., Pomeau Y. and Vidal C. *L'ordre dans le chaos.* Paris, Herman, 1984.

1.5   Birkhoff G. D. *Dynamical Systems.* Providence, A. M. S. Publications, 1927.

1.6   Bishop A. R., Campbell D. K. and Nicolaenko B. (Eds.) *Nonlinear Problems: Present and Future.* Amsterdam, North Holland, 1982.

1.7   Bishop A. R., Gruner G. and Nicolaenko B. (Eds.) *Spatiotemporal Coherence and Chaos in Physical Systems. Physica* **23D** (1986).

1.8   Bogolyubov N. N. and Mitropolskii Yu. A. *Asymptotic Methods in the Theory of Nonlinear Oscillations.* Moscow, Nauka, 1974 (in Russian).

1.9   Butenin N. V., Neimark Yu. I. and Fufayev N. A. *Introduction into Theory of Nonlinear Oscillations.* Moscow, Nauka, 1976 (in Russian).

1.10 Campbell D. and Rose H. (Eds.) *Order in Chaos. Physica* **7D** (1983).

1.11 Cartwright M. L. and Littlewood L. E. On non—linear differential equations of the second order. *J. London Math. Soc.* **20** (1945) 180.

1.12 Christiansen P. L. and Parmentier R. D. (Eds.) *Stucture, Coherence and Chaos in Dynamical Systems.* Manchester, Manchester University Press, 1989.

1.13 Collet P. and Eckmann J.—P. *Iterated Maps on the Interval as Dynamical Systems.* Boston, Birkhauser, 1980.

1.14 Cvitanovic' P. (Ed.) *Universality in Chaos* (reprint collection). Bristol, Adam Hilger, 1984.

1.15 Ebeling W. and Klimontovich Yu. L. *Selforganization and Turbulence in Liquids.* Leipzig, Teubner, 1984.

1.16 Eckmann J.—P. Roads to turbulence in dissipative dynamical systems. *Rev. Mod. Phys.* **53** (1981) 643.

1.17 Farmer J. D. Chaotic attractors of an infinite—dimensional system. *Physica* **4D** (1982) 366.

1.18 Feigenbaum M. J. Universal behaviour in nonlinear systems. *Los Alamos Science* **1** (1980) 4.

1.19 Ford J. How random is a coin toss? *Phys. Today* **36** (1983) 40.

1.20 Gaushus E. V. *Investigation in Dynamical Systems by the Methods of Discrete Mappings.* Moscow, Nauka, 1976 (in Russian).

1.21 Glass L. and Mackey M. C. *From Clocks to Chaos: The Rhythms of Life.* Princeton, Princeton University Press, 1988.

1.22   Guckenheimer J. and Holmes P. *Nonlinear Oscillations, Dynamical Systems, and Bifurcations of Vector Fields.* New York, Springer, 1983, 1986.

1.23   Gumowski I. and Mira C. *Recurrences and Discrete Dynamic Systems.* Lecture Notes in Mathematics **809** (1980).

1.24   Gurel O. and Rössler O. E. (Eds.) *Bifurcation Theory and Applications in Scientific Disciplines. Ann. N.Y. Acad. Sci.* **316** (1979).

1.25   Haken H. (Ed.) *Evolution of Order and Chaos in Physics, Chemistry, and Biology.* Springer Series in Synergetics **17**, Berlin, Springer, 1982.

1.26   Haken H. *Advanced Synergetics.* Berlin, Springer, 1983.

1.27   Hale J. *Oscillations in Nonlinear Systems.* New York, McGraw – Hill, 1963.

1.28   Hale J. *Asymptotic Behavior of Dissipative Systems.* Providence, A. M. S. Publications, 1988.

1.29   Hao Bai – lin (Ed.) *Chaos* (reprint collection). Singapore, World Scientific, 1984.

1.30   Hao Bai – lin (Ed.) *Directions in Chaos.* Singapore, World Scientific, 1987 (Vol.1) and 1988 (Vol.2).

1.31   Hayashi T. *Nonlinear Oscillations in Physical Systems.* New York, McGraw – Hill, 1964.

1.32   Helleman R. H. G. Self – generated chaotic behavior in nonlinear mechanics. In *Fundamental Problems in Stat. Mech.* V, ed. E. G. D. Cohen, Amsterdam, North – Holland, 1980, p. 165.

1.33   Helleman R. H. G. (Ed.) *Nonlinear Dynamics. Ann. N.Y. Acad. Sci.* **357** (1980).

1.34   Hénon M. A two – dimensional mapping with a strange attractor. *Commun. Math. Phys.* **50** (1976) 69.

1.35   Hénon M. and Heiles C. The applicability of the third integral of motion, some numerical experiments. *Astron. J.* **69** (1964) 73.

1.36   Holmes P. (Ed.) *New Approaches to Nonliear Problems in Dynamics.* Philadelphia, SIAM Publications, 1980.

1.37   Iooss G., Helleman R. G. H. and Stora R. (Eds.) *Chaotic Behaviour of Deterministic Systems.* Amsterdam, North – Holland, 1983.

1.38   Kuramoto Y. (Ed.) *Chaos and Statistical Methods.* Berlin, Springer, 1984.

1.39   Landa P. S. *Oscillations in Systems with a Finite Number of Degrees of Freedom.* Moscow, Nauka, 1980 (in Russian).

1.40   Landa P. S. *Oscillations in Distributed Systems.* Moscow, Nauka, 1983 (in Russian).

1.41   Landau L. D. and Lifshitz E. M. *Fluid Mechanics.* Oxford, Pergamon Press, 1959.

1.42   Li T. and Yorke J. A. Period three implies chaos. *Am. Math. Monthly* **82** (1975) 985.

1.43   Lichtenberg A. J. and Lieberman M. A. *Regular and Stochastic Motion.* Appl. Math. Sci. **38**, New York, Heidelberg, Berlin, Springer, 1982.

1.44   Lorenz E. N. Deterministic nonperiodic flow. *J. Atmos. Sci.* **20** (1963) 130.

1.45   Mackey M. G. and Glass L. Oscillations and chaos in physiological control systems. *Science* **197** (1977) 287.

1.46   Mandelbrot B. *The Fractal Geometry of Nature.* San Francisco, Freeman, 1982.

1.47   Mitropolski Yu. A. (Ed.) *Application of the Theory of Nonlinear Oscillations.* Kiev, Naukova Dumka, 1984 (in Russian).

1.48   Monin A. S. On the nature of tubulence. *Sov. Phys. Usp.* **21** (1978) 429.

1.49   Moser J. *Stable and Random Motions in Dynamical Systems.* Princeton, Princeton University Press, 1973.

1.50   Neimark Yu. I. *Dynamical Systems and Controllable Processes.* Moscow, Nauka, 1978 (in Russian).

1.51   Neimark Yu. I. *Method of Discrete Mappings in the Theory of Nonlinear Oscillations.* Moscow, Nauka, 1972 (in Russian).

1.52   Neimark Yu. I. and Landa P. S. *Stochastic and Chaotic Oscillations.* Moscow, Nauka, 1987 (in Russian).

1.53   Poincaré H. *Les méthodes nouvelles de la méchanique celeste.* 3 Vols., Paris, Gauthier – Villars, 1899.

1.54   Rabinovich M. I. Stochastic self−oscillations and turbulence. *Sov. Phys. Usp.* **21** (1978) 443.

1.55   Rabinovich M. I. and Trubeckov D. I. *Introduction to the Theory of Oscillations and Waves.* Moscow, Nauka, 1984 (in Russian).

1.56   Ruelle D. and Takens F. On the nature of turbulence. *Comm. Math. Phys.* **20** (1971) 167.

1.57   Schuster H. G. *Deterministic Chaos. An Introduction.* Weinheim, Physik, 1984.

1.58   Smale S. Differentiable dynamical systems. *Bull. Am. Math. Soc.* **73** (1967) 747.

1.59   Smale S. *The Mathematics of Time: Essays on Dynamical Systems, Economic Processes and Related Topics.* New York, Springer, 1980.

1.60   Sonechkin D. M. *Stochasticity in Models of Atmospheric Circulation.* Moscow, Gidrometeoizdat, 1984 (in Russian).

1.61   Sparrow C. *The Lorenz Equations: Bifurcations, Chaos, and Strange Attractors.* Appl. Math. Sci. **41**, New York, Springer, 1982.

1.62   Swinney H. L. and Gollub J. P. (Eds.) *Hydrodynamical Instabilities and the Transition to Turbulence.* New York, Springer, 1981.

1.63   Temam R. *Infinite-Dimensional Dynamical Systems in Mechanics and Physics.* New York, Springer, 1988.

1.64   Vidal C. and Pacault A. (Eds.) *Nonlinear Phenomena in Chemical Dynamics.* New York, Springer, 1981.

1.65   Wolfram S. (Ed.) *Theory and Applications of Cellular Automata.* Singapore, World Scientific, 1986.

1.66   Zaslavsky G. M. *Stochasticity of Dynamical Systems.* Moscow, Nauka, 1984 (in Russian).

2

Differential equations, maps and asymptotic behaviour

2.1 Time evolution and dynamical systems

Physical, chemical, biological or social phenomena can be seen as systems characterized by a time evolution of their properties. Such *evolution systems* are ubiquitous in nature. Often we are able to express the rate of change of the properties of a considered evolution system in the form of equations, applying and combining the relevant laws of nature. Solutions of the constructed *mathematical model* then mimic the time evolution of the real system.

Our aim is to predict this evolution using a proper mathematical model. An instantaneous *state* of the model system can be given by a finite set of numbers or by a finite set of functions. A set of all states of the system will be called a *state space* (in physics literature it is also called a *phase space*). A system will be considered as *deterministic* if its future and past are fully determined by its current state. In a *semideterministic* system only the future is uniquely determined, while in a *stochastic* system neither the past nor the future is unique (this type of system will not be treated here).

A system of bodies moving according to laws of classical mechanics, electronic circuits or interacting populations in a closed ecological system may be considered as deterministic systems. An isothermal chemical reaction in an ideally stirred (homogeneous) environment is another example of a deterministic system, while the consideration of molecular diffusion makes the system semideterministic.

A substantial difference between the last two examples of evolution systems is in the *dimension* of the corresponding state space. Let X denote the state space and **x** its elements (states of the system). The state of the system is, in the first case, described at a given time by n values of concentrations x_1, ..., x_n of components taking part in n independent reaction steps. Hence **x** is an n-vector with the state space defined on a subset of n-dimensional *Euclidean space* R^n. In the second case, the concentration of each of the n reaction components is dependent on the spatial coordinates and also has to satisfy conditions imposed on the boundaries of the system. The corresponding state space consists of elements formed by n-tuples of concentration profiles satisfying the boundary conditions. As each element can be generally described by an infinite number of real numbers, the corresponding state space is a *functional space* of infinite dimension. In such spaces, we generally cannot determine unambiguously the evolution of the system in the past. Semideterministic systems can be also defined in the finite-dimensional state spaces.

We shall mostly limit ourselves to the description of *finite-dimensional systems* with the state space being a suitable subset of an *n*-dimensional Euclidean space R^n. Such a simplification is often possible in dissipative systems.

Let a one-parameter system $g^t(x)$ map the state space X to itself. The parameter t that denotes time can be continuous or discrete. We shall usually assume a certain smoothness of g^t with respect to x and in the case of continuous time also smoothness with respect to t.

Let us assume that g^t has the following properties:

(i) $g^0(x) = \text{identity}$;

(ii) $g^{t+s}(x) = g^s(g^t(x)) = g^t(g^s(x))$, $t, s > 0$.

Then g^t defines a *semiflow* on the state space X, i.e. the future evolution of the system is determined by the current state of the system. If, moreover, the inverse map $g^{-t}(x)$ is uniquely defined for all $x \in X$ then a *flow* on the state space X is defined and the behaviour both in the past and in the future is unambiguously determined by the current state of the system.

If an initial state x at time $t = 0$ is given, then the flow g^t generates an *orbit*. When the time is continuous, the orbit is a smooth curve in X parametrized by the time t; in the discrete case the orbit consists of the sequence of points in X, see Fig. 2.1. If the map $x \rightarrow g^t(x)$, together with its inverse, is differentiable for

(a)   (b)

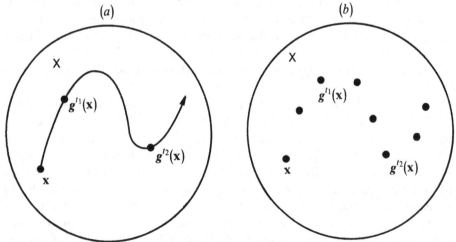

Fig. 2.1. Orbits in the state space X; x is mapped to $g^t(x)$: (*a*) orbit of a continuous time system, $t_2 > t_1$; (*b*) orbit of a discrete time system, $t_1 = 2$, $t_2 = 5$.

any value of t, then g^t is called a *diffeomorphism*. In the case of the semiflow the orbit is defined only in the positive direction of the time. The collection of all orbits in the state space is called a *phase portrait*. We have tacitly assumed that the evolution process is defined for all positive times. Although this assumption need not hold generally (e.g. it does not hold in the theory of explosions, etc.) we shall in the following consider cases belonging to the class of asymptotically well defined processes.

The pair (X, g^t) consisting of the state space and a flow or a semiflow defined on it is called a *dynamical system*. Our requirement of predictability of the future is automatically satisfied if a dynamical system is defined. Unfortunately, there are at least two factors that complicate the situation. First, the laws of nature that form the basis of the model can often be described only locally. Thus, given the instantaneous state of the system, we can determine only the instantaneous velocity of the state point motion along the orbit in the case of continuous time or the next state of the discrete time system; hence we cannot explicitly determine $g^t(x)$. The second factor has initiated a large upsurge in the study of dynamical systems in the last 20 years and also forms the subject of this book. It is now well known that many dynamical systems have such complicated orbits that their effective description requires a statistical approach. This results from the internal instability of the system, reflected in the fact that two orbits starting at two neighbouring points can diverge exponentially fast and after a finite time become practically uncorrelated.

Let us return to the first factor, our uncertainty about the flow caused by the local nature of the mathematical model. In the system with continuous time we know only the velocity $v(x)$ of the motion at the state point x, i.e.

$$v(x) = \left. \frac{\partial g^t(x)}{\partial t} \right|_{t=0}.$$

The vector $v(x)$ is located in the tangent space to X at the point x. The tangent space is Euclidean space and the mapping $x \to v(x)$ is called a *vector field*. The vector field can be defined on n-dimensional manifolds and can also be generalized to infinite-dimensional systems such as the systems of parabolic partial differential equations and equations with time delay.

If we know the velocity of the state point motion at each point of the state space, we can write down the equation

$$\frac{dx}{dt} = v(x), \tag{2.1}$$

which in n-dimensional Euclidean state space R^n represents the system of n *ordinary differential equations* for n variables. A flow g^t, associated with this system, identically satisfies (2.1). If the vector field v is differentiable, then g^t is a diffeomorphism. The equation (2.1) is *autonomous*, i.e. the vector field does not depend explicitly on time. This property reflects a reproducibility of the modelled natural process. For a chosen *initial condition* $x(t_0) = x$, the solution $x(t)$ is uniquely determined and we can always choose $t_0 = 0$. Then any solution of Eq. (2.1) can be written as $x(t) = g^t(x)$. The solution $x(t)$ with the initial condition $x(0) = x$ can also be considered as a curve $t \to x(t)$ passing through the point x at $t = 0$. The image of this curve in the state space is called a *trajectory* and is identical with the corresponding orbit of the flow g^t. The reversibility in time is reflected in the fact that two different trajectories never intersect. When we consider semiflows, then two or more trajectories may merge and this prevents the system from being reversed in time, see Fig. 2.2.

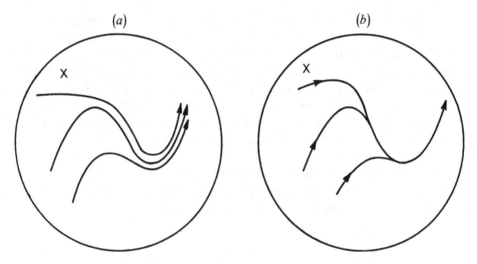

Fig. 2.2 (*a*) orbits of a flow; (*b*) orbits of a semiflow.

Let us mention that while a vector field exists for every flow, the opposite is not generally true. The existence and uniqueness of the solution of Eq. (2.1) is guaranteed only locally, in a small neighbourhood of the point x under consideration. The existence of a global flow is guaranteed only on state spaces which are compact manifolds; in this case orbits cannot escape into infinity.

When the time t can attain only discrete values, we can consider it as an integer valued variable k and write down analogously to Eq. (2.1)

$$x_{k+1} = f(x_k),\tag{2.2}$$

which corresponds in the case of a Euclidean state space R^n to a *system of difference equations* in the vector notation. The trajectory $\{x_k\}$ is generated by the subsequent iterations of f according to (2.2). Using the following notation for the composition of f with itself

$$f^0(\mathbf{x}) = \mathbf{x}, \qquad f^1(\mathbf{x}) = f(\mathbf{x}), \qquad f^2(\mathbf{x}) = f(f(\mathbf{x})), \dots$$

and

$$f^{q+p}(\mathbf{x}) = f^q(f^p(\mathbf{x})),$$

we can identify the semiflow corresponding to Eq. (2.2) with the family of maps $f^k(\mathbf{x})$ parametrized by the time $k = 0, 1, \dots$. The orbits generated by f^k are identical with the trajectories of (2.2), i.e. $\mathbf{x}_k = f^k(\mathbf{x}_0)$. If the map f has a unique inverse f^{-1}, then the orbit is also defined in the negative time direction and f^k defines a flow. Similarly, as in the case of continuous time we call the differentiable mapping $f(\mathbf{x})$ with a differentiable inverse a *diffeomorphism*. The mapping f which is only continuous with continuous inverse is called a *homeomorphism*. In contrast to differential equations, we can here define a semiflow in the Euclidean phase space.

Let us note that the elements $\mathbf{x}$ of the state space associated with Eq. (2.2) can be functions and then f will act in the state space of an infinite dimension. This point of view is used, for example, in renormalization theory which has been successfully applied to the study of the onset of chaos.

Given a dynamical system, we wish to determine first its *asymptotic behaviour*, i.e. the behaviour of orbits for large t. This approach is relevant to systems in nature that are governed by time-independent dynamical laws, i.e. systems that are autonomous or can be transformed to autonomous systems (e.g. periodically forced systems). It involves an idealization because dynamical laws governing many evolution processes (e.g. biological or social phenomena) actually depend on time. However, the changes in time are often very slow and thus the dynamical systems approach is applicable.

We are interested in open *dissipative systems*, where losses of mass and energy caused by dissipation are compensated by their inflows. In such systems a contraction of volumes in the state space is caused by the action of the flow. The volume decreases to zero asymptotically and the resulting motion has fewer degrees of freedom, i.e. the asymptotic behaviour is describable within a subset of the state space with lower dimension than that of the original state space. For example, we can typically expect that the asymptotic behaviour of an infinite-dimensional system will be limited to a subset of the state space with a finite (often low) dimension and hence will be more easily describable in the context of finite-dimensional systems.

The study of dynamical systems may be approached from two directions. First, we can investigate the asymptotic behaviour of individual orbits. In the second approach we can study sets of orbits and follow their distribution in the state space. Both approaches can be unified within the framework of ergodic theory, see Section 2.5.

We shall study those subsets of the state space that are invariant with respect to flow. The set A is *invariant* with respect to the flow g^t if, for each $\mathbf{x} \in A$, $g^t(\mathbf{x}) \in A$ for all t. The most simple example of an invariant set is a unique orbit. It is evident that every invariant set will be, in general, a union of orbits. Because we are interested in a nontransient asymptotic motion we are looking for invariant sets within which the orbits have a recurrent character, i.e. a state point on the orbit based at $\mathbf{x}$ will pass arbitrarily close to $\mathbf{x}$ after a finite time (the orbit need not necessarily be closed). This recurrent property can be defined asymptotically.

Let us consider an orbit φ of the flow $g^t(\mathbf{x})$ (t can be continuous or discrete) passing through the point $\mathbf{x}$. The point $\mathbf{z}$ is an *ω-limit point* of the orbit φ if there exists a sequence of times $t_i \to \infty$ such that $g^{t_i}(\mathbf{x}) \to \mathbf{z}$. The set of all ω-limit points of φ is called an *ω-limit set of the orbit φ*. If the sign of the time is reversed in this definition, we obtain an *α-limit set of the orbit φ*.

In the following section we shall study simple cases of limit sets and then describe more general examples of limit sets possessing a complex dynamical and geometrical structure.

2.2 Stationary points

Let us consider a dynamical system with continuous time defined by a flow $g^t(\mathbf{x})$ on an n-dimensional state space R^n. The simplest case of the limit set is a unique point $\mathbf{z}$. It satisfies $g^t(\mathbf{z}) = \mathbf{z}$ for all t.

The corresponding differential equation

$$\frac{d\mathbf{x}}{dt} \equiv \dot{\mathbf{x}} = v(\mathbf{x}), \qquad \mathbf{x} \in R^n,$$

or equivalently

$$
\begin{aligned}
\dot{x}_1 &= v_1(x_1, ..., x_n) \\
\dot{x}_2 &= v_2(x_1, ..., x_n) \\
&\vdots \\
\dot{x}_n &= v_n(x_1, ..., x_n)
\end{aligned}
\qquad
\mathbf{x} = \begin{bmatrix} x_1 \\ x_2 \\ \vdots \\ x_n \end{bmatrix},
\qquad
v = \begin{bmatrix} v_1 \\ v_2 \\ \vdots \\ v_n \end{bmatrix},
\qquad (2.3)
$$

has a constant solution $x(t) = z$ and the point z can be determined as a solution of $v(x) = 0$. The point z is usually denoted as the *equilibrium, singular or stationary point.*

We are interested in the behaviour of trajectories in the neighbourhood of the stationary point. Because the vector field is generally nonlinear, we can find exact solutions only for a specific and very limited number of cases. However, we can linearize Eqs (2.3) in the neighbourhood of z and obtain a set of linear equations

$$\dot{a} = Aa, \qquad a \in R^n.$$ (2.4)

Here a is an element of the tangent space to the state space at the point z and

$$A = D_x v(x)\big|_{x=z} = \left\{ \frac{\partial v_i}{\partial x_j} \right\}$$

is the *Jacobi matrix* evaluated at the point z. Eq. (2.4) forms a homogeneous system of linear differential equations with constant coefficients and can be solved exactly. A trivial solution $a(t) = 0$ corresponding to zero perturbation of z always exists. A general solution of (2.4) can be expressed as a linear combination of n linearly independent solutions $a_1(t), ..., a_n(t)$;

$$a(t) = \sum_{j=1}^{n} c_j a_j(t) = C(t) c,$$ (2.5)

where the vector $c = (c_1, ..., c_n)$ consisting of n constants can be determined from the initial condition $a(0) = a_0$. We have written a system of linearly independent solutions in the form of the *fundamental matrix*

$$C(t) = \{a_1(t), ..., a_n(t)\}.$$

The flow $g^t(a)$ associated with Eq. (2.4) can be written by means of the exponential matrix e^{At} as

$$g^t(a) = e^{At}a,$$ (2.6)

where $e^{At} = C(t) C^{-1}(0)$. To construct n independent solutions we determine eigenvalues λ and eigenvectors ω of the matrix A. Eigenvalues and eigenvectors satisfy the well known relation

$$(A - \lambda I) \omega = 0.$$

When A has n linearly independent eigenvectors ω_j, $j = 1, ..., n$, then

$$a_j(t) = e^{\lambda_j t}\omega_j,$$ (2.7)

where λ_j is an eigenvalue corresponding to ω_j. When we have a pair of complex conjugate eigenvalues λ_j, λ_{j+1} with the corresponding complex eigenvectors, we can form a pair of linearly independent real solutions

$$\mathbf{a}_j(t) = e^{\mathrm{Re}(\lambda_j)t}[\mathrm{Re}\ (\omega_j) \cos \mathrm{Im}\ (\lambda_j)\ t - \mathrm{Im}\ (\omega_j) \sin \mathrm{Im}\ (\lambda_j)\ t]\ ,$$

$$\mathbf{a}_{j+1}(t) = e^{\mathrm{Re}(\lambda_j)t}[\mathrm{Re}\ (\omega_j) \sin \mathrm{Im}\ (\lambda_j)\ t + \mathrm{Im}(\omega_j) \cos \mathrm{Im}\ (\lambda_j)\ t]\ , \qquad (2.8)$$

where $\mathrm{Re}\ (\ )$ and $\mathrm{Im}\ (\ )$ denote real and imaginary parts of the complex argument.

When multiple eigenvalues and fewer than n linearly independent eigenvectors exist, we can form generalized eigenvectors (solving equations $(A - \lambda I)^k\ \omega = 0$ for some $1 \leq k \leq n$).

It is evident from Eqs $(2.6) - (2.8)$ that the time dependence of the functions $\mathbf{a}_j(t)$ is given by the real parts of the eigenvalues. It follows further that if the initial point is located in the subspace generated by some k-tuple of the (generalized) eigenvectors, $1 \leq k \leq n$, then the entire trajectory is contained in it, i.e. the *eigensubspaces* are invariant with respect to the flow e^{At}.

Let us divide the invariant subspaces generated by (generalized) eigenvectors into three groups, E^s, E^u, E^c, with the corresponding eigenvalues having negative, positive or zero real parts, respectively. The stable subspace E^s is characterized by an exponential decay of $\mathbf{a}(t)$. In the unstable subspace E^u an exponential increase of $\mathbf{a}(t)$ occurs. The solutions in the centre subspace E^c are either constant or periodic or diverge[2.3, 2.31]. Let us denote the number of eigenvalues with negative, positive and zero real part n_s, n_u, n_c, respectively. Then the dimensions of the corresponding subspaces are given by these numbers and

$$n_s + n_u + n_c = n\ .$$

The theory of linear differential equations is essentially complete; the interested reader can find detailed information, for example, in the books by Arnold[2.3] or Hirsch and Smale[2.31].

The knowledge of solutions of linear systems can be used in the study of the nonlinear system (2.3) in the neighbourhood of the stationary point $\mathbf{z}$. When the matrix A does not have an eigenvalue with zero real part, that is $n_c = 0$, then the linear flow e^{At} characterizes uniquely the original flow g^t in a certain neighbourhood of the point $\mathbf{z}$ in the state space R^n. The relation between both flows is called *topological equivalence* and the stationary point is called *hyperbolic*. A hyperbolic point $\mathbf{z}$ is stable if $n_s = n$, otherwise it is unstable. The stable stationary point is also called an *attractor*. There are two kinds of unstable hyperbolic points. If $n_u = n$, then all orbits move away from $\mathbf{z}$ and the stationary point is called a *repellor*. When both n_s and n_u are nonzero then the stationary point z is called a *saddle*. An example of three types of the phase portraits in the neighbourhood of z for $n = 2$ is shown in Fig. 2.3.

The stationary points which are not hyperbolic cannot be unambiguously described by a linear flow e^{At}. If we change the vector field $v(\mathbf{x})$ slightly, then the

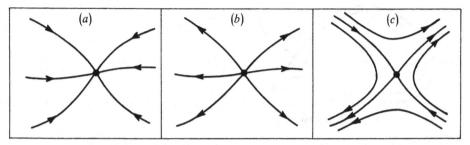

Fig. 2.3. Hyperbolic stationary points: (a) attractor; (b) repellor; (c) saddle.

phase portrait in the neighbourhood of a hyperbolic point does not change qualitatively, while a qualitative change in the arrangement of orbits – a *bifurcation* – occurs when the stationary point is not hyperbolic. Bifurcations will be studied in more detail in the next chapter.

The invariant subspaces spanning the tangent space at the point z have their analogues in the state space. Let us assume first that the point z is hyperbolic. Then in a certain neighbourhood U of the point z *stable* and *unstable manifolds* defined in the following way,

$$W^s(\mathbf{z}) = \{\mathbf{x} \in U; \; g^t(\mathbf{x}) \to (\mathbf{z}) \text{ for } t \to \infty, \; g^t(\mathbf{x}) \in U \text{ for all } t \geq 0\};$$

$$W^u(\mathbf{z}) = \{\mathbf{x} \in U; \; g^t(\mathbf{x}) \to \mathbf{z} \text{ for } t \to -\infty, \text{ and } g^t(\mathbf{x}) \in U \text{ for all }$$

$$t \leq 0\}$$

uniquely exist.

Both $W^s(\mathbf{z})$ and $W^u(\mathbf{z})$ have the same dimension n_s, n_u as the corresponding eigenspaces E^s and E^u and are tangent to them at the point z. The neighbourhood U can be chosen arbitrarily large hence the invariant manifolds can be defined globally. A stable (or unstable) manifold cannot intersect itself; however, the intersection of the stable manifold with the unstable manifold of the same stationary point is possible. Also an intersection of the stable and unstable manifold of two different stationary points can occur. As an example we can take the orbit that has a saddle stationary point as both α- and ω-limit sets. This orbit is contained both in $W^u(\mathbf{z})$ and $W^s(\mathbf{z})$ and it is called *homoclinic* to z or a *saddle loop*, see Fig. 2.4. A similar trajectory connecting two stationary points is called *heteroclinic*. Under certain conditions which will be specified later a very complicated invariant set with *chaotic behaviour* may exist in the neighbourhood of the homoclinic (or heteroclinic) orbit. A complicated phase portrait can also be expected in the case where many stationary points are located close to each other in the state space.

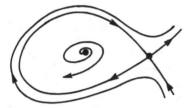

Fig. 2.4. A homoclinic orbit or saddle loop.

When the stationary point z is not hyperbolic, in addition to the stable and unstable manifolds there exists a *centre manifold* with dimension n_c. The centre manifold is tangent to E^c at the point z if the flow g^t is sufficiently smooth. The centre manifold is generally not defined uniquely and to describe the course of orbits located in it we have to approximate the original flow with an accuracy of at least the second order. The centre manifold plays an important role in the study of bifurcation phenomena which are described in Chapter 3.

2.3 Periodic oscillations

A periodic orbit is another simple ω-limit set. Let us have a flow g^t (t continuous or discrete) on a state space R^n. The *periodic orbit* γ with a period $T > 0$ is a set of points $\mathbf{x} \in \gamma$ for which $g^{t+T}(\mathbf{x}) = g^t(\mathbf{x})$ and $g^{t+s}(\mathbf{x}) \neq g^t(\mathbf{x})$ for all $0 < s < T$. Each point of the periodic orbit is a fixed point of the mapping g^T. The orbit γ is a closed curve in the continuous case and the phase point travels around it once per period. In the discrete case γ is formed by a finite number of points.

Let us first discuss systems with discrete time described by Eq. (2.2); we shall denote time k instead of t and the period q instead of T to differentiate them from the continuous case. The flow (or semiflow) connected with the difference equation (2.2) is then f^k. The periodic orbit γ is a set of q points, $\mathbf{x}_0^p, \mathbf{x}_1^p, ..., \mathbf{x}_{q-1}^p$ generated by successive iterations of (2.2). Each point on γ is called a *q-periodic point* of the mapping f and it is a *fixed point* of the mapping f^q. To study the stability of the periodic orbit it is advantageous to linearize f^q in the neighbourhood of some periodic point, say $\mathbf{x}_0^p$. We obtain a linear system

$$\mathbf{a}_{k+1} = U\mathbf{a}_k, \tag{2.9}$$

where the matrix $U = D_x f^q(x)|_{x=x_0^p}$ can be determined according to the chain rule as a product of q matrices,

$$U = D_x f(x_{q-1}^p) \, D_x f(x_{q-2}^p) \, ... \, D_x f(x_0^p) \, .$$

From Eq. (2.9) we can infer that if all eigenvalues σ_i of the matrix U have moduli less than 1, i.e. $|\sigma_i| < 1$, $i = 1, ..., n$, then the norm of $\mathbf{a}_k$ will decrease exponentially with time and γ is stable. If $|\sigma_i| > 1$ for some i then γ is unstable.

Definitions of stable, unstable and centre invariant subspaces E^s, E^u, E^c, can be introduced in way analogous to the case of linear continuous flows. A stable (unstable, respectively) subspace is generated by (generalized) eigenvectors of U with the corresponding eigenvalues with moduli less (greater, respectively) than 1. The dimensions n_s, n_u, n_c of E^s, E^u and E^c are again given by a number of eigenvalues which have the modulus less than, greater than and equal to 1, respectively. The eigenvalues of the matrix U do not depend on the periodic point x_i^p at which the linearization was performed.

A periodic orbit is called hyperbolic if it does not have any eigenvalue with modulus equal to one. A topological equivalence between a linear and a non-linear flow in a neighbourhood of γ exists for hyperbolic orbits. Invariant manifolds $W^u(\gamma)$, $W^s(\gamma)$ of a periodic orbit γ with dimensions n_u and n_s can be defined as for continuous flows with g^t replaced by f^k; they are tangent to invariant subspaces E^u, E^s. If the dimension of the stable (unstable, respectively) manifold of the hyperbolic orbit γ is equal to the dimension of the state space, then γ is an attractor (repellor, respectively). When both n_s and n_u are nonzero then γ is called a saddle. A nonhyperbolic orbit has a centre manifold W^c of dimension n_c in addition to W^s and W^u.

Stable and unstable manifolds of a periodic orbit may intersect and then a homoclinic orbit is located in $W^s(\gamma) \cap W^u(\gamma)$. A heteroclinic orbit is located in the intersection of the stable manifold of one periodic orbit with an unstable manifold of another periodic orbit. It follows from the definition of invariant manifolds that the periodic orbit γ is both the α- and ω-limit set of points in the intersection of $W^u(\gamma)$ and $W^s(\gamma)$, i.e. the forward and backward iterations of the points located in $W^u(\gamma) \cap W^s(\gamma)$ asymptotically approach the periodic orbit γ. The existence of the homoclinic orbit thus guarantees that the unstable (or stable) manifold will oscillate wildly and fold up on itself in a neighbourhood of γ. An example of complex behaviour in two-dimensional state space with $n_s = n_u = 1$ is shown schematically in Fig. 2.5. It implies irregular oscillations of the iterates of the points located on the oscillating manifold or in its neighbourhood. Poincaré[1.53] and Birkhoff[1.5] have shown that in the case of a transverse intersection of $W^s(\gamma)$ and $W^u(\gamma)$ there are infinitely many different homoclinic orbits which approach the periodic orbit γ in a very different way and, moreover, in any neighbourhood of a transverse homoclinic point infinitely

many periodic points exist. The phase portrait in the neighbourhood of homoclinic orbits is thus very complex and orbits of two points located arbitrarily
close to each other may have completely different dynamics. These properties are
the sources of the chaotic behaviour of deterministic dynamical systems.

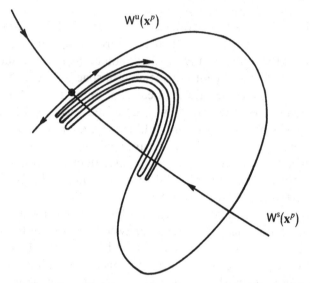

Fig. 2.5. Homoclinic intersection of stable and unstable manifolds (W^s and W^u) of the
1-periodic orbit $\gamma \equiv x^p$.

In the case of dynamical systems with continuous time the linearization of Eq.
(2.1) along γ leads to the nonautonomous system of linear equations with
T-periodic coefficients

$$\dot{a} = A(t)a , \tag{2.10}$$

where the matrix

$$A(t) = D_x v(x(t))\big|_{x(t) \, \in \, \gamma}$$

is periodic in t, $A(t + T) = A(t)$. It is proved in the theory of differential
equations with periodic coefficients (Floquet theory)[2.4] that the fundamental
matrix can be written in the form

$$U(t) = Z(t) \, e^{tR} , \tag{2.11}$$

where R is a constant matrix and $Z(t)$ is a periodic matrix of period T. We can
specifically choose

$$U(0) = Z(0) = I$$

where I is the unit matrix, and then we obtain the monodromy matrix

$$U(T) = Z(T) e^{TR} = e^{TR} . \qquad (2.12)$$

Hence we choose any point $\mathbf{x}^p \in \gamma$ and after one cycle around the closed curve γ the matrix $U(T)$ is obtained by simultaneously integrating Eq. (2.10) with n column vectors of I as initial conditions at $t = 0$. The eigenvalues of the matrix $U(T)$ (called *multipliers* of the orbit γ) will govern an expansion or a contraction of the discrete linear flow $g^k(\mathbf{a}) = U(kT)\mathbf{a}$ in the same manner as in the case of the linear difference equation (2.9). The eigenvalues of the matrix $U(T)$ again do not depend on the choice of the point $\mathbf{x}^p \in \gamma$. However, a difference exists when compared with the systems with discrete time: one eigenvector of the linearization coincides with the direction along the orbit γ. Exponential contraction or expansion does not occur in this direction and hence the matrix $U(T)$ will always have one eigenvalue equal to one. The remaining eigenvalues can be divided according to whether their moduli are greater than, less than or equal to one and we denote their number n_u, n_s and n_c, respectively. Then $n_s + n_u + n_c + 1 = n$. The periodic orbit is called hyperbolic if $n_c = 0$.

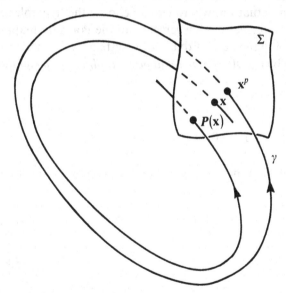

Fig. 2.6 Poincaré mapping.

To obtain full analogy with discrete systems, we choose in the neighbourhood of the point $\mathbf{x}^p \in \gamma$ a plane (or a surface) Σ of *codimension one* (a codimension of a k-dimensional submanifold in $\mathbb{R}^n$ is equal to $n - k$) transverse to γ at the point $\mathbf{x}^p$. This cross-section divides the state space locally into two parts. Orbits in the vicinity of γ will intersect Σ close to the point $\mathbf{x}^p$, see Fig. 2.6. Let us follow a sequence of intersections $\{x_k\}$ of an orbit close to γ with Σ and define the mapping $x \rightarrow P(x)$ from Σ into itself. The mapping P (called the *Poincaré mapping*) can be written down by means of the flow g^t as $x_{k+1} = g^{t_k}(x_k)$, where t_k is the time necessary for the phase point to travel between the intersections x_k and x_{k+1}. The construction of the mapping P has its origin in the works of Poincaré[1.53].

We have thus defined a discrete system with the $(n-1)$-dimensional state space Σ, which is associated with the flow g^t in the neighbourhood of the periodic orbit. Eigenvalues of the linearized mapping $D_x P$ are equal to the eigenvalues of the matrix $U(T)$ except for the eigenvalue corresponding to the direction along the orbit which is equal to 1. The theorem on the topological equivalence between the linear flow $D_x P^k$ and the nonlinear flow P^k and the theorem on stable and unstable manifolds W^s, W^u for hyperbolic periodic points again hold. If necessary, we can extend the state space Σ of the Poincaré mapping and thus follow the behaviour of orbits not only in a small neighbourhood of γ. This approach, often used in numerical analysis of models, has a drawback: certain orbits may be tangent to Σ or after several intersections with Σ may go away into other parts of the state space. Hence the choice of Σ has to be made carefully.

There exists one important situation, where we can choose the hyperplane Σ in such a way that all orbits intersect it transversely and the Poincaré mapping is thus defined globally: namely the case of dynamical systems with an external T-periodic forcing modelled by a system of *nonautonomous differential equations*

$$\dot{x} = v(x, t),$$

$$v(x, t + T) = v(x, t), \qquad T > 0. \tag{2.13}$$

Because of the periodicity of the function $v(x, t)$ in t, we can rewrite the system into the form

$$\dot{x} = v(x, s), \tag{2.14}$$

$$\dot{s} = 1, \tag{2.15}$$

where the newly-introduced dependent variable s is defined on a circle S^1 of length T (the coordinate s along the circle may be understood as a real number modulo T). One cycle around S^1 is realized in the course of one forcing period T. The state space $R^n \times S^1$ of Eqs (2.14) and (2.15) has dimension $n + 1$.

Now we choose a point s_1 on S^1 (e.g. $s_1 = 0$ mod T) and define a hyperplane $\Sigma = R^n \times \{s_1\}$. Then the Poincaré mapping $P: \Sigma \rightarrow \Sigma$ associated with

(2.14) and (2.15) is defined on n-dimensional state space Σ and trajectories of Eqs (2.14) and (2.15) intersect this hyperplane transversely. This construction for a two-dimensional system is shown in Fig. 2.7.

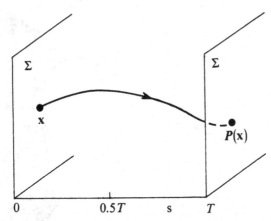

Fig. 2.7. Poincaré mapping for periodically driven systems.

2.4 Asymptotic behaviour and chaotic attractors

The asymptotic behaviour of deterministic systems described so far has been rather simple – stationary points or periodic oscillations were studied. A more complicated dynamics has already been introduced in Chapter 1. Here we briefly discuss two other simple models which produce asymptotic orbits more complicated than stationary points or periodic oscillations.

Let us consider a discrete dynamical system with a one-dimensional state space, the well-known *logistic equation*

$$N_{k+1} = N_k(a - bN_k) . \tag{2.16a}$$

Using transformation $x = bN/a$ this becomes

$$x_{k+1} = f_a(x_k) = ax_k(1 - x_k) . \tag{2.16b}$$

The equation was originally proposed for the description of the dynamics of a population of organisms that appear in discrete generations, such as insects[2.41].

The density N_{k+1} of the population in the next season is dependent on its current density N_k. We can see from Eq. (2.16a) that the density increases geometrically (i.e. exponentially) if $b = 0$ and $a > 1$. The parameter a is the rate of the population growth and the parameter b is the mortality factor which decreases the effective growth rate in the case of too high population densities. When $a \leq 1$ or $a > 4$, then the population density decreases to zero or becomes negative, respectively, which means that the population becomes extinct. May[2.41] presents a number of examples ranging from genetic problems to sociology which are modelled by equations of the type (2.16).

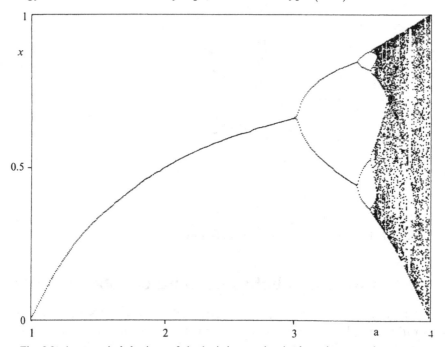

Fig. 2.8. Asymptotic behaviour of the logistic equation in dependence on the parameter a.

The logistic equation is one of the most often discussed prototypes of the complex behaviour of deterministic systems. Even if the equation looks simple, its solutions may be so complex that we need to use statistical methods for their description. From the mathematical point of view, the function $f_a(x)$ maps the interval $I = [0, 1]$ into itself for $0 \leq a \leq 4$. This means that the asymptotic behaviour must be contained in I. Fig. 2.8 shows computer generated orbits for a large number of iterations for a discrete set of values of a, separated by narrow intervals. The orbit generated for fixed a approximates an asymptotic limit set. In the scale of Fig. 2.8 we can observe certain periodic orbits with low periods and also the orbits which seem to be scattered densely in some subintervals of I. These orbits indicate the complex behaviour of Eq. (2.16).

However, we can also observe another structure which appears if we follow the dependence of the limit set on a. This structure results from successively repeated bifurcations which cause a self-similarity of the diagram when the scale is increased. The best visible is a sequence of successive *period doublings*, observable in the interval $3 \lesssim a \lesssim 3.57$. This *bifurcation sequence* will be discussed in more detail in the following chapter.

The next example also stems from biology, although analogous situations occur in a number of other fields. Let us have a continuous time dynamical system with an attracting periodic orbit perturbed by an external periodic force. One of the simplest nonlinear models, containing a repelling stationary point surrounded by an attracting periodic orbit can be described by two ordinary differential equations,

$$\dot{x} = u(x, y) = ax\left(1 - \sqrt{x^2 + y^2}\right) - 2\pi y\left[1 + \alpha\left(1 - \sqrt{x^2 + y^2}\right)\right],$$
(2.17)

$$\dot{y} = v(x, y) = ay\left(1 - \sqrt{x^2 + y^2}\right) + 2\pi y\left[1 + \alpha\left(1 - \sqrt{x^2 + y^2}\right)\right].$$
(2.18)

On rewriting into polar coordinates,

$$r = \sqrt{x^2 + y^2}, \qquad \varphi = \frac{1}{2\pi} \arctan\frac{y}{x},$$

we obtain

$$\dot{\varphi} = 1 + \alpha(1 - r),$$
(2.19)

$$\dot{r} = ar(1 - r).$$
(2.20)

This simple model is often used in the context of biological oscillators, see Winfree[6.58] and Guevara and Glass[6.26]. Solutions of Eqs (2.19) and (2.20) are in the form

$$\varphi(t) = t + \alpha\left[t - a^{-1}\ln|1 + r(e^{at} - 1)|\right] + \varphi_0 \;(\text{mod } 1),$$
(2.21)

$$r(t) = \left(\frac{1 - r_0}{r_0}e^{-at} + 1\right)^{-1},$$
(2.22)

with initial conditions

$$\varphi(0) = \varphi_0, \qquad r(0) = r_0.$$

From (2.17) and (2.18) we can see that for $a > 0$ the stationary state z located at the origin of the Cartesian coordinates is a repellor; the eigenvalues of the linearization of (2.17) and (2.18) in the neighbourhood of z are

$$\lambda_{1,2} = a \pm i\, 2\pi(1 + \alpha).$$

The periodic orbit γ, see Fig. 2.9, is the unit circle S^1 centred at the origin and its period is equal to one. It is an attractor if $a > 0$; then the multipliers of γ are $\sigma_1 = 1$, $\sigma_2 = e^{-a}$. The system (2.19) and (2.20) is conservative when $a = 0$ and dissipative for $a > 0$.

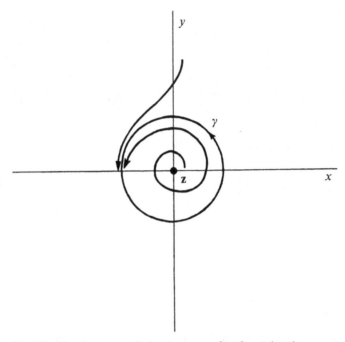

Fig. 2.9. The phase portrait for the system (2.17) and (2.18).

Eq. (2.20) can be considered as a continuous time analogue of the logistic difference equation (2.16). However, its asymptotic behaviour is far simpler. It follows from (2.22) that all initial conditions except for $r(0) = 0$ lead monotonically to $r = 1$ as $t \to \infty$. This behaviour follows from a different number of degrees of freedom. Even if both logistic equations have one-dimensional state spaces, the difference equation (2.16) is a noninvertible discrete dynamical system, while the system (2.20) is an invertible dynamical system with a continuous time variable. In fact, a minimal dimension of a state space of a dynamical system with complex (chaotic) behaviour is:

(1) $n = 1$ for noninvertible discrete time systems,

(2) $n = 2$ for invertible discrete time systems,

(3) $n = 3$ for invertible continuous time systems.

Let us take the system (2.17), (2.18) and perturb it by a discrete jump of an amplitude A in the state space at every integer multiple of T:

$$x(kT+) = x(kT) + A , \qquad (2.23)$$

$$y(kT+) = y(kT) , \qquad k = 0, 1, ... \qquad (2.24)$$

Each pulse will shift the phase point by A in the positive direction of the x axis. Starting from the point $(x_k, y_k) = (x(kT), y(kT))$ we can transform (2.23), (2.24) into polar coordinates and using (2.21) and (2.22) we obtain the difference equations

$$\varphi_{k+1} = T + \alpha[T - a^{-1} \ln |1 + r_0(\varphi_k, r_k) (e^{aT} - 1)|] +$$

$$+ \varphi_0(\varphi_k, r_k) \ (\text{mod } 1) , \qquad (2.25)$$

$$r_{k+1} = \left[\frac{1 - r_0(\varphi_k, r_k)}{r_0(\varphi_k, r_k)} e^{-aT} + 1 \right]^{-1} , \qquad (2.26)$$

where

$$\varphi_0(\varphi_k, r_k) = \frac{1}{2\pi} \arctan\left(\frac{d}{c}\right) , \qquad r_0(\varphi_k, r_k) = \sqrt{c^2 + d^2} ,$$

$$d = r_k \sin 2\pi \, \varphi_k, \qquad c = A + r_k \cos 2\pi \, \varphi_k . \qquad (2.27)$$

Eqs (2.25) and (2.26) can also be written in the condensed form

$$\begin{bmatrix} \varphi_{k+1} \\ r_{k+1} \end{bmatrix} = P \begin{bmatrix} \varphi_k \\ r_k \end{bmatrix} . \qquad (2.28)$$

Eqs (2.17–18) and (2.23–24) form a nonautonomous system, which can be augmented to obtain a three-dimensional autonomous system with the state space $R^2 \times S^1$, as it was discussed in the previous section. The flow connected with this system is discontinuous because of the pulses and the orbits possess discontinuities of magnitude A at the times $t \ (\text{mod } T) = 0$. When we choose a Poincaré section Σ at $t \ (\text{mod } T) = 0$, then we obtain the discrete system (2.28), where P is a diffeomorphism, because the trajectories of the continuous time system are also unique in the negative time direction (even if they are discontinuous).

Several typical computer generated orbits of the mapping (2.28) are shown in Figs. 2.10(*a*)–(*d*). In the first case, Fig. 2.10(*a*), the orbit traces regularly a closed curve (the ends of the curve in Fig. 2.10(*a*) are in fact joined because of the cyclic character of the variable φ). Such an orbit is called *quasiperiodic*. Its continuous

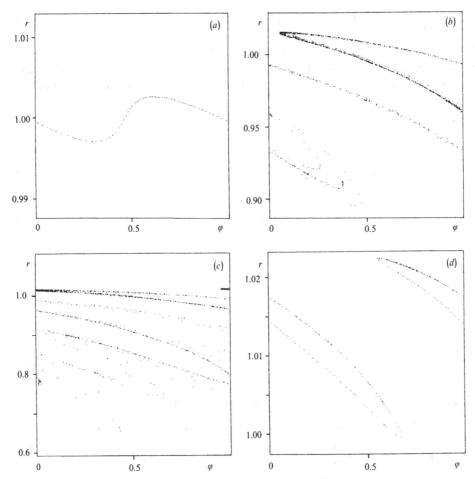

Fig. 2.10. Asymptotic behaviour of the system (2.25)–(2.26) for different values of the amplitude A and the forcing period T, $\alpha = 3$, $a = 3$ in all cases: (*a*) $A = 0.1$, $T = 1.2$; (*b*) $A = 0.7$, $T = 1.1$; (*c*) $A = 0.9$, $T = 1.1$; (*d*) $A = 2$, $T = 1.15$.

time variant fills a two-dimensional torus. Quasiperiodicity usually arises from two weakly coupled periodic phenomena with an irrational ratio of frequencies. Hence it can be expected to occur if the amplitude A of the pulses is sufficiently small. The invariant curve does not deviate very much from the unperturbed periodic orbit γ (which is the unit circle) and tends to it if $A \rightarrow 0$.

When the amplitude of the pulses increases, deviations from γ become larger and the quasiperiodic behaviour is replaced by a more complicated regime, see Fig. 2.10(b). The points of the asymptotic orbit are evidently located along curves oriented in the angular direction; they fold up on themselves and thus

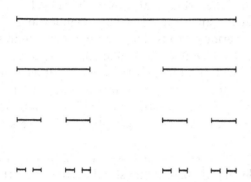

Fig. 2.11. The construction of the 'middle third' Cantor set.

form a complicated structure in the radial direction. These curves can, in principle, be identified with unstable manifolds of hyperbolic periodic points contained in the limit set[2.32]. Complex behaviour of unstable manifolds indicates the presence of homoclinic orbits, which imply an infinity of periodic points in the limit set[2.30]. Hence, it is not surprising that such *chaotic limit sets* show structure in the radial direction after an arbitrary increase of the scale.

Such a structure is typical for *Cantor sets*. A classical construction of such a set proceeds as follows: let us take a closed interval, and take away from it the middle one-third. From the remaining two intervals again take away the middle one-third, see Fig. 2.11 and so on *ad infinitum*. The resulting set C is a closed set that does not contain intervals and every point of C is an accumulation point of C (notice the self-similarity of C). Hence we can view the *local structure of a chaotic limit set* as the Cartesian product of a manifold and a Cantor set. However, unlike C the chaotic set does not need to possess a *self-similar structure*.

We observe that an increase in the amplitude of perturbations causes a transition from quasiperiodic to chaotic behaviour, see Fig. 2.10(a), (b). Chaotic orbits for values of amplitude A comparable to the radius of γ are illustrated in Fig. 2.10(c). The fluctuations become very large in both angular and radial directions. The last example, see Fig. 2.10(d) corresponds to a strong perturbation of the system. The width of the chaotic limit set is very small in the radial direction, even if its internal structure has the same complexity as in the other two cases.

Naturally, the quasiperiodic and the chaotic behaviour of the mapping P are not observable at all values of the parameters A, T, a and α. The mapping P has attracting periodic points in many cases, and often the type of the observed asymptotic behaviour depends on the chosen initial state.

The examples of the logistic equation and the periodically perturbed dissipative oscillator may serve as a motivation for the definition of the chaotic behaviour via the invariant sets and the orbits that approach them. In other words, we need to generalize the idea of an attractor to more complex invariant sets other than stationary points or periodic orbits. The notion of the ω-limit set as defined at the end of Section 2.1 above does not suffice, because it can be expected that a general attractor will contain saddle periodic orbits which themselves are ω-limit sets of orbits located in their stable manifolds. On the other hand, most orbits asymptotic to such an attractor will have far more complex ω-limits. Hence we have to consider entire sets of orbits at the same time.

Analogously to stable stationary points or periodic orbits we can require that an invariant set Λ (with a possible chaotic behaviour) should attract all orbits from a certain neighbourhood U. Hence $g^t(U) \subset U$ for all $t > t_0 > 0$ and $g^t(U) \to \Lambda$ for $t \to \infty$, where g^t is a (semi)flow, discrete or continuous in time t. This means that ω-limit sets of all orbits with the initial state $\mathbf{x} \in U$ are contained in Λ. Such a set is called an *attracting set*[2.50]. To ensure unique dynamic properties for 'most' orbits on Λ we must moreover require the *indecomposability* of Λ. A possible definition of indecomposability[1.22] is that Λ contains an orbit that is dense in Λ but other possibilities exist[2.16, 2.50].

An *attractor* may be defined as an indecomposable attracting set. A set W of all initial states $\mathbf{x}$ whose orbits approach the attractor Λ is called a *basin of attraction* of Λ. An example of an attractor is a stable stationary point or a stable periodic orbit. Another example is an attracting torus (a closed curve in systems with discrete time) with a quasiperiodic orbit.

Howevever, certain problems exist with the above definition of an attractor. For example, a situation may arise that any neighbourhood U of an invariant indecomposable set Λ contains periodic orbits or other invariant sets that do not belong to Λ, i.e. Λ is not an attracting set but 'most' orbits in U do approach Λ. In this case, a condition for Λ to be an attractor is that the set of points in U whose orbits approach Λ has a positive *Lebesgue measure* (i.e. a nonzero volume in the state space)[2.42].

The object of the greatest interest is a *chaotic* or *strange attractor*. An appropriate definition of a chaotic (strange) attractor is based on its dynamical properties such as a positive *metric entropy*, an *algebraic complexity* and *Lyapunov exponents*; these properties will be discussed in the next section. Here we remark that the chaoticity is implied by a *sensitive dependence on initial conditions*[1.56, 2.16], which means that two neighbouring points on the attractor separate locally exponentially fast with time. Such dynamics lead to an unpredictability if we do not know the initial point with an arbitrary precision.

The term 'chaotic' is related to a complex behaviour in time while the term 'strange' rather reflects a complex geometry of the attractor[2.16]. There are attractors with chaotic dynamics and with simple geometric structure but, on the other hand, there are attractors with complicated geometry (Cantor set-like structures) and with dynamics that are not sensitively dependent on initial conditions. Typically, chaotic (strange) attractors of multidimensional flows contain homoclinic orbits[1.22]. The presence of a homoclinic orbit will cause an infinite folding of an unstable manifold, see Fig. 2.5, and this causes both chaotic dynamics and geometric complexity of the attractor.

Chaotic dynamics can also occur on indecomposable invariant sets that are not attractors. Such objects may be called *chaotic repellors* or *chaotic saddles* and play an important role in bifurcations of chaotic attractors (see Chapter 3).

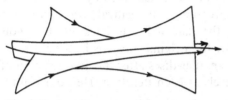

Fig. 2.12. Locally exponential contraction and expansion in the neighbourhood of orbits contained in a hyperbolic invariant set.

Another problem is the requirement of *structural stability*, i.e. the requirement that the phase portrait generated by the flow g^t in a neighbourhood of the attractor will not change its properties qualitatively when g^t is slightly changed. Structurally stable attractors occur in a class of flows generating indecomposable sets with the property of *hyperbolicity*[2.11, 2.12]. The *hyperbolic invariant set* Ω is a direct generalization of a hyperbolic stationary point or a periodic orbit, hence the tangent space to the state space at any point $\mathbf{x} \in \Omega$ can be decomposed into subspaces where an exponential expansion or contraction occurs, see Fig. 2.12. If Ω is indecomposable with a contracting neighbourhood, and if it contains a homoclinic orbit (i.e. Ω is not only a stationary point or a periodic orbit) then Ω is *a hyperbolic chaotic attractor* and the dynamics on it will be very complex and sensitively dependent on initial conditions. Stable and unstable manifolds of periodic orbits contained in a hyperbolic chaotic attractor Ω intersect transversely and the unstable manifolds are contained in Ω. Special cases of hyperbolic attractors are *Axiom A attractors* studied by Smale[1.58] and others[2.11, 2.12]. These attractors are densely filled by (saddle) periodic orbits, which makes their analysis easier.

A classical example of the mapping which generates a hyperbolic set is the *Anosov diffeomorphism*[1.3] defined on a two-dimensional torus $T^2 = S^1 \times S^1$

$$\begin{bmatrix} x_{n+1} \\ y_{n+1} \end{bmatrix} = \begin{bmatrix} 2 & 1 \\ 1 & 1 \end{bmatrix} \begin{bmatrix} x_n \\ y_n \end{bmatrix} \ (\text{mod } 1) . \tag{2.29}$$

Here Ω is formed by the entire torus T^2 and can be understood as a chaotic attractor that covers the whole state space, i.e. it does not have a complex geometrical structure. The eigenvalues of the matrix in (2.29) are independent of the point on T^2,

$$\sigma_1 = (3 + \sqrt{5})/2 > 1 \,, \qquad \sigma_2 = (3 - \sqrt{5})/2 < 1 \,,$$

and thus the expansion and contraction associated with σ_1 and σ_2, respectively, are uniform.

Another example of a hyperbolic attractor with a contracting neighbourhood is a *solenoid*[1.56, 2.49], generated by a diffeomorphism in R^3. Its construction is not defined algebraically but geometrically, see Fig. 2.13 (a)–(c). We take a solid torus $T \subset R^3$. The map f from T into itself is described by first shrinking the torus, then stretching it along the direction of its central core approximately twice, and finally folding and placing the distorted torus back into the undistorted torus T. The transverse section of the original torus is in the form of two discs; each of the discs contains two smaller discs after the first iteration of the mapping and this process repeats itself at each iteration. Hence a Cantor set arises, and the attractor

$$\Omega = \bigcap_{k=1}^{\infty} f^k(T)$$

has a structure of curves in the longitudinal and a Cantor set in the transverse directions.

However, hyperbolic chaotic attractors have found until now only theoretical applications. This follows from the fact that chaotic attractors arising in models of practical interest are frequently structurally unstable unlike hyperbolic attractors. This structural instability causes changes in the structure of an attractor when parameters in the model equation are varied.

The requirement of structural stability is at first sight justifiable, but probably too strong for practical purposes, as discussed by Guckenheimer and Holmes[1.22]; structural instability does not preclude the applicability of the models to the description of real systems. An interesting definition of structural stability based on an application of small unbounded random noise was proposed by Zeeman[2.62].

Considering the problem of structural stability a definition of an attractor that will express its stability with respect to *small bounded random perturbations* would be appropriate. Hyperbolic attractors have this property[2.36], but the definition can be formulated in a more general way[2.50].

The stability against a random noise implies that numerically found chaotic orbits reflect the dynamics on a 'real' chaotic attractor, because the round-off errors may be considered as a small random noise[2.29, 2.50]. However, a computer

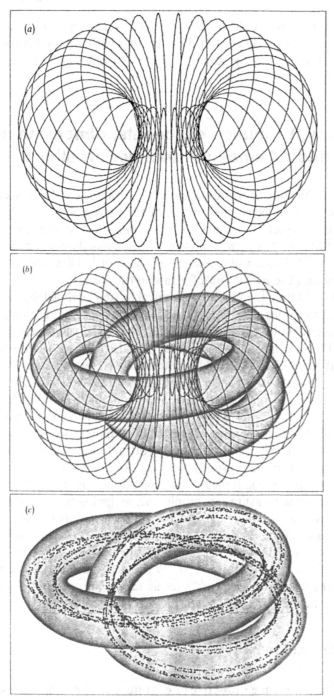

Fig. 2.13 (*a*) The original torus T; (*b*) the image $f(\mathsf{T})$ of the original torus is contained in (T); (*c*) successive iterates of $f(\mathbf{x})$, $\mathbf{x} \in \mathsf{T}$ approach a hyperbolic chaotic attactor – the solenoid.

used in these computations should possess sufficient accuracy to ensure that periodic cycles, which necessarily exist in the computer with a finite number of states, will be far longer than the time scale we are interested in, see e.g. Ref.2.8, 2.38.

2.5 Entropy, Lyapunov exponents, dimension and the effect of small random perturbations

The dynamic behaviour of trajectories on the attractor can be very complex. Different ways of an effective description of the randomness of a chaotic trajectory are the subject of the *ergodic theory* of dynamical systems[2.9, 2.11–2.14, 2.16, 2.47, 2.56, 2.58], which is in many aspects closely connected with probability theory[2.14, 2.46]. In a certain sense there is not much difference between a chaotic trajectory and the realization of a random process.

Let us consider a finite-dimensional dynamical system defined by a pair (X, g^t). In ergodic theory we also need to define a *measure*, i.e. a countably additive nonnegative function μ defined on the σ-algebra Σ of subsets of the state space X[2.46]. The asymptotic behaviour of the dissipative system is concentrated on the attractor. We shall study a statistical description of trajectories located inside the basin of attraction W of the attractor Λ. Let us assume for simplicity that there is a neighbourhood U of Λ contained in W such that $g^t(U) \subset U$, $t \geq 0$ and $\Lambda = \bigcap_{t \geq 0} g^t(U)$. If more attractors coexist, then it is better to study each attractor separately. Nonattracting limit sets or boundaries between the regions of attractions of neighbouring attractors are also of interest.

Theorems of ergodic theory are usually valid not for all $\mathbf{x}$ but for almost all $\mathbf{x}$ with respect to some measure μ. It means that exceptions exist but only on sets of zero measure. Lebesgue measure is a simple one, assigning to every element of Σ its volume. Dissipative dynamical systems contract volumes in the state space and thus the attractor Λ has zero volume. Hence the Lebesgue measure is not useful for its description. A relevant measure must be invariant with respect to g^t, i.e. $\mu(C) = \mu(g^{-t}(C))$, $C \in \Sigma \subset W$, $t > 0$. Such a measure may be constructed in the following way[2.19, 2.48]: let us take a point $\mathbf{x}_0 \in W$ and a subset C of W, then the measure $\mu(C, \mathbf{x}_0)$ of the set C (depending on $\mathbf{x}_0$) can be written as

$$\mu(C, \mathbf{x}_0) = \lim_{\tau \to \infty} \mu_\tau(C, \mathbf{x}_0), \tag{2.30}$$

where μ_τ is a fraction of the time τ (discrete or continuous) which the state point moving along the orbit with the initial state $\mathbf{x}_0$ spends in C. Because all orbits inside W asymptotically approach $\varLambda$, the measure is concentrated on $\varLambda$. When the attractor is a stationary point or a periodic orbit, the measure is determined uniquely; a large number of different invariant measures can be defined by Eq. (2.30) in the general case.

For the sake of simplicity we shall, however, assume that almost all points x_0 (with respect to the Lebesgue measure) lead to a unique measure $\mu(C)$, called the *asymptotic* or *natural measure*. This assumption is supported by numerical experiments, which show that asymptotic behaviour of orbits is not dependent on the initial state. Because all orbits in the basin of attraction W are asymptotic to the attractor, the measure of any set $C \subset W$ *such that* $\varLambda \subset C$ is $\mu(C) = = 1$, in particular $\mu(\varLambda) = 1$. Hence we can interpret $\mu(C)$ as a probability that the phase point lies in C.

A measure μ is called *ergodic* if either $\mu(C) = 0$ or $\mu(C) = 1$ for any invariant subset $C \subset \varLambda$. Ergodic measure reflects the indecomposability of $\varLambda$. For such a measure a well-known *ergodic theorem* holds: for μ-almost all $x \in W$ and for any continuous real function φ time averages are equal to averages taken over the state space,

$$\lim_{T \to \infty} \frac{1}{T} \int_0^T \varphi(g^t(\mathbf{x})) \, \mathrm{d}t = \int_\varLambda \varphi(\mathbf{x}) \, \mathrm{d}\mu(\mathbf{x}) \,. \tag{2.31}$$

A special case of the ergodic theorem states the equality of the *autocorrelation function* averaged over time and over the state space

$$b(\tau) = \lim_{T \to \infty} \frac{1}{T} \int_0^T \varphi(g^{t+\tau}(\mathbf{x})) \, \varphi(g^t(\mathbf{x})) \, \mathrm{d}t = \int_\varLambda \varphi(g^\tau(\mathbf{x})) \, \varphi(\mathbf{x}) \, \mathrm{d}\mu(\mathbf{x}) \,.$$

In the case of discrete time the integrals over time are replaced by sums.

It is evident that ergodicity itself is not sufficient for the description of chaotic attractors. Stationary, periodic, or quasiperiodic attractors are also ergodic. The autocorrelation function of a periodic or a quasiperiodic orbit will be periodic or quasiperiodic, respectively. To correspond to a chaotic orbit the autocorrelation function has to decay,

$$\lim_{\tau \to \infty} b(\tau) - \left[\int_\varLambda \varphi(\mathbf{x}) \, \mathrm{d}\mu(\mathbf{x}) \right]^2 = 0 \,. \tag{2.32}$$

This property is called mixing and guarantees that $\varLambda$ has a more complex structure than that of a periodic or of a quasiperiodic attractor. The same differentiation is possible by means of the *power spectrum*

$$P(\omega) = \int_{-\infty}^{\infty} b(\tau) \, \mathrm{e}^{-\mathrm{i}\omega\tau} \, \mathrm{d}\tau \tag{2.33}$$

which consists of δ-functions for periodic or quasiperiodic attractors and which is continuous in the presence of a mixing mechanism. A chaotic attractor may be defined as an attractor with the mixing property. However, a typical chaotic attractor should have an exponentially decaying autocorrelation function. This property is difficult to verify and has been proved rigorously only for certain systems such as diffeomorphisms satisfying Axiom A[2.11, 2.12].

Hence we need other criteria of chaoticity. One of them is a *measure theoretic (Kolmogorov, metric) entropy* [2.1, 2.9, 2.16]. Let us cover an attractor by a net of m cubes $E_1, .., E_m$ of side ε and let us choose a time interval of length $\tau > 0$ (in discrete time systems it suffices to choose $\tau = 1$). The state space is 'coarse grained' and we can uniquely assign a symbol to each cube E_i, e.g. the numbers $1, 2, ..., m$. This formulation expresses the fact that experimental measurements have a finite accuracy and we are able to measure only discrete sequences of values recorded at discrete times. We would like to determine the probability that the orbit will pass subsequently through N specified cubes $E_{i_1}, E_{i_2}, ..., E_{i_N}$, i.e. we assume that in the observations we record the sequence of numbers $i_1, i_2 ... i_N$. This sequence is called a 'word' of length N; the method based on the above concept is called a *symbolic dynamics*[2.1]. Even if the indices i_k, $k = 1, ...,$ N, can attain any values from the set $\{1, ..., m\}$ not all 'words' are admissible, because the orbits cannot pass through the state space arbitrarily.

Let $C_{i_1... i_N}$ be the set of all initial points of orbits, which are described in the first N measurements by a given 'word' $i_1 ... i_N$, *and* C^N be a set of all $C_{i_1 ... i_N}$ such that $i_1 ... i_N$ is any admissible 'word'. In particular $C_{i_1} = E_{i_1}$ and the number of elements (*cardinality*) of the set C^1 is m. It can be expected that the cardinality of C^N will increase at most geometrically with increasing N. Considering the (semi)flow g^t we can write

$$C^N = \{C_{i_1 ... i_N} = \bigcap_{k=1}^{N} g^{(1-k)\tau}(C_{i_k}) ;$$

such that $i_1 ... i_N$ is an admissible 'word'} .

Making use of the invariant measure μ we can define the probability of the orbit passing through subsequent specified cubes as

$$P(i_1 ... i_N) = \mu(C_{i_1 ... i_N}) .$$

A complexity of the dynamic behaviour of orbits on the set C^N can be expressed by means of the entropy H,

$$H(C^N) = -\sum_{C^N} P(i_1 ... i_N) \log P(i_1 ... i_N) . \tag{2.34}$$

The entropy $H(C^N)$ has an information theoretic interpretation: $H(C^N)$ is equal to an average amount of information obtained after N successive measurements, if we observe a 'word' of length N. If the base of the logarithm in (2.34) is two then H is measured in bits.

The metric entropy h_μ of the attractor is now defined as

$$h_\mu = \frac{1}{\tau} \lim_{\varepsilon \to 0} \lim_{N \to \infty} \frac{H(C^N)}{N} \qquad [\text{bits per time unit}] . \qquad (2.35)$$

From the point of view of *information theory*[2.53, 2.54] h_μ is equal to the mean flow of the information produced by the system. The metric entropy depends on μ. It is commonly assumed that μ is the natural measure but this measure generally does not give a maximal entropy. A maximal value of h_μ (over all invariant ergodic measures),

$$h = \sup_\mu h_\mu , \qquad (2.36)$$

is called a *topological entropy* (h can be defined independently of h_μ); generally $h_\mu \leqq h$. We can define the attractor to be chaotic when $h_\mu > 0$.

The metric entropy is defined for the attractor as a whole. There exists a concept which assigns to an orbit starting at $\mathbf{x}$ a nonnegative number $K(\mathbf{x})$ called an *algebraic complexity* of the orbit. This number is related to the length of the shortest program which is able to generate the orbit based at $\mathbf{x}$ on a computer. The exact definition is given, for example, in Ref. 2.1. In general $K(\mathbf{x}) \leqq h_\mu$; if μ is an ergodic invariant measure on the attractor then $K(\mathbf{x}) = h_\mu$ for almost all $\mathbf{x}$ with respect to μ. In particular, if the metric entropy is positive then almost all orbits exhibit behaviour as complicated as the entire dynamical system on the attractor.

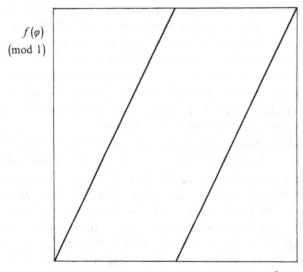

Fig. 2.14. The Bernoulli shift mapping.

Ford[2.21] illustrated the concept of the algebraic complexity on a simple dynamical system

$$\varphi_{k+1} = f(\varphi_k) = 2\varphi_k \pmod{1}. \tag{2.37}$$

The mapping $f(\varphi)$, see Fig. 2.14, is a noninvertible 'linear' mapping of the circle S^1 into itself such that if φ circulates S^1 once, then $f(\varphi)$ circulates it twice. This circle map is at the same time a one-dimensional limit of the solenoid discussed in Section 2.4, see Fig. 2.13 (a)–(e). The mapping $f(\varphi)$ preserves the Lebesgue measure and the metric entropy (in this simple case also the topological entropy) is equal to log 2. If we choose $\varphi_0 \pmod{1}$, then Eq. (2.37) generates a sequence of numbers which in the binary representation arise if we shift the binary point in φ_0 by one position to the right and neglect the leftmost digit. This procedure is called a *Bernoulli shift*[1.49].

The complexity of the orbit thus depends on the properties of the initial state φ_0. If φ_0 is given with a finite accuracy, say N bits, then after N iterations an information not contained explicitly in the initial data is obtained. The system thus acts as a noise amplifier. The flow of information through the system is equal to one bit per iteration, in agreement with the value of the metric entropy $(\log_2 2 = 1)$.

Let us assume now that the initial state φ_0 is known exactly. If φ_0 is rational, i.e. with periodic binary expansion, then the length of the program generating the orbit will be proportional to the length of the period and the complexity of the orbit will be zero[2.21]. It can be shown that for an irrational initial state φ_0 the length of the program which generates the first N bits of φ_0 can be proportional either to log N or to N. If $N \to \infty$ the first case leads again to the zero complexity of the orbit, however, the second case leads to the complexity equal to log 2. Almost all irrational choices of φ_0 lead to the positive complexity of orbits[2.21], hence it is evident that the source of the deterministic chaos is hidden just in these numbers in the real axis. In fact, the sequence of binary digits representing most of the numbers of the continuum cannot in any reasonable way be distinguished from the sequence of 'heads' and 'tails' in coin tossing. In this sense deterministic chaos is equivalent to randomness.

Unfortunately, the computations of the entropy and the complexity are in most practical applications relatively difficult. Hence we shall turn our attention to quantities that can be computed more easily on a sufficiently fast computer.

As discussed above, almost all orbits inside the attractor move away from one another exponentially fast. This expansion, however, can occur only in certain directions and must be balanced by an exponential contraction of orbits in other directions because the orbits cannot leave the attractor. If we follow the directions of contraction and expansion along a chosen orbit, we observe that they are changing in a complex way. To characterize the orbit, however, we need to know only the mean velocities of the expansion and contraction along the individual directions. The mean velocities can be described on introducing the notion of *Lyapunov exponents*[2.43, 2.48].

Let us assume that we have a dynamical system (X, g^t) and that the dimension of the state space X containing an attractor Λ with a basin of attraction W is n. For simplicity we set $X = R^n$. Let us choose an initial state $x_0 \in W$ and follow the time evolution of the moduli of the eigenvalues $\sigma_i(t)$ of the linear mapping $U(t) = D_x\, g^t(x)|_{x=x_0}$. If x_0 lies on a periodic orbit with period T, then the eigenvalues of $U(T)$ are just the multipliers of the orbit, see Section 2.3.

In a general case, $\sigma_i(t)$ express local expansion or contraction in the neighbourhood of the orbit with the initial state x_0. The Lyapunov exponents λ_i are defined as time averages of the logarithms of the moduli of the eigenvalues $\sigma_i(t)$[2.19]

$$\lambda_i = \lim_{t \to \infty} \frac{1}{t} \log |\sigma_i(t)|, \qquad i = 1, ..., n. \tag{2.38}$$

We assume here that the λ_is are ordered with respect to their magnitudes, i.e. $\lambda_1 \geq \lambda_2 \geq ... \geq \lambda_n$. As the metric entropy, the Lyapunov exponents are invariant with respect to coordinate transformations in the state space. For the original definition of the Lyapunov exponents see Ref. 2.43, and a generalization to infinite-dimensional systems can be found in Ref. 2.51. If μ is an ergodic measure on the attractor Λ, then the spectrum of Lyapunov exponents is constant for μ-almost all $x_0 \in \Lambda$. Our assumptions on the properties of the natural measure imply that the λ_is are constant for almost all x_0 (with respect to Lebesgue measure) from the basin of attraction W.

The *spectrum of Lyapunov exponents* can be schematically written as $(+, ..., +, 0, ..., 0, -, ..., -)$, where the symbols $+, 0, -$ denote positive, zero and negative exponents, respectively. It is evident that if at least one λ_i is positive, an exponential expansion inside Λ occurs and such an attractor can be denoted as chaotic. Continuous time orbits which are not stationary points have at least one $\lambda_i = 0$ because neither exponential expansion nor contraction occurs along the orbit. According to the symbolic spectrum the attractors can be classified in the following way: $(-, ..., -)$ – stationary point, $(0, -, ..., -)$ – periodic orbit, $(0, 0, -, ..., -)$ – two-dimensional torus with a quasiperiodic orbit (tori of higher dimension will contain an appropriate number of zeros in the spectrum), $(+, 0, -, ..., -)$ – chaotic attractor with a single direction of expansion, etc. ...

The identical zero is absent in systems with discrete time. Let us stress that the spectrum is constant only for the orbits typical with respect to the measure μ. For example, unstable periodic orbits with a different spectrum of Lyapunov exponents (some of them can also be positive) can exist inside of the chaotic attractor. However, these orbits are exceptional in the sense of the measure μ.

It can be shown that under certain assumptions the spectrum of exponents of an infinite-dimensional system is discrete and the number of positive exponents is finite[2.51]; hence, the dynamics on an attractor of an infinite dimensional system can be described as a finite-dimensional system.

There exists a close relation between the exponential expansion expressed by positive Lyapunov exponents and the metric entropy of the attractor Λ. For an ergodic measure

$$h_\mu \leqq \sum_{\lambda_i > 0} \lambda_i(\mathbf{x}) \,, \tag{2.39}$$

the independence of λ_i on $\mathbf{x}$ and the equality in (2.39) are expected in typical cases[2.48] and hold rigorously for the attractors satisfying Axiom A[2.12]. As will be discussed in more detail in Chapter 4 dealing with numerical techniques, Lyapunov exponents can be determined numerically from model equations and also from experimental data (at least several positive λ_is). Some procedures are applicable also to systems of an infinite dimension. The metric entropy (and thus also the complexity of trajectories) can be computed as the sum of positive Lyapunov exponents.

Another characteristic which can be extremely useful for a description of chaotic sets is *dimension*. Simple objects, such as points, lines, rectangles or cubes have an integer dimension. This idea of dimension is not useful if the studied objects have a complex structure. Cantor sets can serve as examples of complex objects. It is then convenient to define a noninteger dimension. Chaotic attractors have a local structure that looks like the Cartesian product of a Cantor set with a manifold hence they will be characterized by a *noninteger* dimension. In general, sets with such a complex geometry are called *fractals*[1.46, 2.10, 2.52]

Before we present various definitions of the noninteger dimension we consider an integer dimension called an *embedding dimension*. It is a minimum dimension m of a Euclidean space $\mathbf{R}^m$ such that there is a smooth transformation from the original state space to $\mathbf{R}^m$ which is one-to-one at every point of the attractor. If the studied dynamical system is defined on a Euclidean state space $\mathbf{R}^n$, then the embedding dimension m of the attractor is at most n, however, it can be substantially less. More generally, if the state space is an n-dimensional compact manifold, then the Whitney embedding theorem[2.32] guarantees that $m \leqq 2n + 1$. Let us mention one interesting case, where the embedding dimension is lower than the dimension of the state space. The state space is of infinite dimension but there exists a compact manifold with finite dimension n containing the attractor. The embedding dimension $m \leqq 2n + 1$ is finite and the motion on the attractor is described by a finite-dimensional dynamical system[2.20, 2.39, 2.40].

The procedures for the estimation of the embedding dimension are based on Takens' idea about the reconstruction of an asymptotic orbit in the state space from the knowledge of a single dynamical variable[4.65], see Chapter 4.

The following definitions of the dimensions of attractors[2.19] can be divided into two groups. In the first group are the definitions that consider only the geometric complexity of the attractor, while the definitions in the second group respect the frequency of occurrence of the phase point in the individual parts of the attractor (hence they include the natural measure).

The most simple definition from the first group is the definition of the *capacity*. Let us cover the attractor A by a net of n-dimensional cubes of side ε (n is the dimension of the state space) and denote $M(\varepsilon)$ the number of cubes. (In the terminology used in the definition of the metric entropy $M(\varepsilon)$ is equal to the cardinality of the set C^1). Then the capacity D_C is

$$D_C = - \lim_{\varepsilon \to 0} \frac{\log M(\varepsilon)}{\log \varepsilon} . \tag{2.40}$$

Although the classical 'middle third' Cantor set (described in Section 2.4) is not a model of an actual attractor, it is often presented as an example of the set with a noninteger dimension. Starting from the unit interval, after the n-th deletion 2^n intervals remain, each of length $\varepsilon = 3^{-n}$, see Fig. 2.11; hence

$$D_C = - \lim_{n \to \infty} \frac{\log 2^n}{\log 3^{-n}} = \frac{\log 2}{\log 3} . \tag{2.41}$$

Another concept is that of the *Hausdorff dimension* D_H[2.16]. Generally $D_H \leq D_C$, but it is conjectured that for typical attractors both dimensions acquire the same value D_F called a *fractal dimension*[2.19].

The dimensions related to the natural measure are defined in an analogous way as the capacity, making use of the entropy defined by (2.34). Let us denote the i-th cube C_i; the probability P_i of finding the phase point in C_i is $P_i = \mu(C_i)$ and the entropy is then

$$H(\varepsilon) = - \sum_{i=1}^{n(\varepsilon)} P_i \log P_i$$

(in the notation used in definition (2.34) $H(\varepsilon) = H(C^1)$). An *information dimension*[2.17, 2.46] is then defined by analogy with (2.40) as

$$D_I = - \lim_{\varepsilon \to 0} \frac{H(\varepsilon)}{\log \varepsilon} , \tag{2.42}$$

where D_I expresses the rate of increase of the mean amount of information obtained in one measurement with an accuracy ε as ε decreases. In general, $D_I \leq D_F$.

The dimensions defined in (2.40) and (2.42) can be computed numerically, see Chapter 4. The computations, however, require both a large memory and sufficient computer time. Computations in more than three-dimensional state space require the use of large computers or supercomputers. The computation of the information dimension is easier, because we can omit the cubes which are rarely 'visited' by a typical orbit on the attractor.

Another definition of dimension uses Lyapunov exponents. The *Lyapunov dimension* D_L introduced by Kaplan and Yorke[2.34] is defined as

$$D_L = k + \sum_{i=1}^{k} \lambda_i / |\lambda_{k+1}|, \tag{2.43}$$

where k is a number such that

$$\sum_{i}^{k} \lambda_i \geq 0 \quad \text{and} \quad \sum_{i}^{k+1} \lambda_i < 0.$$

This means that i-dimensional volumes in the tangent space expand along the chaotic orbit if $i \leq k$, and contract if $i > k$. Generally, $D_I \leq D_L$ but it is conjectured that the Lyapunov dimension is equal to the information dimension in typical cases[2.22, 2.61].

Shtern[2.55] and others[2.23, 2.16] have shown that the information dimension can be decomposed into a sum,

$$D_I = \sum_{i=1}^{n} D_I^i, \tag{2.44}$$

where the *partial dimensions* D_I^i describe a geometric complexity of the attractor in n directions of the state space, $D_I^i \in [0, 1]$, and the information flow is balanced if

$$\sum_{i=1}^{n} \lambda_i D_I^i = 0. \tag{2.45}$$

The Lyapunov dimension D_L will be equal to D_I under the assumption that the attractor is (1) a manifold in the first k directions, i.e. $D_I^i = 1$, $i = 1, ..., k$; (2) a Cantor set in the $(k+1)$st direction, i.e. $0 < D_I^{k+1} < 1$ and, finally, (3) the remaining (contracting) directions do not contribute to the dimension of the attractor, i.e. $D_I^i = 0$, $i = k + 2, ..., n$. Then (2.45) implies that

$$D_I^{k+1} = \sum_{i=1}^{k} \lambda_i / |\lambda_{k+1}|$$

and Eqs (2.43) and (2.44) give the same result.

The definition of dimension can be further generalized by introducing entropy of order q[2.25]

$$H_q(\varepsilon) = \frac{1}{1-q} \log \sum_i^{M(\varepsilon)} P_i^q .$$ (2.46)

The *generalized dimension* of order q is then

$$D_q = -\lim_{\varepsilon \to 0} \frac{H_q(\varepsilon)}{\log \varepsilon} .$$ (2.47)

Similarly the *generalized metric entropy* h_q is obtained by introducing the generalized entropy $H_q(C^N)$ of the set C^N (see Eq. (2.34))

$$H_q(C^N) = \frac{1}{1-q} \log \sum_{C^N} P_{i_1 \dots i_N}^q$$ (2.48)

and after taking the limit

$$h_q = \frac{1}{\tau} \lim_{\varepsilon \to 0} \lim_{N \to \infty} \frac{H_q(C^N)}{N} .$$ (2.49)

By comparison with (2.40) and (2.42) we have $D_0 = D_C$ and $D_1 = D_I$ (it follows from the limit for $q \to 1$ using L'Hospital's rule), similarly $h_0 = h$ ($\equiv$ topological entropy), $h_1 = h_\mu$ ($\equiv$ metric entropy). Grassberger and Procaccia[2.25] show that these quantities are easily computable from the experimental data for $q = 2$. They also show that D_q and h_q are decreasing functions of q and therefore D_2 and h_2 are lower limits for the information dimension D_I and the metric entropy h_μ, respectively. The description of the numerical computations of D_2 and h_2 is discussed in Chapter 4. The parameter q in Eqs (2.47) and (2.49) can be any real number and thus D_q and h_q can be seen as functions of q. Then D_q is called a *dimension function* and can be also computed numerically[2.6].

Concepts such as measure, entropy, Lyapunov exponents or dimension can in principle also be applied to invariant indecomposable sets other than attractors. In particular, repellors or generalized saddles (semiattractors) with chaotic behaviour are of interest. These invariant sets are generally contained in the boundaries of basins of attraction between different attractors. In numerical computations we very often observe orbits, which are chaotic for a long time but finally become more simple. This so called transient chaos[2.35] gives evidence about the presence of chaotic repellors or semiattractors[2.33]. These sets can often be considered as hyperbolic. Their importance lies in the fact that they can keep orbits in their neighbourhood for a very long time, even if the asymptotic behaviour is different. Nonattracting chaotic sets and the dynamics of trajectories in their neighbourhood were studied by Eckman and Ruelle[2.16] and Kantz and Grassberger[2.33]. We shall also discuss them in Chapter 3.

Asymptotic motion in dissipative systems takes place on sets which usually have zero volume in state space (i.e. zero Lebesgue measure). If such a set is chaotic, then it usually has a noninteger dimension, smaller than the dimension of the state space. There are, however, sets which have fractal dimension equal to that of the embedding space (i.e. an integer) and a nonzero Lebesgue measure. Such sets are called *fat fractals*. They do not form attractors in dissipative systems but they are important in different contexts[2.18, 2.26, 2.57], for example, are observed in the logistic equation (2.16b) as a set of parameter values for which chaotic attractors occur, see Section 3.4.

A simple example of a fat fractal is a Cantor set constructed similarly to the 'middle third' Cantor set described previously (see Fig. 2.11) but instead of deleting 1/3 at each step of the construction we delete successively 1/3, 1/9, 1/27, etc., from each interval.

The usual definition of the capacity dimension does not characterize the fractal properties of fat fractals because it is an integer. However, it can be modified as described by Grebogi et al[2.26]. Let us consider a closed bounded set C contained in n-dimensional space. A definition equivalent to that given by (2.40) is

$$D_C = n - \lim_{\varepsilon \to 0} \frac{\log V[C(\varepsilon)]}{\log \varepsilon}. \tag{2.50}$$

Here $V[C(\varepsilon)]$ is the n-dimensional volume of the set $C(\varepsilon)$ which contains all points within a distance ε from C. Now the *exterior capacity dimension* is defined as

$$D_E = n - \lim_{\varepsilon \to 0} \frac{\log V[\bar{C}(\varepsilon)]}{\log \varepsilon} \tag{2.51}$$

where $\bar{C}(\varepsilon)$ arises by removing the original set C from the set $C(\varepsilon)$. For ordinary fractals which have zero volume, $D_E = D_C$, while for fat fractals the exterior capacity is different from the usual capacity. Moreover, D_E is useful in characterizing sets which have nonzero volumes and a fractal boundary[2.7, 2.27].

The idea of dimension can be used in the description of the effect of small random fluctuations on the attractor[2.44]. Let us consider a discrete dynamical system with a small additive stochastic term

$$\mathbf{x}_{k+1} = f(\mathbf{x}_k) + \varepsilon \mathbf{u}_k, \tag{2.52}$$

where $\mathbf{x}_k$ is an n-dimensional vector and $\mathbf{u}_k$ is an n-dimensional vector generated by a stationary random process with zero mean value $\langle \mathbf{u}_k \rangle = 0$ and with the correlation function $\langle \mathbf{u}_k \mathbf{u}_l \rangle = 0$ for $k \neq l$ (the so-called δ-correlation).

It is assumed that the probability density distribution function $p(\mathbf{u})$ exists, i.e. that

$$\langle \varphi(\mathbf{u}) \rangle = \int \varphi(\mathbf{u}) \, p(\mathbf{u}) \, \mathrm{d}\mathbf{u} \, ,$$

and ε specifies the level of the noise.

We can say approximately that a bounded noise of magnitude ε will wash out the structure of the attractor on the scale ε and cause the noisy attractor to have nonzero volume in state space. Because the dimension of the noiseless attractor determines the scaling of the structures in the attractor we can expect a relation to exist between the volume of the noisy attractor and the dimension in the deterministic limit $\varepsilon = 0$. Ott, Yorke and Yorke[2.44] have shown that if the noise is bounded, then the volume V of the noisy attractor of the system (2.52) scales with ε as

$$V \sim \varepsilon^{n - D_C} \tag{2.53}$$

where D_C is the capacity for the system with zero noise. For example, a stationary point changes in the presence of noise into an n-dimensional ellipsoid with volume of order ε^n but relation (2.53) holds also for a noninteger dimension.

REFERENCES

2.1   Alekseev V. M. and Yakobson M. V. Symbolic dynamics and hyperbolic dynamic systems. *Phys. Reports* **75** (1981) 287.

2.2   Andronov A. A., Vitt A. A. and Chaikin S. E. *Theory of Oscillations.* Moscow, Nauka, 1981 (in Russian).

2.3   Arnold V. I. *Ordinary Differential Equations.* Cambridge, MA, M. I. T. Press, 1973 (Russian original, Moscow, 1971).

2.4   Arnold V. I. *Geometrical Methods in the Theory of Ordinary Differential Equations.* New York, Springer, 1982 (Russian original, Moscow, 1978).

2.5   Arnold V. I., Varchenko A. N. and Gusein–Zade S. M. *Singularities of Differentiable Mappings.* Vols. I and II. Moscow, Nauka, 1982, 1984 (in Russian).

2.6   Badii R. and Broggi G. Measurement of the dimension spectrum $f(\alpha)$: Fixed–mass approach. *Phys. Lett.* **131A** (1988) 339.

2.7   Battelino P. M., Grebogi C., Ott E., Yorke J. A. and Yorke E. D. Multiple coexisting attractors, basin boundaries and basic sets. *Physica* **32D** (1988) 296.

2.8   Beck C. and Roepstorff G. Effects of phase space discretization on the long–time behavior of dynamical systems. *Physica* **25D** (1987) 173.

2.9   Billingsley P. *Ergodic Theory and Information.* New York, Wiley, 1965.

2.10   Bohr T. and Tél T. The thermodynamics of fractals. In *Directions in Chaos* Vol.2, ed. Hao Bai–lin, Singapore, World Scientific, 1988, p. 194.

2.11   Bowen R. *Equilibrium State and the Ergodic Theory of Anosov Diffeomorphisms.* Lecture Notes in Mathematics **470**, Berlin, Springer, 1975.

2.12   Bowen R. and Ruelle D. The ergodic theory of axiom A flows. *Inventiones Math.* **29** (1975) 181.

2.13   Bylov B. F., Vinograd R. E., Grobman D. N. and Nemyckii V. V. *Theory of Lyapunov Exponents and Their Applications to Stability Problems.* Moscow, Nauka, 1966 (in Russian).

2.14   Cornfeld I. P., Fomin S. V. and Sinai Ya. G. *Ergodic Theory.* New York, Springer, 1982.

2.15   Čenys A. and Pyragas K. Estimation of the number of degrees of freedom from chaotic time series. *Phys. Lett.* **129A** (1988) 227.

2.16   Eckman J.–P. and Ruelle D. Ergodic theory of chaos and strange attractors. *Rev. Mod. Phys.* **57** (1985) 617.

2.17   Farmer J. D. Information dimension and the probabilistic structure of chaos. *Z. Naturforsch.* **37a** (1982) 1304.

2.18   Farmer J. D. Sensitive dependence on parameters in nonlinear dynamics. *Phys. Rev. Lett.* **55** (1985) 351.

2.19   Farmer J. D., Ott E. and Yorke J. A. The dimension of chaotic attractors. *Physica* **7D** (1983) 153.

2.20 Foias C. and Temam R. The algebraic approximation of attractors : The finite dimensional case. *Physica* **32D** (1988) 163.

2.21 Ford J. Chaos: Solving the unsolvable, predicting the unpredictable. In *Chaotic Dynamics and Fractals*, ed. M. F. Barnsley and S. G. Demko, New York, Academic Press, 1985.

2.22 Frederickson P., Kaplan J. L., Yorke E. D. and Yorke J. A. The Lyapunov dimension of strange attractors. *J. Diff. Eqs.* **49** (1983) 185.

2.23 Grassberger P. Information aspects of strange attractors. Preprint, Wuppertal, Univ. of Wuppertal, 1984.

2.24 Grassberger P. Generalizations of the Hausdorff dimension of fractal measures. *Phys. Lett.* **107A** (1985) 101.

2.25 Grassberger P. and Procaccia I. Measuring the strangeness of strange attractors. *Physica* **9D** (1983) 189.

2.26 Grebogi C., McDonald S. W., Ott E. and Yorke J. A. Exterior dimension of fat fractals. *Phys. Lett.* **110A** (1985) 1.

2.27 Grebogi C., Ott E. and Yorke J. A. Chaos, strange attractors, and fractal basin boundaries in nonlinear dynamics. *Science* **238** (1987) 632.

2.28 Guckenheimer J. Noise in chaotic systems. *Nature* **298** (1982) 358.

2.29 Hammer S. M., Yorke J. A. and Grebogi C. Do numerical orbits of chaotic dynamical processes represent true orbits? Preprint, Maryland, Univ. of Maryland, 1987.

2.30 Hirsch M. W. The dynamical systems approach to differential equations. *Bull. (New) Am. Math. Soc.* **11** (1984).

2.31 Hirsch M. W. and Smale S. *Differential Equations, Dynamical Systems and Linear Algebra.* New York, Academic Press, 1974.

2.32 Hirsch M. W., Pugh C. C. and Shub M. *Invariant Manifolds.* Lecture Notes in Mathematics **583**, New York, 1977.

2.33 Kantz H. and Grassberger P. Repellers, semi – attractors, and long – lived chaotic transients. *Physica* **17D** (1985) 75.

2.34 Kaplan J. L. and Yorke J. A. Chaotic behaviour of multidimensional difference equations. In *Functional Differential Equations and Approximation of Fixed Points*, ed. H. – O. Peitgen and H. – O. Walther, Lecture Notes in Mathematics **730**, Berlin, Springer, 1979, p. 204.

2.35 Kaplan J. L. and Yorke J. A. Preturbulence: A regime observed in a fluid flow model of Lorenz. *Commun. Math. Phys.* **67** (1979) 93.

2.36 Kifer Yu. I. On small random perturbations of some smooth dynamical systems. *Izv. Akad. Nauk SSSR, Ser. Mat.* **38** (1974) 1091 (in Russian).

2.37 Ledrappier F. Some relations between dimension and Liapunov exponents. *Commun. Math. Phys.* **81** (1981) 229.

2.38 Levy Y. E. Some remarks about computer studies of dynamical systems. *Phys. Lett.* **88A** (1982) 1.

2.39 Mallet – Paret J. Negatively invariant sets of compact maps – an extension of a theorem of Cartwright. *J. Diff. Eq.* **22** (1976) 331.

2.40 Mañé R. On the dimension of the compact invariant sets of certain nonlinear maps. In *Dynamical Systems and Turbulence.* Lecture Notes in Mathematics **898**, New York, Springer, 1981, p.230.

2.41 May R. M. Simple mathematical models with very complicated dynamics. *Nature* **261** (1976) 459.

2.42 Milnor J. On the concept of attractor. *Commun. Math. Phys.* **99** (1985) 177.

2.43 Oseledec V. I. A multiplicative ergodic theorem. Lyapunov characteristic numbers for dynamical systems. *Trudy Mosk. Mat. Obsc.* **19** (1968) 179 (in Russian).

2.44 Ott E., Yorke E. D. and Yorke J. A. A scaling law: How an attractor's volume depends on noise level. *Physica* **16D** (1985) 62.

2.45   Palis J. and de Melo W. *Geometric Theory of Dynamical Systems: An Introduction.* New York, Springer, 1982.
2.46   Renyi A. *Probability Theory.* Budapest, Academiai Kiado, 1970.
2.47   Ruelle D. Ergodic theory of differentiable dynamical systems. *Publ. Math. IHES* **50** (1979) 275.
2.48   Ruelle D. Measures describing a turbulent flow. *Ann. N. Y. Acad. Sci.* **357** (1980) 1.
2.49   Ruelle D. Strange attractors. *Math. Intelligencer* **2** (1980) 126 (*La Recherche* **108** (1980) 132).
2.50   Ruelle D. Small random perturbations of dynamical systems and the definition of attractors. *Commun. Math. Phys.* **82** (1981) 137.
2.51   Ruelle D. Characteristic exponents and invariant manifolds in Hilbert space. *Am. Math.* **115** (1982) 243.
2.52   Röhricht B., Metzler W., Parisi J., Peinke J., Beau W., Rössler O. E. The classes of fractals. In Springer Series in Synergetics **37**, Berlin, Springer, 1987, p. 275.
2.53   Shannon C. E. and Weaver W. *The Mathematical Theory of Information.* Urbana, Univ. of Illinois Press, 1949.
2.54   Shaw R. S. Strange attractors, chaotic behavior, and information flow. *Z. Naturforsch.* **36a** (1981) 80.
2.55   Shtern V. N. Arrangement and dimension of turbulent motion attractors. Preprint, Novosibirsk 1982.
2.56   Sinai Ya. G. *Introduction to Ergodic Theory.* Princeton, Princeton University Press, 1976.
2.57   Umberger D. K. and Farmer J. D. Fat fractals on the energy surface. *Phys. Rev. Lett.* **55** (1985) 661.
2.58   Walters P. Ergodic Theory — Introductory Lectures. Lecture Notes in Mathematics **458**, Berlin, Springer, 1975.
2.59   Yamaguchi Y. Analytical approach to the homoclinic tangency in a two–dimensional map. *Phys. Lett.* **133A** (1988) 201.
2.60   Yamaguchi Y. and Sakai K. Structure change of basins by crisis in a two–dimensional map. *Phys. Lett.* **131A** (1988) 499.
2.61   Young L.–S. Dimension, entropy and Liapunov exponents in differentiable dynamical systems. *Physica* **124A** (1984) 639.
2.62   Zeeman E. C. Stability of dynamical systems. *Nonlinearity* **1** (1988) 115.

3

Transition from order to chaos

3.1 Bifurcations of dynamical systems

Until now we have assumed that the dynamical system is given by fixed relations. However, real processes are often described by models which depend on variable parameters. The purpose of the modelling can often be formulated in such a way that we seek a set of parameter values for which the orbits generated by the dynamical system have required properties.

Properties of dynamical systems will generally vary with the change of parameters. Qualitative changes in phase portraits are studied by the bifurcation theory. Let us take a one-parameter system of differential equations

$$\dot{\mathbf{x}} = v_\alpha(\mathbf{x}), \qquad \mathbf{x} \in \mathsf{R}^n, \qquad \alpha \in \mathsf{R}^d \tag{3.1}$$

where α is a vector valued parameter. We are interested in situations where the structure of orbits in the state space will change qualitatively with the variation of α. Such a qualitative change is called a *bifurcation* and a critical value α_c, where the bifurcation occurs, is the *bifurcation value* of the parameter α. The dynamical system is *structurally unstable* at this parameter value[3.74]. We shall be particularly interested in the sequences of bifurcations which lead to successively more complicated behaviour of orbits in the state space, and, finally, to chaotic behaviour.

A large number of different kinds of changes of phase portraits exists and therefore it is difficult to classify them hierarchically. The simplest qualitative changes in the dynamics occur in the close neighbourhood of stationary points or periodic orbits and they are usually denoted as *local bifurcations*. On the other hand, *global bifurcations* include such changes in the structure of orbits, which occur at many locations of the state space simultaneously. The formation of chaotic sets is often connected with bifurcations of homoclinic orbits characterized by a nontransverse intersection of stable and unstable manifolds[1.22].

Local bifurcations can be classified according to a minimum dimension of the parameter space necessary for the bifurcation to occur as a typical (generic) case. Let us consider a stationary solution $\mathbf{x}(t) = \mathbf{z}$ of Eq. (3.1) for a fixed value α_0. This solution is represented by a point in the state space.

According to the implicit function theorem z is a function of α in a neighbourhood of α_0 as far as det $[D_x \, v_{\alpha_0}(z)] \neq 0$. The graph of this function forms a
d-dimensional hypersurface in the product space $R^n \times R^d$ of all pairs (x, α).

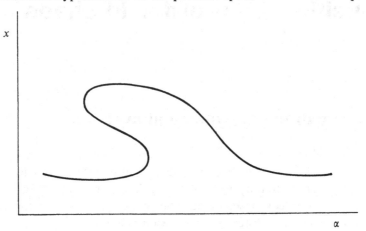

Fig. 3.1. Solution curve in the product space R × R.

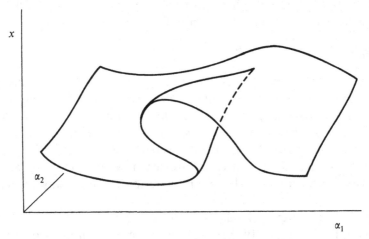

Fig. 3.2. Two-dimensional solution surface in the product space R × R².

Hence we can expect that stationary points of Eq. (3.1) for all values of α will
lie on d-dimensional surfaces in the product space, specifically on the curves if
$d = 1$ and on two-dimensional sheets if $d = 2$, cf. Figs 3.1 and 3.2.

The local bifurcation of the stationary point z occurs if the matrix $D_x v(x)|_{x=z}$
has some eigenvalues with zero real parts. Hence z will have a nonempty centre
manifold, see Chapter 2. If α_c is a bifurcation value of the parameter, then the
pair (z, α_c) is called a bifurcation point. All such points usualy form bifurcation
surfaces located in the d-dimensional surfaces of solutions and their dimension

is less than d. Various types of bifurcations differ according to how many eigenvalues have zero real parts and how the eigenvalues change as the pair $(\mathbf{z}, \boldsymbol{\alpha})$ passes through the bifurcation point. Exact formulation requires the study of transverse intersections of certain submanifolds of the space of all n by n matrices[1.22]. Here we shall only review results for the simplest cases.

Let us consider first one-dimensional parameter space. Local bifurcations occur in *generic cases* at isolated bifurcation points for isolated values of the parameter $\boldsymbol{\alpha}$. These bifurcations are characterized by the existence of either a unique zero real eigenvalue or a pair of purely imaginary eigenvalues of the matrix $D_{\mathbf{x}}v_\alpha$ at the bifurcation point $(\mathbf{z}, \boldsymbol{\alpha}_c)$; certain transversality conditions must be satisfied simultaneously (they are, generally, in the form of inequalities, e.g. the eigenvalue must cross the imaginary axis with a non-zero speed, etc.).

If the dimension of the parameter space is two, then the surface of stationary solutions is two-dimensional and the bifurcation points observed for $d = 1$ will be replaced by bifurcation curves. However, additional bifurcations that are not typical for $d = 1$ may occur at certain points along these curves. These bifurcations can occur when either the matrix $D_{\mathbf{x}}v_\alpha$ is singly degenerate and a special case of transversality conditions is satisfied or $D_{\mathbf{x}}v_\alpha$ is doubly degenerate with following possibilities:

(1) double zero eigenvalue;
(2) two pairs of purely imaginary eigenvalues;
(3) one zero eigenvalue and a pair of purely imaginary eigenvalues.

The number of possible types of bifurcations increases rapidly with the increasing dimension of the parameter space. It is useful to introduce the notion of *codimension of bifurcation*. It is the lowest dimension of a parameter space which is necessary to observe a given bifurcation phenomenon[1.22]. The bifurcations are organized hierarchically with increasing codimension. The case of codimension two includes in a natural way the case of codimension one. However, more degenerate points must be included. Local bifurcations of codimension one and two are treated, e.g. in Refs 1.22 and 3.9. This subject is also connected with *catastrophe theory*[3.67, 3.74, 3.78] and with more general *singularity theory*[3.32, A.10].

Local bifurcations can be studied using centre manifolds. If $\mathbf{z}$ is not hyperbolic then the centre manifold theorem guarantees that there exists an invariant manifold W^{c} tangent to the centre subspace E^{c} and its dimension n_{c} is equal to the number of eigenvalues of the matrix $D_{\mathbf{x}}v|_{\mathbf{x}=\mathbf{z}}$ having zero real parts, see Chapter 2. In contrast to the stable or unstable manifolds the centre manifold need not be determined uniquely and the linear approximation of the flow close to the nonhyperbolic point $\mathbf{z}$ is not sufficient. However, a knowledge of the dynamics in W^{c} enables us to study bifurcations in the neighbourhood of the bifurcation point $(\mathbf{z}, \boldsymbol{\alpha}_c)$.

The centre manifold of the point $\mathbf{z}$ with dimension n_{c} can be unfolded in the neighbourhood of $(\mathbf{z}, \boldsymbol{\alpha}_c)$ into the system of manifolds of dimension $n_{\mathrm{c}} + d$. It can be shown that all qualitative changes in the behaviour of orbits for small perturbations of the point $(\mathbf{z}, \boldsymbol{\alpha}_c)$ occur in this *unfolded centre manifold*. A systematic approach to the construction of canonical models of dynamics on unfolded centre manifolds in the form of a set of n_c differential equations with d parameters gives the *method of normal forms*[1.22, 2.4, 3.21]. Normal forms play an essential role in bifurcation theory because they provide the simplest system of equations that describe the dynamics of the original system close to bifurcation points. Basic information on normal forms including a list of relevant equations and typical phase portraits can be found in Appendix A.

The same method of approach can be used for periodic points of discrete time systems and via the Poincaré maps also for periodic orbits of systems with continuous time.

The importance of the study of local bifurcations for the description of chaotic behaviour stems from the fact that formation of chaotic sets is often accompanied by local bifurcations, which accumulate and form a 'prelude' to chaotic behaviour. Certain normal forms that describe local bifurcations of codimension two (and higher) directly include global bifurcations leading to chaotic sets.

The chaotic attractor just after its formation is often a low-dimensional object. When a control parameter is varied then further (global) bifurcations may occur and the dimension of the attractor may increase. For example, in hydrodynamical systems motion becomes more and more complex and, finally, a *developed turbulence* appears. This process, however, still cannot be described in the frame of bifurcation theory and thus research focuses on the study of the *onset of turbulence* (i.e. on a low dimensional chaos).

The evolution of chaotic behaviour is usually studied in dependence on a single parameter, but the phenomena observed in experiments can often be explained in a more appropriate way if the dimension of the parameter space is increased. Such investigations are difficult and challenging both from the experimental and theoretical points of view.

3.2 Local bifurcations and dependence of solutions on a parameter

We shall describe typical local bifurcations in the neighbourhood of stationary points or periodic orbits of a family of differential equations

$$\dot{\mathbf{x}} = v_{\alpha}(\mathbf{x}), \qquad \mathbf{x} \in \mathsf{R}^n, \qquad \boldsymbol{\alpha} \in \mathsf{R}^d . \tag{3.2}$$

The local bifurcations of periodic orbits in dynamical systems with discrete time described by a family of difference equations

$$x_{k+1} = f_\alpha(x_k), \qquad x \in R^n, \qquad \alpha \in R^d \tag{3.3}$$

will be also of interest. A dynamical system with continuous time can be transformed into a simpler system of the form of Eq. (3.3) by means of a properly chosen Poincaré section.

The number of different types of local bifurcations of a given codimension depends on the class of the considered vector fields v (or of the mappings f). Model equations often possess various symmetries or other special properties specific to the studied problem. If we consider the simplest symmetries and special properties, then there exist four typical local bifurcations of stationary points of codimension one, see Appendix A for their normal forms. Here we limit ourselves to a qualitative description. Figs 3.3–3.6 show the dependence of a chosen norm of the bifurcating solutions on the scalar parameter α. Such dependences are called *solution diagrams*[4.41]. The cases in Figs 3.3–3.5 correspond to the passing of a real eigenvalue through the imaginary axis; only the first case is not connected with special properties or with symmetry of the vector field.

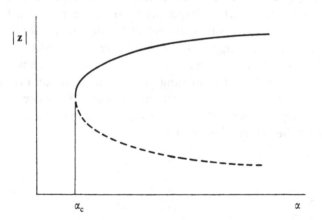

Fig. 3.3. Saddle-node bifurcation.

(1) *Saddle-node bifurcation* – Fig. 3.3. The bifurcation point, called a limit or a turning point arises by joining two branches of the curve of solutions; solutions on one branch are locally unstable (saddles), the second branch consists of locally stable solutions which are approached by nonoscillating orbits (nodes). Let us mention that the stability of stationary points on the curve of solutions is considered here only with respect to the orbits inside the unfolded centre manifold (which has in this case dimension $n_c + d = 2$). Hence, if the unstable

manifold exists at the bifurcation point (i.e. some eigenvalues lie to the right of the imaginary axis), then all solutions in the neighbourhood of the bifurcation point are unstable. This holds also for all following cases.

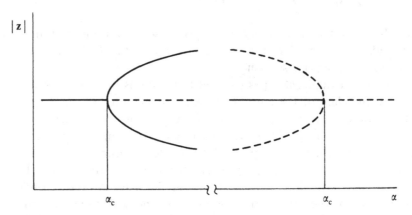

Fig. 3.4. Symmetry breaking bifurcation; supercritical (left) and subcritical (right) cases are shown.

(2) *Symmetry breaking bifurcation* – Fig. 3.4. This case arises in systems which possess symmetry. As an example of a simple symmetry we can take a two-dimensional dynamical system invariant with respect to the coordinate transformation $(x, y) \rightarrow (y, x)$. The original solution changes its stability at the bifurcation point and a pair of new, mutually symmetric solutions arises. Two situations differing in the direction of branching of new solutions can occur, a *subcritical* bifurcation (the bifurcating pair of solutions is unstable close to the bifurcation point) and a *supercritical* bifurcation (the bifurcating pair is stable in the neighbourhood of the bifurcation point).

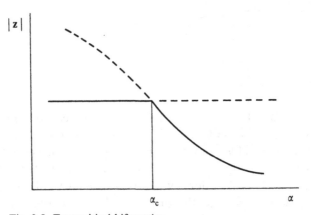

Fig. 3.5. Transcritical bifurcation.

(3) *Transcritical bifurcation* – Fig. 3.5. The third type of bifurcation occurs in systems that admit the existence of an elementary solution which is independent of λ. If the bifurcation occurs then the new branch exists for values both above and below the bifurcation value α_c and the exchange of stability occurs on the curves of solutions.

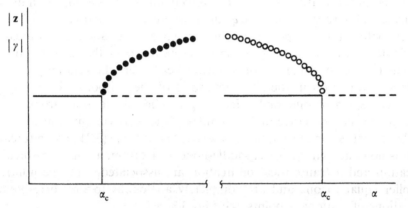

Fig. 3.6. Hopf bifurcation; supercritical (left) and subcritical (right) cases are shown.

(4) *Hopf bifurcation* – Fig. 3.6. The last case of bifurcation of codimension one from the stationary point is connected with the pair of purely imaginary eigenvalues $i\omega$, $-i\omega$. The family of periodic orbits branches off the curve of stationary solutions[3.43], which changes stability at the bifurcation point. The period at the bifurcation point is $T = 2\pi/\omega$ and the amplitude of periodic orbits approximately increases as $|\alpha - \alpha_c|^{1/2}$ in the neighbourhood of the bifurcation point, see Fig. 3.7. Both subcritical and supercritical bifurcations can occur.

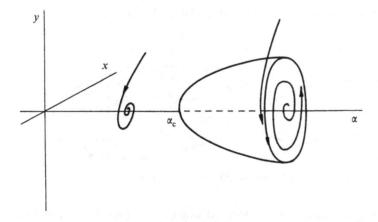

Fig. 3.7. Hopf bifurcation in the product space $R^2 \times R$. The one-parameter family of periodic orbits lies on a paraboloid.

There exists extensive literature dealing with the problem of Hopf bifurcation, including infinite-dimensional systems and numerical computations[3.40, 3.56].

Hopf bifurcation couples stationary and periodic behaviour. One-parameter systems of periodic orbits of Eq. (3.2) form 'tubes' of variable diameter in the (x, α) product space. As is shown in Fig. 3.7 such a 'tube' can arise just at the point of Hopf bifurcation where it has zero diameter. In systems with discrete time instead of a 'tube' we observe q curves where q is the period of the followed orbit. As with stationary points, parameter dependences and local bifurcations of periodic solutions can be graphically represented in the form of solution diagrams (when we assign a proper norm to each periodic solution).

A local bifurcation in the neighbourhood of the periodic orbit γ with period T occurs, when some multipliers – eigenvalues of the monodromy matrix $U(T)$ – lie on the unit circle in the complex plane. A trivial multiplier equal to one always exists for autonomous systems such as Eq. (3.2). The bifurcation 'point' is now the pair (γ, α_c). A saddle-node bifurcation, a symmetry-breaking bifurcation and a transcritical bifurcation are associated with one nontrivial multiplier equal to one and are completely analogous to the corresponding bicurcations of stationary points, see Figs 3.3–3.5.

The case of two complex conjugate multipliers passing through the unit circle, however, is more complicated than the analogous Hopf bifurcation from the stationary point. If two complex conjugate multipliers σ, $\bar{\sigma}$, passing through the unit circle with a nonzero speed fulfil the condition

$$|\sigma|^m = |\bar{\sigma}|^m \neq 1 , \qquad m = 1, 2, 3, 4,$$

at the bifurcation point, then in the neighbourhood of the bifurcation point (γ, α_c) a system of two-dimensional invariant tori arises, see Fig. 3.8 (in the discrete case it is a system of closed curves). As in the case of the Hopf

$$\alpha < \alpha_c \qquad\qquad\qquad \alpha > \alpha_c$$

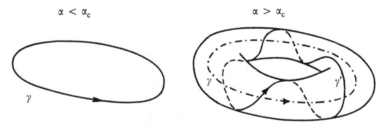

Fig. 3.8. Torus bifurcation; the bifurcated torus contains a periodic orbit.

bifurcation the dimension of the formed object is higher by 1. The tori are attracting in the supercritical case and repelling in the subcritical one; the corresponding solution diagram is similar to that in Fig. 3.6.

Depending on the parameter α, the bifurcated tori can contain either quasiperiodic orbits which cover the torus densely or periodic orbits[3.46]. The quasiperiodic and periodic dynamics can be described by means of a *rotation number* which is generally a nontrivial function of the parameter[3.41] as will be described in the Sect. 3.3.6. The case of strong resonance, where the condition for σ is not satisfied, is described, for example in Ref. 3.46.

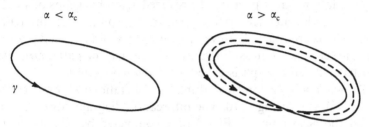

Fig. 3.9. Period-doubling bifurcation.

There exists still another type of bifurcation, which has no analogy in bifurcations of stationary points. It is the case where a single multiplier passes through -1. In this case a new system of periodic orbits with double period at the bifurcation point branches off the original periodic orbit, see Fig. 3.9. In the discrete case a system of orbits with period $2q$ branches off the system of orbits with period q. Again two cases of the change of stability can occur and the

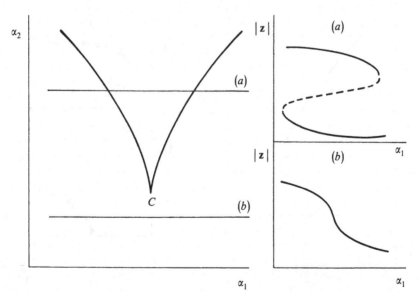

Fig. 3.10. Cusp bifurcation; corresponding solution diagrams along the straight lines (a) and (b) are shown in the right part of the figure; full lines represent stable stationary points and dashed lines unstable ones.

corresponding solution diagram is similar to that in Fig. 3.6. The *period-dou-bling bifurcation* often occurs repeatedly ad infinitum and accompanies a forma-tion of chaotic attractors.

The description of local bifurcations of codimension two is substantially more complicated and often includes the description of bifurcations which have nonlocal character[1.22]. We shall limit ourselves to several simple examples. A graphical representation can in principle be realized again by means of solution diagrams, i.e. by the dependence of the norm of the solution on the pair of parameters α_1, α_2. This approach, however, requires three-dimensional plots and therefore only the set of all bifurcation values $(\alpha_{1c}, \alpha_{2c})$ in the parametric plane is usually constructed. This graph is called a *bifurcation diagram*[4.41]

The simplest local codimension two bifurcation of stationary points called a *cusp bifurcation* is shown in Fig. 3.10. The bifurcation diagram consists of two curves joined at the cusp point C. Fig. 3.10 is completed by several solution diagrams for constant values of one of the parameters. Three stationary points exist inside the cusp region, two of them are stable and one unstable; only one stationary point exists outside. The system (3.2) linearized at the point C has a simple zero eigenvalue, but the corresponding normal form is degenerate in the higher order terms, see Appendix A. An analogous bifurcation occurs also for periodic solutions.

The second example generalizes Hopf bifurcation and describes the change of the supercritical bifurcation into a subcritical one, see Fig. 3.11. Analogous

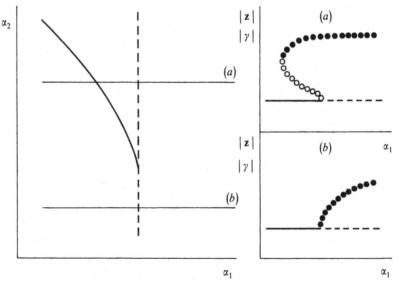

Fig. 3.11. Codimension two Hopf bifurcation; corresponding solution diagrams contain both stationary points (lines) and periodic orbits (circles: solid circles – stable periodic orbits; open circles – unstable periodic orbits).

bifurcation diagrams have all types of bifurcations where the supercritical bifurcation changes into a subcritical one, i.e. symmetry breaking bifurcations both on stationary and periodic solutions and period doubling and tori bifurcations.

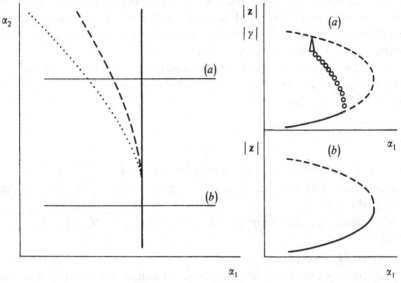

Fig. 3.12. Bifurcation with double zero eigenvalue at the cusp point. The dotted line in the bifurcation diagram corresponds to a family of homoclinic orbits (see the triangle in the solution diagram (a)). The dashed line corresponds to the Hopf bifurcation.

The last example we shall mention here is a bifurcation from the stationary point with a double zero eigenvalue[3.8], see Fig. 3.12. The periodic orbits arising via Hopf bifurcation disappear after collision with the stationary point, i.e. at a homoclinic orbit (a saddle-loop, see Fig. 2.4). This bifurcation does not occur in the small neighbourhood of the stationary point only and can therefore be understood as a global one.

3.3 Bifurcations and chaos in one-dimensional maps

It is well known that one-dimensional discrete dynamical systems can possess chaotic behaviour if they are noninvertible. On the other hand, dynamical systems connected with differential equations are invertible in time, hence there is not an immediate relation to one-dimensional noninvertible mappings. However, sufficiently strong dissipation causes such contraction of volumes in the state space that an approximation by means of one-dimensional mappings is often satisfactory. This is possible particularly in cases where the chaotic attractor generated by a continuous time system has a sheet-like structure with only small width, i.e. its fractal dimension is only slightly larger than two. In this case the Poincaré section through the attractor is an approximately one-dimensional curve. When reducing the problem to a one-dimensional map, we have to sacrifice the fine Cantor set-like structure of the chaotic attractor observed in the Poincaré cross-section. This simplification is reflected in the noninvertibility of the one-dimensional model. Moreover, discrete maps or difference equations arise naturally as models of many real processes[3.68].

We shall now describe several simple one-dimensional maps whose iterates yield chaotic behaviour. We are interested in the asymptotic behaviour of orbits of the difference equation

$$x_{k+1} = f_a(x_k) \,, \tag{3.4}$$

where the corresponding state space may be either an interval I on the real line R, or a circle S^1. The function f_a is dependent on the parameter a (generally multidimensional).

Let us consider first the mapping of the interval into itself, $f: I \rightarrow I$, $I = [0, 1]$ for which

(1) $f(0) = f(1) = 0$,
(2) f has only one extreme, let us say a maximum, at the point c located inside I (c is called a *'critical point'*)
(3) f is strictly increasing on $[0, c)$ and strictly decreasing on $(c, 1]$.
    Such a map is called a *unimodal map*. If, moreover,
(4) $f'(0) > 1$ and f has negative *Schwarzian derivative*, i.e. f is three times differentiable and

$$S(f) = \frac{f'''(x)}{f'(x)} - \frac{3}{2}\left(\frac{f''(x)}{f'(x)}\right)^2 < 0 \tag{3.5}$$

for all $x \in I$ such that $f'(x) \neq 0$, then f has at most one attracting periodic orbit. More generally, if the mapping $f: I \rightarrow I$ is m-modal on I and $S(f) < 0$ outside the m critical points, then f has at most m attracting periodic orbits[3.72]. It can be expected that when the conditions $(1) - (4)$ are satisfied, then f has just one attractor in the interval I.

Let us consider now a one-parameter system of mappings, $f_a: I \rightarrow I$, satisfying conditions $(1) - (4)$ for all a from a certain interval. The bifurcation structure connected with the change of the parameter a in principle does not depend on the concrete chosen one-parameter system. As an example, let us consider the already mentioned logistic difference equation (Eq. (2.16))

$$x_{k+1} = f_a(x_k) = a x_k(1 - x_k) \,, \tag{3.6}$$

which maps the unit interval into itself for $a \in [0, 4]$. The critical point for the logistic function is $c = 1/2$; the graph of the function $f_a(x)$ is given in Fig. 3.13 for several values of the parameter a. Any orbit can be constructed by means of

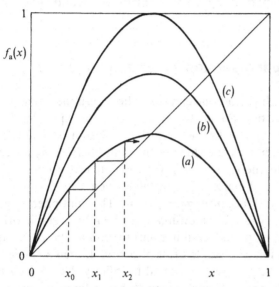

Fig. 3.13. The graph of the logistic function: curve (a) $a = 2$; curve (b) $a = 3$; curve (c) $a = 4$. The construction of an orbit is shown for $a = 2$.

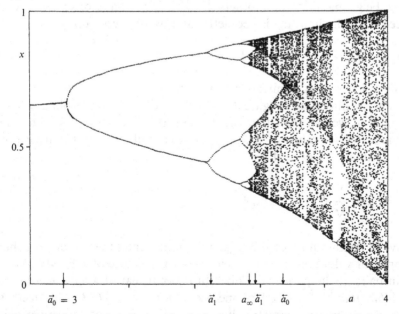

Fig. 3.14. Asymptotic behaviour for the logistic map in the region of period-doubling bifurcations and chaos.

the diagonal, as it is shown in the figure. The state space is one-dimensional and therefore the solution diagram can be constructed in such a way that we directly plot a numerically generated asymptotic orbit dependent on the parameter a, see Fig. 3.14. This is a magnified part of Fig. 2.8 showing the bifurcation phenomena and chaos.

3.3.1 Period-doubling sequences and universality

The simplest orbits with period one can easily be determined analytically. When $a > 1$, then the logistic map has two fixed points (or, equivalently, two different orbits of period 1), $x_{f1} = 0$ and $x_{f2} = 1 - 1/a$. Local stability of the periodic orbits is determined by the value of the multipliers σ_1, σ_2 which are in this simple case equal to the derivative $f'_a(x)$ evaluated at points x_{f1}, x_{f2}, respectively. While $\sigma_1 = a$ and hence x_{f1} is unstable when $a > 1$, the second point x_{f2} is stable for $a \in (1, 3)$ because $\sigma_2 = 2 - a$. The multiplier σ_2 equals to -1 at $a = 3$ and a period-doubling bifurcation generates a stable orbit of period 2, see Fig. 3.14. A new period doubling bifurcation at the point $\check{a}_1 \approx$ ≈ 3.45 leads to the formation of a stable orbit of period 4, then a stable orbit of period 8 bifurcates and the process is repeated to infinity. The sequence $\{\check{a}_i\}$ of the bifurcation values of the parameter approaches a finite value $a_\infty \approx$ $\approx 3.570^{3.23}$. This is a famous cascade of *period-doubling bifurcations* studied in detail by Feigenbaum[3.23] and others[3.12, 3.18, 3.50, 3.76]. The convergence of the sequence $\{\check{a}_i\}$ to the limit a_∞ is geometric and we can write down

$$(a_\infty - \check{a}_i) \sim \delta^{-i}, \qquad i \gg 1 \tag{3.7}$$

where $\delta = 4.669 \dots$ is a universal constant.

Now we are interested in global features of the period-doubling sequence. From Fig. 3.14 we can infer that the dynamical behaviour of Eq. (3.6) will be very complicated for values of $a > a_\infty$. More careful numerical analysis reveals that the Lyapunov exponent

$$\lambda = \lim_{N \to \infty} \frac{1}{N} \sum_{k=0}^{N} \log |f'_a(x_k)|$$

is positive for many values of $a > a_\infty$ and thus indicates the existence of chaotic attractors. For values of $a > a_\infty$ there exists a structure of 2^i bands which are joined in pairs at the points $\check{a}_i$. This bifurcation is called a *band merging* or a *reverse bifurcation*[3.51, 3.52], the last one occurs for $\check{a}_0 \approx 3.678$. The decreasing sequence $\{\check{a}_i\}$ converges geometrically to a_∞ with the rate of convergence given by δ, in a way analogous to the period-doubling sequence. In fact, these two complementary phenomena occur repeatedly in a complicated hierarchy in the interval $[a_\infty, 4]$ and cause the complex structure observed in Fig. 3.14.

The mechanism of the period-doubling (and of the band merging, respectively) is general and it is best observable when studying the q-th iterate $f_a^q(x)$, where

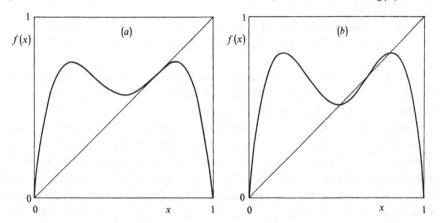

Fig. 3.15. The graph of the second iterate of the logistic function, (a) at, and (b) slightly above the critical point $\check{a}_0 = 3$; an orbit with double period is created.

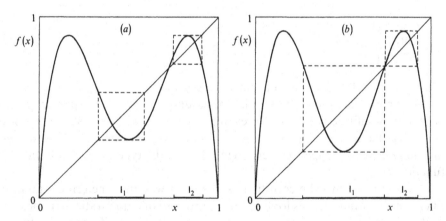

Fig. 3.16. The graph of the second iterate of the logistic function, (a) slightly below, and (b) at the critical point $\bar{a}_0 \approx 3.678$; invariant intervals l_1 and l_2 merge.

q is the period of the orbit created by the period-doubling. For example, bifurcations at the points $\check{a}_0$ and $\bar{a}_0$ are well observable in the graph of the function $f_a^2(x)$, see Figs. 3.15(a), (b), and 3.16 (a), (b), respectively. In the case of the period-doubling the graph of $f_{\check{a}_0}^2(x)$ has an inflection point at $x = x_{f2}$, see Fig. 3.15(a). When a exceeds the critical value $\check{a}_0$, two fixed points of $f_a^2(x)$ in the neighbourhood of the now unstable point x_{f2} appear, see Fig. 3.15(b); these two points form a period two orbit of $f_a(x)$. The situation is analogous at the band merging point $\check{a}_0$, only the points are substituted by the intervals I_1, I_2. These intervals are invariant under $f_{\check{a}}^2(x)$. The end points of I_1 (I_2) are given by the first and second iterate of the critical point contained in $I_1(I_2)$. When a is slightly less than $\check{a}_0$, then I_1 and I_2 are disjoint and the unstable point x_{f2} lies between them, see Fig. 3.16(a).

When $a = \check{a}_0$, then critical points contained in I_1 and I_2 are mapped into x_{f2} after two iterations of $f_a^2(x)$; x_{f2} is the common point of I_1 and I_2, see Fig. 3.16(b).

The rate of convergence of the period doublings and of the band mergings given by the constant δ is independent of the details of a one-parameter system of unimodal maps with a negative Schwarzian derivative[3.12].

Another important universal constant α appears when we follow scaling properties of periodic orbits of period 2^i in the state space. It is convenient to follow an orbit of period 2^i such that one point of the orbit coincides with the critical point c. Such an orbit is called *superstable* and its multiplier is equal to zero. Let us denote $\hat{a}_i$ the corresponding value of the parameter a; just one such point exists in the interval ($\hat{a}_i$, $\hat{a}_{i+1}$). Let d_i denote the distance between the critical point c and the point on the superstable orbit of period 2^i which is closest to c. Then

$$\lim_{i \to \infty} \frac{d_i}{d_{i+1}} = \alpha \tag{3.8}$$

where $\alpha = 2.502\,9...$ is also a universal constant.

Feigenbaum has shown that under the assumption that the sequence of period doublings is infinite (which is, for example, guaranteed by the above conditions $(1)-(4)$) the universal properties are given by the existence of a *universal function* $g(x)$ with properties dependent only on the type of critical point of the function $f(x)$.

It is useful to shift the coordinates in such a way that the critical point c of $f(x)$ is at the origin. Let us look at the typical case of a quadratic extreme, where in the neighbourhood of the critical point $c = 0$ we can describe the function $f(x)$ by means of the quadratic parabola

$$f(x) \approx |x|^z, \quad z = 2 \,.$$

It was demonstrated first heuristically[3.23] and then rigorously[3.12, 3.50] that the universal function $g(x)$ is a fixed point of a *doubling operator* T, which is defined as

$$T(f(x)) = -\alpha f^2(-x/\alpha) \,, \tag{3.9}$$

where α is the scaling parameter from (3.8). Hence

$$g(x) = T(g(x)) = -\alpha g^2(-x/\alpha) . \qquad (3.10)$$

It follows from Eq. (3.10) that $\beta g(x/\beta)$, $\beta \neq 1$, is also a fixed point of T and thus the universal function does not have an absolute scale. A conventional choice of scale is

$$g(0) = 1 , \qquad (3.11)$$

which implies

$$\alpha = -g(0)/g^2(0) = -1/g^2(0) . \qquad (3.12)$$

T operates in a suitably chosen functional space and can be seen as a discrete dynamical system with an infinite-dimensional state space. Repeated iterations of T generate a sequence of functions $\{[f(x)]_k\}$ with the initial state $[f(x)]_0$. The action of T implies two iterations of f and the change of coordinates $x \to -x/\alpha$.

We are now interested in initial states of orbits $\{[f(x)]_k\}$ converging to $g(x)$. It can be shown that they are just those functions, which correspond to accumulation points in one-parameter systems $f_a(x)$. For example, subsequent iterates $T^k(f_{a_\infty}(x))$ of the logistic equation at $a = a_\infty$ lead to $g(x)$. However, this function is not normalized with respect to condition (3.11). To normalize it, we

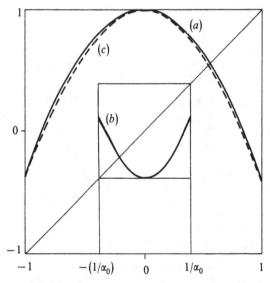

Fig. 3.17. Logistic function at $a_\infty \approx 3.570$ under a doubling operator: curve (a) the graph of the original (normalized) function $f_{a_\infty}(x)$; curve (b) the graph of $f^2_{a_\infty}(-x/\alpha_0)$; curve (c) the graph of $-\alpha_0 f^2_{a_\infty}(-x/\alpha_0)$.

can at every iteration of T rescale $[f(x)]_k$, $k = 0, 1, 2, ...$, in such a way that $[f(0)]_k = 1$. This means that α will depend on $[f(x)]_k$ according to

$$\alpha_k = 1/[f^2(0)]_k, \qquad k = 0, 1, 2 ... \tag{3.13}$$

and $\alpha_k \to \alpha$. This procedure is shown for $k = 0, 1$ in Fig. 3.17. The similarity of the functions before and after the rescaling is already evident after the first iteration.

However, this iteration scheme is not stable. When the initial function is not chosen exactly at the accumulation point a_∞, then the subsequent iterations $[f(x)]_k$ are finally repelled from $g(x)$, i.e. $g(x)$ behaves like a saddle point. The stable manifold of $g(x)$ has codimension one and the complementary unstable manifold is one-dimensional[3.23].

The local dynamics on the unstable manifold close to $g(x)$ defines the universal number δ. Let us consider a one-parameter system $f_a(x)$ and the sequence of functions $\{f_{\hat{a}_i}(x)\}$ which have a superstable cycle of period 2^i at $a = \hat{a}_i$. A sequence of universal functions $g_r(x)$ is defined by

$$g_r(x) = \lim_{k \to \infty} (-\alpha)^k f_{\hat{a}_{k+r}}^{2^k}(x/(-\alpha)^k), \qquad r = 0, 1, ... \tag{3.14}$$

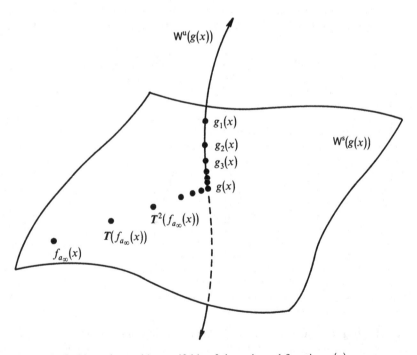

Fig. 3.18. Stable and unstable manifolds of the universal function $g(x)$.

It is evident that the limit of this sequence is just the universal function $g(x)$

$$\lim_{r \to \infty} g_r(x) = g(x) . \tag{3.15}$$

It can be shown that

$$g_{r-1}(x) = T(g_r(x)) \tag{3.16}$$

and thus for sufficiently large r the functions $g_r(x)$ are repelled from T with a speed that is just equal to δ; functions $g_r(x)$ lie in the unstable manifold of $g(x)$ and δ is the only eigenvalue of the linearization of T which has modulus greater than one. The situation is schematically shown in Fig. 3.18.

There is a large number of universal features associated with the existence of the universal function $g(x)$. Let us mention one aspect which forms an important connection with experimental observations, i.e. power spectra.

A periodic orbit $\{x_1^p, x_2^p, ..., x_q^p\}$ with period $q = 2^p$ can be represented by its Fourier expansion as

$$x_k^p = \sum_j A_j^p \exp (i\omega_j k) , \tag{3.17}$$

where the frequency $\omega_j = 2\pi_j/2^p$, $j = 0, 1, ..., 2^{p-1}$. Thus the power spectrum will consist of δ-functions at the frequencies ω_j with amplitudes determined by

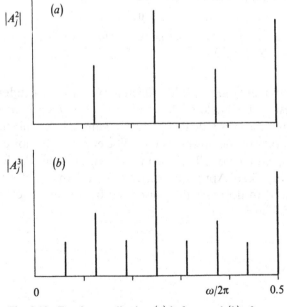

Fig. 3.19. Fourier amplitudes, (a) before, and (b) after a period-doubling bifurcation.

$|A_j^p|^2$. The number of components of the spectrum will double after each period doubling, $p \to p + 1$; new components will appear at odd multiples of the subharmonic frequency $\omega_j/2^{p+1}$, $j = 1, 3, 5, \ldots$. The old odd components will become even harmonics after the bifurcation and remain practically unchanged, see Fig. 3.19. Feigenbaum[3.24, 3.25] has shown that averaged odd components which appear after the period doubling are approximately equal to odd components before the period-doubling bifurcation, multiplied by the factor μ^{-1}:

$$|A_{\text{odd}}^{p+1}| \approx \mu^{-1}|A_{\text{odd}}^p| ,$$

where $\mu = 4\alpha(2(1 + 1/\alpha^2))^{-1/2}$. Using the known value of α, see Eq. (3.8), we have $\mu \approx 6.6$. The amplitudes of the power spectrum are proportional to $|A_j^p|^2$ and their average drop is then 10 log $\mu^2 \approx 16.4$ dB.

Another way of averaging the amplitudes of the power spectrum gives the value 13.5 dB and holds for the low-frequency components of the spectrum[3.59].

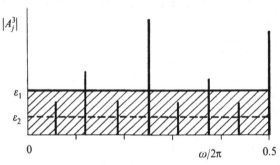

Fig. 3.20. The effect of noise on the period-doubling. New subharmonics emerge after a decrease of the noise level ε by the factor $\varkappa$.

The experimental measurements are usually affected by a noise of different origin. Because the smallest length scale of doubled orbits becomes finer with increasing period, a noise with a subsequently decreasing amplitude will suffice to wash out the above structure of the power spectra. The effect of the noise will manifest itself by the disappearance of all subharmonics with amplitudes lower than that of the noise, see Fig. 3.20. Approximately, to observe the next period-doubling bifurcation we need to decrease the noise level by a value given by a universal constant[3.14, 3.15, 3.16] $\varkappa \approx 6.65$.

3.3.2 The U-sequence

One parameter systems of unimodal mappings with negative Schwarzian derivative such as Eq. (3.6), exhibit an ordered structure of superstable periodic orbits called a *universal sequence* or *U-sequence*. The superstable cycles can best be described by methods of symbolic dynamics. The state space $I = [0, 1]$ of a unimodal mapping f satisfying the condition (3.5) can be divided into three parts: $I_L = [0, c), c, I_R = (c, 1]$ ($c = 1/2$ for the logistic map). Any orbit generated by f can be in principle described by a sequence of symbols assigned to I_L, c, I_R; we consider, for example, symbols $L, 0, R$, respectively. The method of symbolic dynamics applied to unimodal mappings was fully developed by Milnor and Thurston[3.58], see also Refs. 3.38 and 1.22.

The superstable periodic orbit always contains the point c. If its period is p, then the orbit is uniquely determined by a sequence consisting of the symbols R and L attached to the $p - 1$ successive iterates of c. The first and the second iterate of c must lie in I_R and I_L, respectively, and hence the first two symbols are RL; the entire symbolic sequence can be written as $RL^{p_1} R^{p_2} L^{p_3}$ ... where p_1 denotes the number of successive iterates falling in I_L after the first iteration, etc.

Table 3.1 U-sequence in the logistic mapping

	p	Pattern	a
1	2	R	3.236 068 0
2	4	RLR	3.498 561 7
3	6	RLR^3	3.627 557 5
4	7	RLR^3	3.701 769 2
5	5	RLR^2	3.738 914 9
6	7	RLR^2LR	3.774 214 2
7	3	RL	3.831 874 1
8	6	RL^2RL	3.844 568 8
9	7	RL^2RLR	3.886 045 9
10	5	RL^2R	3.905 706 5
11	7	RL^2R^3	3.922 193 4
12	6	RL^2R^2	3.937 536 4
13	7	RL^2R^2L	3.951 032 2
14	4	RL^2	3.960 270 1
15	7	RL^3RL	3.968 976 9
16	6	RL^3R	3.977 766 4
17	7	RL^3R^2	3.984 747 6
18	5	RL^3	3.990 267 0
19	7	RL^4R	3.994 537 8
20	6	RL^4	3.997 583 1
21	7	RL^5	3.999 397 1

Let us consider now a one-parameter system of unimodal mappings which satisfy condition (3.5), $f_a: I \to I$, $a \in [a_-, a_+]$ such that

$$f_{a_-}(c) = c \quad \text{and} \quad f_{a_+}(c) = 1 . \tag{3.18}$$

In the case of the logistic map this condition is satisfied for $a \in [2, 4]$. A superstable cycle with a given symbolic pattern will occur for a certain value of a. Metropolis, Stein and Stein[3.57] were the first who observed that all superstable orbits with the period p, $1 < p \leq q$ (q arbitrary) form a sequence ordered along the parameter axis with respect to the symbolic representation of orbits. This sequence is called the U-sequence.

An example of the U-sequence for the logistic mapping and $q = 7$ is given in Table 3.1. When the value of q increases, then new terms are inserted between the original terms of the sequence, this procedure can be defined algorithmically. Combinatorics enables us to enumerate all admissible symbolic sequences of the same length and thus to determine the number of superstable orbits of a given period. These numbers for $q \leq 15$ are given in Table 3.2. The properties of the

Table 3.2 Number of superstable orbits in the logistic mapping for the period q

q	2	3	4	5	6	7	8	9	10	11	12	13	14	15
Number of solutions		1	1	2	3	5	9	16	28	51	93	170	315	585	1 091

U-sequence and corresponding symbolic dynamics were further developed in Refs. 3.18, 3.38, 3.58. There is a connection between the period-doubling sequences and the U-sequence. Every term of the U-sequence is a superstable periodic orbit with the multiplier $\sigma = 0$. Continuation of any superstable cycle to the right along the parameter axis leads to the passing of σ through -1, i.e. the period-doubling occurs. When we continue in the opposite direction, σ approaches 1 and two cases can occur:

(1) the curve of orbits terminates at the period-doubling bifurcation point; a new branch of periodic orbits with half period exists and is stable to the left of the bifurcation value of the parameter,

(2) the curve of orbits passes through the limit point, where the orbits change their stability.

The entire picture is shown in Fig. 3.21. The orbit with the basic period q arises via a saddle-node bifurcation and both branches exist to the right of the critical value of the parameter. The stable branch loses its stability at the period-doubling bifurcation point and a new stable orbit with the period $2q$ arises, this orbit again undergoes the period-doubling, until an accumulation point is reached. The saddle-node and the accumulation point define an interval on the parameter axis called a *q-window* corresponding to periodic attractors that subsequently arise with period $q \times 2^i$, $i = 0, 1, \ldots$, from the left to the right.

When we increase the value of the parameter from $a = a_-$ then at first the basic sequence of period doubling bifurcations ending at $a = a_\infty$ occurs and

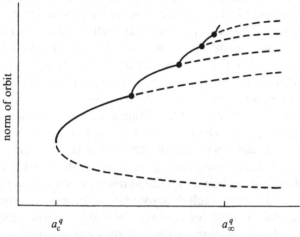

Fig. 3.21. Structure of periodic orbits in a window.

only after that the saddle-nodes and the associated period-doublings appear. The sequences of period doublings in each window have universal properties connected with the existence of the universal function $g(x)$. The U-sequence implies that an infinite number of (nonoverlapping) windows exists in the interval $[a_\infty, a_+]$. Numerical computations show that the width of the window decreases very fast with the increasing basic period q.

The U-sequence in principle determines all periodic orbits and their order of appearance depending on increasing a. A different order is given by the well-known theorem by Sharkovskii[3.69, 3.73]. Consider the following ordering of all positive integers

$$1 \prec 2 \prec 4 \prec \ldots \prec 2^k \prec \ldots$$

$$\prec 2^k(2m + 1) \prec 2^k(2m - 1) \prec \ldots \prec 2^k . 5 \prec 2^k . 3 \prec \ldots$$

$$\prec 2(2m + 1) \prec 2(2m - 1) \prec \ldots \prec 2 . 5 \prec 2 . 3 \prec \ldots$$

$$\prec 2m + 1 \prec 2m - 1 \prec \ldots \prec 5 \prec 3 .$$

If f is a continuous map of an interval to itself with a periodic orbit of period p and $q \prec p$ in this ordering, then f has a periodic orbit of period q.

When applied to unimodal family f_a with negative Schwarzian derivative, the theorem gives the order of appearance of orbits with distinct periods as a is increased.

3.3.3 Chaotic behaviour

We can ask now whether a one-parameter family f_a exhibiting the U-sequence also contains mappings generating chaotic orbits and chaotic attractors. The direct consequence of the Sharkovkii theorem is that if a continuous mapping f of the interval to itself has a periodic orbit with period different from 2^i, then f has positive topological entropy. This fact includes the well known statement 'period three implies chaos' of Li and Yorke[1.42] (see also Ref. 3.63). This in principle means that even if there exists an attracting periodic orbit of period $\neq 2^i$, then there also exist orbits which are asymptotically aperiodic and exponentially unstable. However, ω-limit sets of such orbits need not be attractors. For example, the logistic map may have chaotic attractors for parameter values within the interval $(a_\infty, a_+]$, where $a_\infty \approx 3.57$ and $a_+ = 4$, as there are orbits of period $\neq 2^i$. In fact, the periodic windows form a dense set in $(a_\infty, a_+]$ and parameter values corresponding to chaotic attractors form a Cantor set[3.47]. Chaotic attractors of one-dimensional maps do not have complex geometric structure as is usual in multidimensional state spaces. Guckenheimer[3.39] shows that the chaotic attractor is simply formed by a finite union of closed intervals. It has both information and fractal dimensions equal to one, but its Lyapunov exponent and metric entropy are positive. Grassberger[3.34] has shown numerically in several examples that nonattracting chaotic sets coexisting with periodic and/or chaotic attractors satisfy the inequalities $D_I < D_F < 1$. Hence these chaotic 'repellors' are Cantor sets.

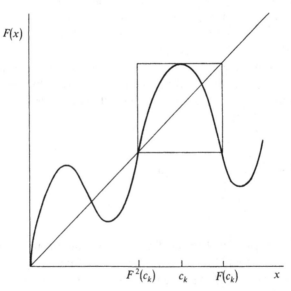

Fig. 3.22. The invariant box structure of the graph of $F(x) = f^q(x)$ indicating the presence of a chaotic attractor.

We shall describe the structure of chaotic attractors on the example of the logistic equation (3.6). A common feature of all chaotic attractors is that they are formed by q intervals $I_1, ..., I_q$ which are invariant with respect to f_a^q for a certain value of a. Each interval I_k contains just one critical point $c_k = f^{k-1}(c_1)$, $k = 1, ..., q$ where $c_q \equiv c$ is the critical point of f, see Fig. 3.22. The first and second iterates, $f^q(c_k), f^{2q}(c_k)$, are the endpoints of I_k and $\{f^{2q}(c_1), ..., f^{2q}(c_q)\}$ is an unstable periodic orbit of period q (or of period $q/2$, see below).

The most simple case occurs when $a = 4$, where $I = [0, 1]$ forms the chaotic attractor. Here $q = 1$, $c_1 = c = \frac{1}{2}$ and $f(f(\frac{1}{2})) = x_{f1} = 0$ is an unstable periodic point of period 1. In fact, the dynamics of the mapping f_a for $a = 4$ is equivalent to the dynamics of the Bernoulli shift mapping $2x$ (mod 1), described in Section 2.5. In particular, both mappings have the same topological and metric entropy, $h = h_\mu = \log 2$, which is positive and ensures that almost every orbit is chaotic in the sense of the algebraic complexity of the orbit. The corresponding power spectrum then contains a continuous broad-band noise.

A similar situation is met whenever the chaotic attractor consists of q intervals, $q > 1$. However, here an additional periodic structure will occur, because though a typical orbit fills out each of the q intervals completely irregularly, successive iterates visit them periodically. This property gives the result that the power spectrum will contain periodic components (δ-functions) on the frequency $1/k$ and its harmonics in addition to broad-band noise. This phenomenon is also called *periodic chaos*.

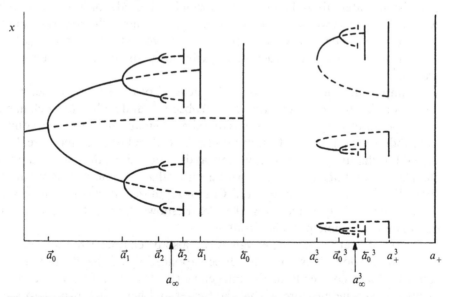

Fig. 3.23. Bifurcation structure of periodic orbits and chaotic attractors in the logistic map, schematically. Only the period-doubling and the reverse bifurcation sequences associated with period 1 and period 3 orbits are shown.

A specific situation occurs at reverse bifurcation points $\tilde{a}_i$, $i = 0, 1, \ldots$. For example the attractor corresponding to $\tilde{a}_0$ is formed by the union of two intervals I_1, I_2 having a single common point of period 1, see Fig. 3. 16(b). Each of the intervals is invariant under $f_{\tilde{a}_0}^2$ and at the same time $f_{\tilde{a}_0}(I_1) = I_2$, $f_{\tilde{a}_0}(I_2) = I_1$. In general, attractors occuring at the band merging points $\tilde{a}_i$, $i = 0, 1, \ldots$ contain 2^{i+1} intervals invariant under $f^{2^{i+1}}$ which have pairwise common points. These points are just the unstable periodic points of period 2^i from the corresponding period doubling sequence. Thus the attractor consists of 2^i disjoint closed intervals, see Fig. 3.23, each of them containing two subintervals separated by a periodic point of period 2^i. The power spectra exhibit universal properties[3.59] both in the periodic components and in the broad-band noise if $i \to \infty$ because of existence of the universal function $g(x)$.

The reverse chaotic attractors occurring at the parameter values $\tilde{a}_i$, $i = 0, 1$, together with the chaotic attractor at $a = a_+$ form a structure of attractors that contain unstable orbits of period 2^i, $i = 0, 1, \ldots$, see Fig. 3.23. The unimodality together with (3.5) implies that the same structure exists repeatedly to the right of any q-periodic window.

Denoting a_∞^q the accumulation point of a periodic window of basic period q (there exists more than one q-periodic window if $q > 3$, see Tab. 3.2), the corresponding reverse bifurcation sequence $\{\tilde{a}_i^q\}$ tends to a_∞^q from the right. The chaotic attractor at $a = \tilde{a}_i^q$ consists of $q \times 2^i$ disjoint intervals each of them containing an unstable $q \times 2^i$- periodic point. The attractor at $a = a_+^q$ consists of q disjoint intervals and contains an unstable q-periodic orbit in its boundary. An example of the window of period 3 in Fig. 3.23 shows the structure and how the $q \times 2^i$- periodic points are created as a is increased. The intervals $[a_\infty^q, a_+^q]$ form a 'box-within-box' structure on arbitrarily small scales when q is increased[3.18].

An important question is: what is the measure of the set of parameter values corresponding to a chaotic attractor? Jakobson[3.47] and Collet and Eckmann[3.10] show that for one-parameter systems of unimodal mappings satisfying condition (3.5) and (3.18) this set is a Cantor set with a positive Lebesgue measure. Hence, even if the chaotic attractors occur only at discrete points, there exists a positive chance of finding a chaotic parameter value. Numerical computations show[2.18, 2.26] that the measure of this Cantor set in the interval $[a_\infty, a_+]$ is $\mu \approx 0.85$ for the logistic mapping; this clearly shows that stable chaotic motions dominate over stable periodic orbits.

The set of chaotic parameter values can also be characterized by means of a fractal dimension. However, the direct application of the definition of the capacity dimension leads to the integer value $D_C = 1$. This is caused by a positive Lebesgue measure of this Cantor set and hence it is proper to use the definition of the exterior capacity dimension D_E, see Section 2.5. Numerical results[2.18, 2.26] give the value $D_E \approx 0.587$ for the logistic mapping.

Unimodal mappings with a negative Schwarzian derivative have besides chaotic and periodic attractors also another type of attractor which exists for values of the parameter a corresponding to accumulation points of period-doubling sequences. The structure of this attractor is suggested by the sequence of reverse chaotic attractors approaching the accumulation point from above. The number of bands is doubled at every step and their width decreases at the same time. Thus a Cantor set, similar to the classical middle-third Cantor set, is formed at the accumulation point. This attractor does not exhibit sensitive dependence on initial conditions, i.e. its Lyapunov exponent is zero. It was found that the fractal dimension of this attractor $D_F \approx 0.538\,04$ is universal[3.33] in the same sense as the sequence of the period-doubling bifurcations. This so called *Feigenbaum attractor* is interesting mathematically; repelling periodic orbits (from the forward bifurcation sequence) which do not belong to this attractor exist arbitrarily close to every point of the attractor[2.16]. The set of parameter values corresponding to this type of attractor has a zero Lebesgue measure and therefore is negligibly small in comparison with the parameter set yielding a stable periodic and/or chaotic behaviour.

3.3.4 Intermittency

Until now we have discussed the transition from periodicity to chaos via the period-doubling sequence. The chaotic set (not necessarily attracting) is formed after the first accumulation point $(a_\infty \approx 3.570$ for the logistic mapping) is reached. In the chaotic region of the logistic map the periodicity re-emerges in periodic windows which are bounded by the accumulation point from the right and by the saddle-node bifurcation from the left. A reverse bifurcation sequence occurs above the accumulation point. Below the saddle-node bifurcation occurs the phenomenon of *intermittency* which we now describe.

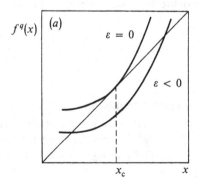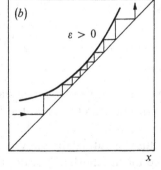

Fig. 3.24. The graph of the function $F_\varepsilon(x)$ at the onset of intermittency. (*a*) Two fixed points for $\varepsilon < 0$ collapse at $\varepsilon = 0$, and (*b*) a narrow channel arises for $\varepsilon > 0$.

Let us consider a window of basic period q, which is bounded from the left by a saddle-node bifurcation occuring at $a = a_c^q$. Let us denote $\varepsilon = a_c^q - a$, and let x_c be one of the q periodic points of the critical map $f_{a_c^q}$. The function $F_\varepsilon(x) = f_a^q(x)$ in the neighbourhood of the bifurcation point (x_c, a_c^q) is shown in Fig. 3.24.

If $\varepsilon < 0$, then the graph of the function intersects the diagonal line at two points corresponding to stable and unstable q-periodic points. If $\varepsilon > 0$, the graph does not intersect the diagonal line and a narrow channel is formed, through which the orbit must pass. When ε decreases, then the number of iterations necessary for passing through the channel increases. The case $\varepsilon = 0$ corresponds to the existence of a semistable periodic point x_c. For a sufficiently small positive ε the dynamics of the mapping consists of two distinct phases. A *laminar phase* with an almost periodic behaviour corresponds to the passing of the orbit through the channel, while the *turbulent phase* is aperiodic and corresponds to the dynamics of the mapping outside the channel. Both phases alternate irregularly. The intermittency was first studied by Manneville and Pomeau[3.55, 3.66]. Their idea was to describe the laminar phase via a simple difference equation

$$y_{k+1} = h_\varepsilon(y_k) = \varepsilon + y_k + y_k^2, \tag{3.19}$$

which is a normal form for a saddle-node bifurcation in dynamical systems with discrete time. Eq. (3.19) has a critical fixed point $y_c = 0$ at $\varepsilon = 0$ and approximates the dynamics of $f_a^q(x)$ in the neighbourhood of (x_c, a_c^q). If $0 < \varepsilon \ll 1$ then the number of iterations l necessary for the orbit to pass through the narrow channel (i.e. the length of the laminar phase) can be obtained as a solution of the differential equation

$$dy/dl = \varepsilon + y^2, \tag{3.20}$$

which approximates Eq. (3.19). Solving the initial value problem (3.20) with $y_{min} = y(0)$ the time l required for the orbit to pass through the interval $[y_{min}, y_{max}]$ is

$$l = \frac{1}{\sqrt{\varepsilon}} \left(\arctan \frac{y_{max}}{\sqrt{\varepsilon}} - \arctan \frac{y_{min}}{\sqrt{\varepsilon}} \right). \tag{3.21}$$

The distribution of lengths of the laminar phases depends on the properties of the original mapping $F_\varepsilon(x)$ outside the laminar channel and cannot be determined from (3.20) without additional assumptions[3.66]. However, the following two main conclusions can be drawn: there exists a finite upper bound of laminar lengths and the mean length $\langle l \rangle$ scales as

$$\langle l \rangle \sim \varepsilon^{-1/2}. \tag{3.22}$$

An alternative approach to the description of the intermittency gives the renormalization group method[3.42, 3.45]. Similarly as in the case of the period-doubling we can apply the doubling operator

$$T(h(x)) = \alpha \, h(h(x/\alpha)) \,. \tag{3.23}$$

The function h does not vary much due to the action of T, only the number of iterations necessary for the transition through the channel is approximately two-fold. The fixed point $g(x)$ of T can be normalized via the conditions $g(0) = 0$, $g'(0) = 1$ and can be determined analytically

$$g(x) = (1/x - 1)^{-1} \,. \tag{3.24}$$

The scaling factor is $\alpha = 2$ and the linearization of T has a unique unstable eigenvalue $\delta = 4$. These relations then determine a scaling law for $\langle l \rangle$, which is equivalent to Eq. (3.22).

One interesting property related to the intermittency is the so called $1/f$ *noise*, i.e., the power spectrum $P(f)$ is approximately scaled as $1/f$ for low frequencies[3.47]. Similar scaling is observed in the spectra of many experimentally studied systems[3.42], however, the intermittency is apparently only one of many sources of $1/f$ noise. The presence of an external noise causes that the intermittent dynamics occurs on both sides of the critical value of a and this phenomenon is again associated with universal scaling laws[3.20].

3.3.5 Crises and transient chaos

Chaotic attractors of unimodal mappings with a negative Schwarzian derivative are structurally unstable. Thus small variations of a characteristic parameter cause qualitative topological changes of the attractor. These changes may often occur on a scale which is very small compared with the size of the attractor. For example the unit interval $I = [0, 1]$ forms the attractor at $a = a_+$. Slightly below a_+ the computer generated picture of the attractor does not differ very much from that for $a = a_+$, see Fig. 3.14. When the parameter passes through a_+ from left to right the attractor disappears because f_a does not map I to itself for $a > a_+$. Above a_+ the function $f_a(x) > 1$ in the neighbourhood of c. Thus a gap arises through which almost all orbits starting in I eventually escape, see Fig. 3.25, and tend to $-\infty$, which can be regarded as an attractor. Hence, at the point a_+ a bifurcation occurs at which the chaotic attractor disappears. This change can be seen as a collision of the attractor with the unstable fixed point $x_{f_1} = 0$ which is located in the boundary of the basin of attraction and hence it is called a *boundary crisis*[3.35]. A very similar phenomenon can also be observed at every chaotic parameter value. For example, the chaotic attractor of the logistic map at $a_+^3 \approx 3.858$ consists of three intervals. At slightly subcritical values of the parameter the attractor is contained in three bands, while it expands for supercritical values of the parameter

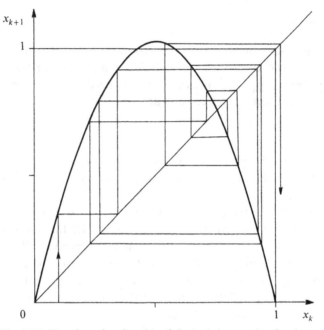

Fig. 3.25. Transient chaotic orbit of the logistic equation for the parameter a slightly above $a_+ = 4$.

and covers a far greater part of the state space, see Fig. 3.14. This type of behaviour is called an *interior crisis*.

When the parameter a passes through the critical value a_+^3 a collision of the chaotic attractor with an unstable periodic orbit of period 3 created at the saddle-node bifurcation occurs, see Fig. 3.23. The same phenomenon arises at all critical values a_+^q; however, the crises cease to be observable with increasing q. The dynamics of the time evolution behind the point of a crisis was investigated by Yorke with coworkers[3.35, 3.65]. They studied a typical case of the logistic equation in the neighbourhood of the point a_+. For values of a slightly greater than a_+ the orbits with the initial point in I remain in I for a certain finite time (i.e. for a finite number of iterations) and behave chaotically, see Fig. 3.25. Only after that they do leave I and tend to $-\infty$. This phenomenon is called a *transient chaos* or a *preturbulence*. For a smooth distribution of initial conditions in I, the length l of a chaotic transient is exponentially distributed

$$\varrho(l) = \frac{1}{\langle l \rangle} e^{-(1/\langle l \rangle)} \tag{3.25}$$

where $\langle l \rangle$ is the average value of transient times. In addition, the scaling of $\langle l \rangle$ with the parameter a is given by

$$\langle l \rangle \sim (a - a_+)^{-1/2} . \tag{3.26}$$

What actually happens for $a > a_+$ is that the chaotic attractor is replaced by a chaotic invariant Cantor set contained in I which is a repellor. This set governs the dynamics of all transient orbits which are sufficiently close to it and the decay rate of the exponential distribution (3.25) is given by its dynamical properties[2.16].

After discussing the notion of the attractor crises we shall shortly mention one global feature arising in such mappings as the logistic mapping[3.77]. Let us choose any periodic window and as above denote a_c^q, $\tilde{a}_0^q$ and a_+^q the bifurcation values of the parameter a where the saddle-node, first period-doubling and the associated crisis occur, respectively. It has been shown that the ratio

$$\alpha = \frac{a_+^q - a_c^q}{\tilde{a}_0^q - a_c^q} \tag{3.27}$$

is close to 9/4 for most windows and differences decrease as the period q is increased. This observation leads to the conjecture that the following scaling law is valid: the fraction of windows of period $\leq q$ for which $\alpha \to 9/4$ approach 1 as $q \to \infty$. The ratio α can be close to 9/4 even for small q but exceptions may exist.

3.3.6 Circle maps

If we replace the interval by a circle we obtain a discrete dynamical system for which we can define the rotation around the circle. This property has the result that in contrast to mappings of the interval, continuous monotonic mappings on the circle may exhibit very complex dynamics. If they are non-monotonic then even in the simplest case they must have two extremes and hence the dynamics of noninvertible circle maps will be more complex than that of unimodal maps of the interval.

Usual mathematical description of the coordinate on the circle S^1 is realized by means of the real numbers modulo the length of the circle (we shall consider unit length, i.e. $\varphi = x(\text{mod } 1)$). The discrete time evolution for the mapping $f: S^1 \to S^1$ is given by

$$\varphi_{k+1} = f(\varphi_k) . \tag{3.28}$$

Continuous circle mappings can be divided into distinct classes according to their *topological degree*, which is an integer determining how many times $f(\varphi)$

revolves around S^1 when φ revolves once. A continuous map on the real axis $x \to F(x)$ can be associated with the continuous circle map f of degree d such that $f = F \pmod 1$, $\varphi = x \pmod 1$, and

$$F(x + 1) = F(x) + d \, . \tag{3.29}$$

The map F is called a *lifting of f* and is defined uniquely up to an additive integer constant (which may be determined by a given model problem).

We describe here the most important case $d = 1$, but let us mention that the cases with $d \neq 1$ are also important. For example, the Bernoulli shift mapping described above, $f(\varphi) = 2\varphi \pmod 1$, is a circle map of degree 2. Though it is linearly increasing, it is noninvertible (two to one) because of its topological degree. Other mappings of the circle with $d \neq 1$ are described in Chapter 6.

If $d = 1$, then the rotation of the orbit $\{f^k(\varphi)\}$ around the circle can be expressed by means of the corresponding orbit $\{F^k(x)\}$ of a given lifting F in the following way

$$\varrho = \lim_{k \to \infty} \frac{F^k(x) - x}{k} \, , \tag{3.30}$$

provided the limit exists. Here ϱ is a *rotation number*, expressing the mean asymptotic velocity of rotation around S^1. Generally, ϱ depends on φ (i.e. different orbits may possess different rotation numbers) and the rotation numbers for all φ fill out a closed interval called the *rotation interval*[3.61]. If Eq. (3.28) has a periodic orbit of period q then for any point $\varphi = x \pmod 1$ on this orbit

$$x + p = F^q(x) \, , \tag{3.31}$$

where p is an integer. Hence the rotation number of a periodic orbit is rational, $\varrho = p/q$. Orbits which are not asymptotically periodic may have both rational and irrational rotation numbers.

If f is a homeomorphism (i.e. f is continuous and invertible with a continuous inverse) then ϱ is independent of φ and the rotation interval degenerates to a single point. The same holds for diffeomorphisms. The rotation number in this case describes the dynamics of a continuous time flow on a two-dimensional torus. Its Poincaré map is just the diffeomorphism of the circle. Rational rotation numbers correspond to periodic orbits and irrational ones to quasi-periodic orbits (the entire circle is an attractor in this case).

Both in physical and computer model experiments we often observe a transition to chaos which seems to be associated with a loss of differentiability of the torus. If the dynamics on the torus is quasiperiodic, then this case is described by iterated circle maps which just become noninvertible. To be able to follow the nature of such a bifurcation, we shall study the circle mapping which is dependent on a parameter (parameters). Let us consider a simple evolution equation

$$\varphi_{k+1} = f_{T,A}(\varphi_k) = T + g_A(\varphi_k) \,(\text{mod } 1), \qquad (3.32)$$

where $T > 0$, $A \geqq 0$ are parameters, g_A is of degree one and becomes identity as $A \to 0$. Eq. (3.32) approximates the Poincaré map of a continuous time system with an attracting periodic orbit, periodically forced by discrete jumps under the assumptions that the amplitude of jumps, A, is sufficiently low and the period of the pulses, T, is sufficiently high. For example, the periodically forced two-dimensional oscillator described in Chapter 2, Eqs (2.23) and (2.24), can be reduced to the model (3.32) if $0 \leq A < 1$.

We shall discuss here the *sine map*, a classical example of a circle map

$$\varphi_{k+1} = f_{T,A}(\varphi_k) = F_{T,A}(x_k) \,(\text{mod } 1) \qquad (3.33)$$

with $F_{T,A}(x) = T + x + (A/2\pi) \sin (2\pi x)$.

Actually we have a two-parameter family of circle maps (of degree 1) which can have two extremes. The character of the nonlinearity depends on the value of the parameter A. We shall consider only $A \geqq 0$, as the mappings $f_{T,A}$ and $f_{T,-A}$ are coupled by the coordinate transformation $\varphi \to \varphi - 1/2$. The graph of the function $f_{T,A}$ for three typical cases is shown in Fig. 3.26. The sine map

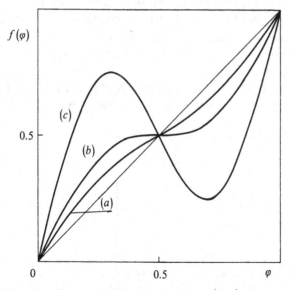

Fig. 3.26. The graph of the sine circle map (3.33) with $T = 0$: curve (a) $A = 0.8$; curve (b) $A = 1$; curve (c) $A = 1.5$.

is a diffeomorphism for $0 \leq A < 1$ and can be seen as a Poincaré map of a continuous time system on a two-dimensional torus. The mapping $f_{T,A}$ has for $A = 1$ a cubic inflection point $\varphi = \frac{1}{2}$ (mod 1) with the derivative $f'(\frac{1}{2}) = 0$. Such mappings are called *critical* and were studied in connection with the transition from quasiperiodicity to chaos[3.26, 3.64]. The critical mappings correspond to the loss of differentiability of a quasiperiodic torus while no direct relation to the dynamics of a more-dimensional system arising after the destruction of the torus exists for $A > 1$. Still the chaotic dynamics originating after the torus destruction can often be described approximately by a nonivertible circle map. This holds particularly for oscillators periodically forced by short pulses.

When $A < 1$, then the rotation number does not depend on φ and fully determines the asymptotic dynamics of the system described by Eq. (3.33). The attractor is either a periodic orbit (with a rational rotation number) or an entire circle (with an irrational rotation number). The rotation number exhibits an interesting dependence on the parameters A and T. This can be seen most easily if we follow the function $\varrho(T)$ parametrized by A. It suffices to consider T on the unit interval, as $\varrho(T + 1) = \varrho(T) + 1$. For $A = 0$, $\varrho(T) = T$ and all values of T with an irrational ϱ form a set with the Lebesgue measure equal to one. If $0 < A < 1$, then $\varrho(T)$ is a continuous nondecreasing function constant on intervals where ϱ is rational and strictly increasing at points where ϱ is irrational[3.41]. The intervals of T values corresponding to different rational rotation numbers are ordered along the T axis according to the *Farey tree*[3.41]: between two intervals with the rotation numbers p/q and m/n there is another interval with the rotation number $(p+m)/(q+n)$ and this construction can be repeated indefinitely. The width of the intervals rapidly decreases with increasing denominators. The graph of $\varrho(T)$ contains a peculiar structure of arbitrarily small steps called the *devil's staircase*, see Fig. 3.27.

The set of T values $(0 < A < 1, A$ fixed$)$ for which $\varrho(T)$ is irrational forms a Cantor set of positive Lebesgue measure, i.e. a fat fractal with the exterior

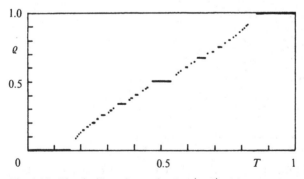

Fig. 3.27. The devil's staircase for Eq. (3.33) with $A = 1$.

capacity dimension $D_E = 0^{2.26}$. The Lebesgue measure of this set decreases as A increases and vanishes for $A = 1$. The fractal dimension of the critical case is $D_F \approx 0.868$ and seems to be universal for one-parameter families of critical maps with a cubic inflection point[3.48].

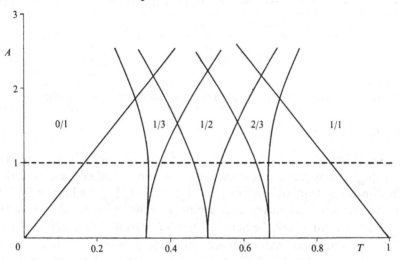

Fig. 3.28. Several basic Arnold tongues for the sine map. There exist an infinite number of other tongues which are ordered according to Farey sequences.

The set of all points in the (A, T) plane $(0 \leq A \leq 1)$ for which ϱ is a given irrational number is a smooth curve while all points for which ϱ is a given rational number form a cusp-shaped region called the *Arnold tongue*[2.4], see Fig. 3.28. These sets for different ϱ's do not intersect if $A < 1$. The plateaus of the devil's staircase correspond to intersections of Arnold tongues with the line $A = $ const. Arnold tongues continue to exist if $A > 1$ and successively begin to overlap as A increases.

3.3.7 Transition from quasiperiodicity to chaos

The transition from quasiperiodicity to chaos occurs when the critical mapping has an irrational rotation number. As described above, one-parameter systems of critical mappings have irrational rotation numbers only on a (Cantor) set of values of a parameter of zero Lebesgue measure. This means that if we follow, for example, the transition to chaos when A increases and T is fixed, then the transition to chaos via quasiperiodicity is not likely. Hence, we have to change both A and T in such a way that we reach a prescribed irrational rotation number at $A = 1$. Under these conditions we can use a renormalization group approach[3.26, 3.64] as in the period-doubling case discussed earlier.

Any irrational rotation number ϱ can be written as a *continued fraction expansion*

$$\varrho = \cfrac{1}{n_1 + \cfrac{1}{n_2 + \cfrac{1}{n_3 + ...}}} = [n_1, n_2, n_3, ...] . \tag{3.34}$$

If we retain only the first m numbers in the continued fraction expansion we have

$$p_m/q_m = [n_1, n_2, ..., n_m] ,$$

and the sequence of rationals, $\{p_m/q_m\}$, converges to ϱ with increasing m. The sequence $\{n_m\}_1^{\infty}$ may be periodic or not. The most studied case is that of a periodic continued fraction, i.e. $n_{m+s} = n_m$, $m = 1, 2, ...$, where $s \geq 1$ is an integer constant. Several universal constants can then be defined. Let us denote T_{∞} the value of T for which the map $f_{T, A}$ has an irrational rotation number with an s-periodic continued fraction representation for a fixed A such that $0 \leq \leq A \leq 1$ and consider orbits starting, for example, at the inflection point $x = \frac{1}{2}$. Then a universal constant

$$\alpha = \lim_{m \to \infty} \left(F_{T_{\infty}, A}^{q_m}(1/2) - p_m\right)/\left(F_{T_{\infty}, A}^{q_{m+s}} (1/2) - p_{m+s}\right) \tag{3.35}$$

exists and takes on two different values for A in the range $0 \leq A < 1$ and for $A = 1$. In fact, α is universal within a certain class of maps of the circle which are diffeomorphisms or critical maps with a cubic inflection point, respectively.

Another universal constant δ can be defined for maps such as (3.33) by

$$\delta = \lim_{m \to \infty} (T_{m+s} - T_m)/(T_m - T_{m-s}) , \tag{3.36}$$

where, for a given value of A (A is either in the range $0 \leq A < 1$ or $A = 1$), it is convenient to consider T_m equal to such a value of T that the point $1/2$ is contained in the periodic orbit with the rotation number p_m/q_m.

In particular, if $s = 1$, $n_m = 1$, then Eq. (3.34) leads to the golden mean rotation number $\varrho_G = (\sqrt{5} - 1)/2$ with p_m/q_m given by the Fibonacci numbers

$$q_1 = 1, \qquad q_2 = 2, \qquad q_{m+1} = q_m + q_{m-1}, \qquad p_m = q_{m-1} , \tag{3.37}$$

and the universal constants are

$$\alpha = -\varrho_G^{-1} \approx -1.618 , \qquad \delta = -\varrho_G^{-2} \approx -2.618\,03 \tag{3.38}$$

for $0 \leq A < 1$ and

$$\alpha \approx -1.288\,62 \,, \qquad \delta \approx -2.833\,61 \tag{3.39}$$

for $A = 1$.

The idea of the renormalization group analysis is to find an operator T_n acting on cubic critical circle maps (or diffeomorphisms) such that if f is a particular map with the rotation number

$$\varrho(f) = \cfrac{1}{n_1 + \cfrac{1}{n_2 + \dots}} \,, \tag{3.40}$$

then the rotation number of $T_{n_1}(f)$ will be

$$\varrho(T_{n_1}(f)) = \cfrac{1}{n_2 + \cfrac{1}{n_3 + \dots}} \,. \tag{3.41}$$

Combining (3.40) and (3.41) we see that the rotation number of $T_{n_1}(f)$ can be obtained from the rotation number of f using the so called *Gauss map*,

$$G(\varrho) = 1/\varrho - [1/\varrho] \,, \tag{3.42}$$

where $[1/\varrho] = n_1$ is the integer part of the argument. This map has an infinite number of unstable periodic points corresponding to different periodic continued fractions; but almost all orbits generated by (3.42) are chaotic, which corresponds to nonperiodic continued fractions.

When the sequence $\{n_m\}$ is periodic of period s, then

$$T = T_{n_{s-1}} \circ T_{n_{s-2}} \circ \dots \circ T_{n_1} \tag{3.43}$$

will have an unstable fixed point whose stability properties determine the universal constants α and δ. The operator T has a nontrivial fixed point for critical maps and δ is the unstable eigenvalue of T defining how fast the critical maps move away from the fixed point. For a precise formulation of the renormalization group transformation we refer the reader to Ref. 3.64.

It is clear that different periodic continued fractions will yield different fixed points of T and that there are an infinite number of them, each having its own α and δ. Moreover, the vast majority of irrational numbers have an aperiodic continued fraction expansion. In such a case the sequence of points in the space of critical circle maps generated iteratively by T_n (with $\{n_m\}$ being almost every sequence generated by the Gauss map (3.42)) is nonperiodic and asymptotically tends to an attractor which is chaotic. This attractor contains all the fixed points of T and thus comprises all local universal properties of these points[3.17, 3.22, 3.75]. It has a single positive Lyapunov exponent $\lambda_1 \approx 15.5$ and its information dimension is $D_I \approx 1.8$. Hence, this is an example of a low-dimensional chaotic attractor in an infinite dimensional state space.

Important features of the transition from quasiperiodicity to chaos are reflected in the power spectra. Their low-frequency part is universal for a given irrational rotation number and, moreover, self-similar if ϱ has a periodic continued fraction expansion[3.64].

3.3.8 Transition from phase-locked dynamics to chaos

Even if the described features of the quasiperiodic transition to chaos are interesting, the one-parameter systems of circle maps exhibit mostly the transition to chaos via the bifurcations from phase-locked periodic orbits. For example, if we increase A for fixed T, than at $A = 1$ we can expect a periodic dynamics since the quasiperiodicity of critical maps is an exceptional case. Instead, the transition to chaos via the Feigenbaum period-doublings occurs when the value of A is increased further.

However, this transition may be complicated by a hysteresis, i.e. by the coexistence of more attractors. The mapping $f_{T,A}$ has negative Schwarzian derivative and for $A > 1$ it has two extremes, hence it can have at most two attracting periodic orbits[3.72]. At the onset of chaos the simultaneous existence of the Feigenbaum attractor and a periodic attractor is likely and the initial state will decide which regime will be observed. Further evolution of (possibly multiple) attractors includes a structure with periodic windows which possesses no universal features if observed along some one-parameter family (i.e. along some chosen path in the parameter plane (T, A)). Hence, if we want to observe some reproducible structure we have to consider a variation of both parameters at the same time.

Let us choose the rotation number $\varrho = p/q$ (p, q relatively prime) and consider the corresponding Arnold tongue denoted by $I_{p/q}$. It continues to exist for $A > 1$ but two (or more) different tongues may overlap, see Fig. 3.28. If this is the case, then the rotation number is no longer independent of φ and a rotation interval $R = [\varrho^-, \varrho^+]$ exists (ϱ^+ and ϱ^- denote the maximal and minimal rotation numbers, respectively). Given the point (T, A) in $I_{p/q}$, there are four possibilities of occurrence of $\varrho = p/q$ in the rotation interval R[3.53].

(1) R is degenerate, i.e. $\varrho = \varrho^- = \varrho^+$,
(2) ϱ lies inside R ,
(3) $\varrho = \varrho^-$,
(4) $\varrho = \varrho^+$.   (3.44)

The first possibility clearly holds if $A \leq 1$, but may occur for $A > 1$ as well. As an example we chose $I_{1/2}$, but the same situation occurs in any other Arnold tongue. The conditions $(1) - (4)$ define four regions in $I_{p/q}$ denoted I–IV, respectively, see Fig. 3.29. We use the lifting $F_{T,A}(x)$ to decribe the situation. For the q-periodic point x_* with $\varrho = p/q$, $F^q(x_*) = x_* + p$ holds and thus the map $F^q(x) - p$ will have a fixed point x_* with $\varrho = 0/1$. The graphs of $F^q(x) - p$ for four typical situations are shown in Figs. $3.30(a)$–(d).

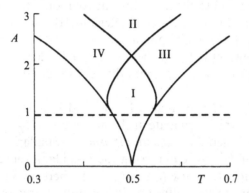

Fig. 3.29. Separation of the Arnold tongue $I_{1/2}$ into distinct regions according to conditions (3.44).

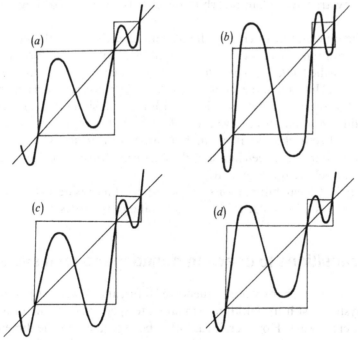

Fig. 3.30. (a)–(d) Typical graphs of the function $F^2(x) - 1$ corresponding to the four regions in Fig. 3.29.

In the first case, Fig. 3.30(*a*) which corresponds to region I of Fig. 3.29, the graph decomposes into invariant boxes and so orbits starting in any box cannot leave it. Hence the rotation number is zero for every orbit of $F^q(x) - p$ and $\varrho = \varrho^+ = \varrho^- = p/q$. Let us note that each invariant box contains two extremes if $A > 1$ and thus the dynamics in region I is governed by a mapping of the interval with two critical points.

The boxes are no longer invariant for (T, A) from region II, Fig. 3.30(*b*). The orbits may diffuse both to the right and to the left. Thus the rotation interval R is not degenerate and ϱ is contained strictly inside R. Regions III and IV are complementary. The orbits of $F^q(x) - p$ corresponding to region III can either stay in the box or diffuse to the right but not to the left, Fig. 3.30(*d*). Hence $\varrho = = \varrho^-$. The orbits corresponding to region IV cannot diffuse to the right, Fig. 3.30(*c*), and thus $\varrho = \varrho^+$.

The crossing of the boundary between the region I $\cup$ IV and II $\cup$ III from left to right implies that the orbits which originate at the maximum of $f_{T,A}$ will have $\varrho > p/q$; this phenomenon is called *deterministic diffusion*[3.29]. Similarly, the orbits starting at the minimum of $f_{T,A}$, when crossing the boundary between I $\cup$ III and II $\cup$ IV from right to left diffuse (here $\varrho < p/q$). There are many scalings associated with the deterministic diffusion. If a chaotic attractor is present at the onset of the deterministic diffusion, a boundary crisis occurs, accompanied by transient chaotic orbits of lengths scaled according to Eq. (3.26).

Another phenomenon of interest is the intermittency. As with the boundary crises, the intermittency can similarly be combined with the diffusion. This phenomenon is called *intermittent diffusion*[3.30] and occurs when crossing the boundaries of $I_{p/q}$. Here again the length of the laminar phase is scaled according to Eq. (3.22), but other scalings associated with the diffusion are also present. Intermittent diffusion may occur even for $A < 1$, but here the intermittency cannot be related to a chaotic attractor but rather to a torus. The main difference is that the bursts between laminar phases are not 'turbulent' (i.e. chaotic) for $A < 1$ and behave quite regularly.

The structure of local bifurcations of periodic orbits inside every Arnold tongue also exhibits self-similarity[3.31, 3.70] and scaling properties[3.2, 3.3, 3.7, 3.19].

3.4 Transitions to chaos in multidimensional systems

The transition to chaos via a sequence of bifurcations when a parameter is varied in systems with a multidimensional state space may occur in a vast number of different ways. However, it can often be expected that the dimension of the just formed chaotic attractor will be close to a minimal possible dimension. Then the chaos formation can be modelled by a low-dimensional dynamical system dependent on a parameter (parameters). The further evolution of the real chaotic system may, of course, lead to an increase in the dimension of attractors (either systematically or jumpwise) and the low-modal approximate description will no longer suffice. It can be expected to occur, for example, in systems with fully developed turbulence (e.g. in hydrodynamics, plasmas, etc.).

The formation of chaotic sets is generally connected with the formation of homoclinic or heteroclinic orbits, i.e. stable and unstable manifolds of periodic orbits become tangent and then intersect transversely (presumably almost everywhere) as a parameter is varied. This global bifurcation gives rise to a chaotic set which need not be an attractor at first, but becomes attractive later, when a parameter is varied further. The homoclinic tangency can be accompanied by sequences of bifurcations similar to those in one-dimensional maps. All bifurcation phenomena described for one-dimensional mappings have their direct counterparts in multidimensional systems. However, the global changes of orbits in the state space are evidently more complex. We shall now present several typical transitions to chaos (called *scenarios*) described in the literature[1.16], using the knowledge of bifurcation phenomena in one-dimensional mappings.

The period-doubling transition to chaos

Sequences of period-doublings observed in one-parameter families of one-dimensional mappings are characterized by the geometric convergence of the bifurcation values of the parameter to the accumulation point. This property is connected with the existence of a universal constant δ with a value dependent only on the type of the extreme of the mapping. In the typical case of the quadratic extreme $\delta = 4.669...$.

Collet, Eckmann and Koch[3.11] (see also Ref. 1.13) have shown that the same behaviour also occurs in the multidimensional case and that the transition is completely analogous.

There exist a large number of numerical studies and experimental data (see Chapter 5), where the bifurcation phenomena known from the one-dimensional mappings are observed: repeated cascades of period-doublings with the window structure of reappearing periodic attractors, the intermittency and various crises of chaotic attractors. Also structures similar to U-sequences known in unimodal mappings occur in multidimensional systems. These properties reflect the strongly dissipative character of the systems which then asymptotically behave as a one-dimensional mapping.

However, the evolution of the dynamics of the system can be quite different from that of a one-dimensional mapping after the accumulation point is reached. The situation varies from case to case and generally we cannot transfer such statements as the Sharkovskii theorem to multidimensional systems. Even in one-dimensional mappings possessing several extremes the period-doubling sequences may occur competitively in distinct regions of the state space and thus a hysteresis arises. The same type of hysteresis may occur in multidimensional cases. An important difference is that the chaotic attractor formed after the accumulation point has a homoclinic structure and it is a fractal in contrast to the situation observed for one-dimensional mappings. Such an attractor then possesses not only complex dynamics but also complex geometric structure.

Intermittency

This phenomenon in the one-dimensional case precedes the formation of a periodic attractor via the saddle-node bifurcation. The mapping already contains an invariant chaotic set (formed after the first accumulation point), which is not an attractor if the periodic attractor is present but a chaotic attractor emerges when the periodic orbits disappear at the saddle-node. The chaotic attractor inherits the basin of attraction after the disappearing periodic attractor.

A completely analogous situation (with the same scaling properties) occurs in multidimensional systems. In fact, the intermittency may appear also in situations where a more regular set (e.g. torus) is substituted for the chaotic set. A transition from periodicity to quasiperiodicity occurs in this case.

The saddle-node bifurcation is just one of several typical local bifurcations of codimension one; two others which may also lead to intermittency are period-doubling and torus bifurcation. Hence, we can classify intermittency according to which type of local bifurcation is involved. Pomeau and Manneville[3.66] describe three typical cases.

(1) Type I intermittency is connected with the saddle-node bifurcation, this case was discussed for the logistic mapping; the average length $\langle l \rangle$ of the laminar phase is scaled with the distance ε from the bifurcation point as $\langle l \rangle \sim \varepsilon^{-1/2}$.

(2) Type II intermittency involves torus bifurcation and thus it cannot occur in one-dimensional mappings. The torus bifurcation has to be subcritical and at the same time a proper chaotic set has to be present such that it becomes an intermittent attractor behind the bifurcation point. The scaling of the laminar phase is now $\langle l \rangle \sim \varepsilon^{-1}$.

(3) Type III intermittency involves a subcritical period-doubling bifurcation, hence it can appear also in one-dimensional mappings. However, the mappings with negative Schwarzian derivative do not exhibit this type of bifurcation and thus we cannot observe it, for example, in the logistic mapping. The scaling of the laminar phase is $\langle l \rangle \sim \varepsilon^{-1}$.

Crises

The interior and boundary crises described for the logistic mapping again have direct analogies in multidimensional systems. The crisis can be seen as a global bifurcation phenomenon, where the chaotic set changes its character from an attracting to a nonattracting set or when more chaotic sets (some of them possibly attractors) mutually interact to form a larger chaotic attractor. Such bifurcations can be quite complex in multidimensional state spaces. The chaotic attractors may come into contact with a boundary of their own basins of attraction, in particular with saddle periodic points and thus they can be indicated in numerical experiments.

In the close neighbourhood of the points of crises we can observe a dynamic behaviour called *transient chaos*. Transient chaos occurs if there are large regions of the state space with the property that the orbits which start in these regions are attracted close to a chaotic set which is not an attractor but rather a (generalized) saddle. Then for a certain (possibly very large) time period these orbits reflect the dynamics on the chaotic set, until they are repelled to an attractor which is located in another region of the state space.

The rate α of repulsion of the chaotic motion close to the nonattracting set can be expressed in terms of Lyapunov exponents λ_i, partial dimensions D_I^i and the metric entropy h_μ of this set. It was shown rigorously in special cases[2.16] and conjectured in general[2.16, 2.33] that

$$\alpha = \sum_{\lambda_i > 0} \lambda_i - h_\mu = \sum_{\lambda_i > 0} \lambda_i \left(1 - D_I^i\right). \tag{3.45}$$

The formula (3.45) yields $\alpha = 0$ for a chaotic attractor because of the nonfractal structure along the unstable directions, i.e. $D_I^i = 1$ for all positive λ_i. In this case the metric entropy is the sum of the positive exponents (see Eq. (2.39)) hence $\alpha = 0$ and the orbits never leave the attractor. In the case of nonattracting chaotic sets, the fractal structure is also present in unstable directions and α becomes positive. The mean length $\langle 1 \rangle$ of the chaotic transient is therefore inversely proportional to α (see also Eq. (3.25)). The value of α is small just close to the points of crises and thus the transient chaos is well observable here.

Homoclinic orbits of stationary points

Certain relatively simple situations in the state space may serve as an indication of the presence of chaotic dynamics. The presence of an orbit which asymptotically approaches the stationary point both in positive and negative time directions belongs among them. Such a homoclinic orbit is substantially simpler than the homoclinic orbit which exists in the intersection of the stable and unstable manifolds of a periodic orbit, but nevertheless this simpler situation may under certain circumstances lead to chaotic behaviour. The homoclinic orbit cannot imply chaos in two-dimensional state spaces (see Fig. 2.4). However, it is already possible in a three-dimensional state space, as it is illustrated by the following example due to Shilnikov[3.71].

Let us consider a vector field on R^3 having a stationary point z and a homoclinic trajectory γ which approaches z in an oscillatory way, see Fig. 3.31. The stationary point z will thus have one real eigenvalue λ and two complex conjugate eigenvalues $\alpha \pm i\beta$. If $|\alpha| < |\lambda|$ and $\beta \neq 0$ then there exist chaotic invariant sets close to γ which persist if the vector field is perturbed. Hence the chaos does not arise at the point of existence of the homoclinic orbit. If we follow the evolution of the system in dependence on a parameter, then chaotic

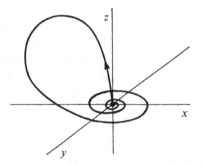

Fig. 3.31. A homoclinic orbit with oscillatory approach to a stationary point at the origin.

attractors should be observable within a certain distance from the critical value of the parameter at which the homoclinic orbit exists. The formation of chaos is, as usually, accompanied by sequences of local bifurcations of periodic orbits, notably by period-doublings and saddle-node bifurcations. Examples of such transitions are in Refs. 3.5 and 3.28.

Under certain special assumptions, such as the presence of symmetry, the chaotic set may arise at the point of a homoclinic bifurcation and may exist only in one direction from the critical value of the parameter. Such a situation arise in the Lorenz model[2.35]. Heteroclinic orbits (i.e. orbits connecting two different stationary states) can under certain circumstances play the same role as the homoclinic ones.

Transitions from torus to chaos

A large class of transitions to chaos includes formation of a two-dimensional torus from the periodic orbit. If this bifurcation is supercritical, then the created torus is attracting and can be either resonant, i.e. it contains an attracting periodic orbit, or quasiperiodic, which means that the entire torus is the attractor covered densely by a single orbit. In multidimensional state spaces a bifurcation of the torus which is resonant at the point of bifurcation may be different from that of a quasiperiodic torus. One possible type of bifurcations corresponds to the loss of the differentiability of the torus. We have argued in Section 3.3.6 that this bifurcation is much more likely on a resonant torus. Another situation arises when a differentiable two-dimensional torus undergoes a further bifurcation. Here the bifurcation of a quasiperiodic three-dimensional torus from the quasiperiodic two-torus is likely to occur but this situation cannot be modelled by the circle maps.

First considerations about the formation of complex dynamics which should describe real turbulence belong to Landau[3.49] and independently to Hopf[3.44]. They considered a sequence of bifurcations of quasiperiodic tori of successively larger dimensions. An N-dimensional torus is filled out by a quasiperiodic orbit generated by N rationally independent (or incommensurate) frequencies ω_1, ..., ..., ω_N. The rational independence means that

$$n_1\omega_1 + ... + n_N\omega_N \neq 0 \qquad (3.45)$$

for any integer valued N-tuple $(n_1, ..., n_N)$. If the successive N-tori remain quasiperiodic for arbitrary N, then the resulting motion will certainly be very complicated.

This hypothesis was discussed by Ruelle and Takens in 1971[1.56], they argued that a model of turbulence should display two principal features: (1) sensitive dependence on initial conditions; and (2) finite dimensionality. Neither of them is satisfied by the Landau–Hopf model. The proposed new model of turbulence was just the strange (or chaotic) attractor. Newhouse, Ruelle and Takens[3.62] proved that a dynamical system on an N-torus, $N \geq 3$, displaying a quasiperiodic motion, can be perturbed to a dynamical system which possesses an Axiom A chaotic attractor embedded in the N-torus. In practice it means that the sequence of bifurcations is as follows: from a periodic orbit to a quasiperiodic two-torus, from a quasiperiodic two-torus to a quasiperiodic three-torus and then an immediate formation of a chaotic attractor.

However, it is not clear a priori how frequently the points corresponding to quasiperiodic, chaotic or other motions occur in some parameter space. Grebogi, Ott and Yorke[3.36] carried out numerical experiments to elucidate this point. They found that small and moderate nonlinear perturbations of a quasiperiodic motion on an N-torus, $N = 3$ and 4, lead to the following frequency of occurrence of points corresponding to quasiperiodic motions in a parameter space: N-frequency quasiperiodic attractors are most common, followed by $(N-1)$-frequency quasiperiodic attractors, ..., followed by periodic attractors. The chaotic attractors are relatively rare but their frequency of occurrence increases considerably with increasing magnitude of perturbations.

These results suggest that the transition from the two-torus to the three-torus and then to the chaotic attractor as a parameter is varied may be very complicated. After the first occurence of chaos a complicated sequence of two- and three-frequency quasiperiodic, chaotic and periodic motions is expected to occur.

An alternative way of transition from a torus to chaos may occur via the loss of the differentiability of the two-torus and its subsequent destruction. The simplest model which displays this behaviour (and at the same time excludes the bifurcation of a three-torus) is mapping of the annulus, i.e. of a two-dimensional manifold which is the Cartesian product of an interval and a circle. The transition from a quasiperiodic torus to chaos in such a model is described by Feigenbaum *et al.*[3.26] and by Ostlund *et al.*[3.64]. If the contraction in the radial direction on the annulus is sufficiently strong then the mapping of the annulus is reduced to the mapping on the circle and the loss of the differentiability of a quasiperiodic motion corresponds to a critical circle map with an irrational rotation number as was described in Section 3.3.7. This transition has universal features which are expected to persist in multidimensional systems[3.64]. The transitions from a resonant torus to chaos in mappings of the annulus are described by Aronson *et al.*[3.6] and Ostlund *et al.*[3.64]. The loss of differentiability of the torus can proceed in many different ways and the further transition to chaotic attractors occurs via routes already described, i.e. via period-doublings, crises and intermittency.

A general mathematical theorem that deals with the loss of differentiability and a subsequent desintegration of a two-dimensional resonant torus is due to Afraimovich and Shilnikov[3.1]. This result is as follows: Let a continuous time dynamical system be given via the differential equation

$$\dot{\mathbf{x}} = v_{\mathbf{a}}(\mathbf{x}), \qquad \mathbf{x} \in \mathbb{R}^n, \qquad n \geq 3, \qquad \mathbf{a} \in \mathbb{R}^d, \qquad d \geq 2, \qquad (3.46)$$

where v is sufficiently smooth with respect to state space variable $\mathbf{x}$ and parameter variable $\mathbf{a}$. Suppose that this dynamical system has an attracting bounded resonant two-dimensional torus $T^2(\mathbf{a}) \subset \mathbb{R}^n$ at $\mathbf{a} = \mathbf{a}_0$. This means that $T^2(\mathbf{a}_0)$ contains a stable-unstable pair of periodic orbits γ_s and γ_u, characterized by a rational rotation number. In fact, the torus T^2 is the union of γ_s, γ_u and the unstable manifold $W^u_{\gamma_u}$ of the orbit γ_u. The multiplicator σ of the stable orbit γ_s which is closest to the unit circle is supposed to be real and without multiplicities. Consider a set $\{\mathbf{a}(r)\}$ of continuous curves in the parameter space $\mathbb{R}^d$ of Eq. (3.46), $0 \leq r \leq 1$, such that the system has a smooth bounded attracting resonant torus $T^2(\mathbf{a})$ at $\mathbf{a} = \mathbf{a}(0) = \mathbf{a}_0$ while there is no such torus at $\mathbf{a} = \mathbf{a}(1)$. Then the theorem on the destruction of $T^2(\mathbf{a})$ states:

(1) There exists an intermediate parameter value r_*, $0 < r_* < 1$, such that $T^2(\mathbf{a}(r))$ loses its smoothness at $r = r_*$. The multiplicator σ becomes complex or the unstable manifold $W^u_{\gamma_u}$ loses smoothness in the vicinity of γ_s at the critical parameter value.

(2) There exists a second critical parameter value r_{**}, $r_{**} > r_*$, such that the dynamical system has no torus $T^2(\mathbf{a}(r))$ for $r > r_{**}$; the torus is destroyed in one of the following ways:

(a) The stable periodic orbit γ_s on T^2 loses its stability at $r = r_{**}$ via a local bifurcation (period-doubling or torus bifurcation).

(b) The stable-unstable pair, γ_s and γ_u, merge together at $r = r_{**}$ (saddle-node bifurcation).

(c) There arises a homoclinic tangency near γ_u at $r = r_{**}$.

The same conclusions hold for discrete time systems with the two-torus replaced by a closed invariant curve. Here the requirement for the dimension of the state space is $n \geq 2$. The sketch in Fig. 3.32 displays the bifurcation diagram

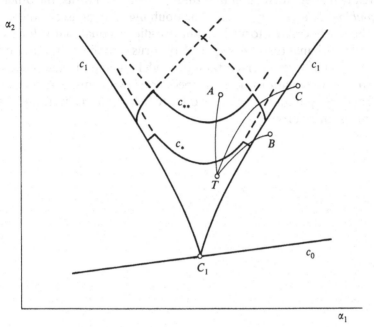

Fig. 3.32. Bifurcation diagram illustrating the possible ways of torus destruction.

in a parameter plane. As in the circle maps, the parameter region corresponding to a resonant torus is bounded by a curve c_1 having a cusp point C_1. The points on c_1 correspond to merging of γ_s and γ_u (saddle-node bifurcation) and the point C_1 is located on a torus bifurcation curve c_0. The curve c_* corresponds to the loss of smoothness, the curve c_{**} corresponds to the torus destruction and the point T corresponds to a smooth resonant torus $T^2(\alpha(0))$. The path TA is described by the case (a) of the theorem: σ becomes complex at the crossing of c_* and period-doubling (or torus bifurcation, alternatively) occurs at the intersection of TA with c_{**}. Subsequent period-doublings may lead to the appearence of a chaotic attractor. The path TB reflects the case (b): $T^2(r)$ loses smoothness and then a saddle-node occurs when TB crosses c_1. The torus disappears and an intermittent route to chaos is observed. The last path TC corresponds to the case (c): After the loss of differentiability a homoclinic tangency arises when TC crosses c_{**}. As a result, a nonattracting chaotic set is formed which may become a multiple attractor (in addition to γ_s) via a crisis bifurcation or may contribute to intermittency behind the crossing of c_1.

The above discussed bifurcation patterns are essentially complete for differential equations systems with $n = 3$ (or for invertible mappings with $n = 2$).

Finally, let us mention some possible bifurcation behaviours when the dimension of the state space exceeds the lower limits. It may be that prior to the destruction of the two-torus a different kind of bifurcation occurs. One possibility is the already discussed creation of a three-dimensional torus, the other one is a *torus-doubling bifurcation*[3.4, 3.27]. Torus-doubling is reminiscent of period-doubling – the newly created torus is again two-dimensional, but it forms two loops around the original torus. A sequence of torus-doublings may also occur. All these phenomena were observed both in modelling and in experiments but mathematical results remain to be developed. In view of this, a lot of careful experiments and computations should be done to help in formulating and finally solving the problem rigorously.

REFERENCES

3.1 Afraimovich V. S. and Shilnikov L. P. Invariant two−dimensional tori, their destruction and stochasticity. In *Methods of Qualitative Theory of Differential Equations*, Gorki, Univ. Press Gorki, 1983 (in Russian).

3.2 Alstrøm P. and Levinsen M. T. Phase locking structure of integrate−and−fire models with threshold modulation. *Phys. Lett.* **128A** (1988) 187.

3.3 Alstrøm P., Christiansen B. and Levinsen M. T. Nonchaotic transition from quasiperiodicity to complete phase locking. *Phys. Rev. Lett.* **61** (1988) 1679.

3.4 Anishchenko V. S. *Dynamical Chaos in Physical Systems − Experimental Investigations of Self−Oscillating Circuits.* Leipzig, Teubner, 1988 (Russian original, Saratov, 1985).

3.5 Arneodo A., Coullet P. H., Spiegel E. A. and Tresser C. Asymptotic chaos. *Physica* **14D** (1985) 327.

3.6 Aronson D. G., Chory M. A., Hall G. R. and McGehee R. P. Bifurcations from an invariant circle for two−parameter families of maps of the plane: A computer−assisted study. *Commun. Math. Phys.* **83** (1982) 303.

3.7 Bak P., Bohr T. and Jensen M. H. Circle maps, mode−locking and chaos. In *Directions in Chaos* Vol.2, ed. Hao Bai−lin, Singapore, World Scientific, 1988, p. 16.

3.8 Bogdanov R. I.: Bifurcation of the limit cycle of a family of plane vector fields. *Sel. Math. Sov.* **1** (1981) 373 ; Versal deformation of a singularity of a vector field on the plane in the case of zero eigenvalues. *Sel. Math. Sov.* **1** (1981) 389.

3.9 Chow S. N. and Hale J. K. *Methods of Bifurcation Theory.* New York, Springer, 1982.

3.10 Collet P. and Eckmann J.−P. On the abundance of aperiodic behavior for maps on the interval. *Commun. Math. Phys.* **73** (1980) 115.

3.11 Collet P., Eckmann J.−P. and Koch H. Period doubling bifurcations for families of maps on R^n . *J. Stat. Phys.* **25** (1981) 1.

3.12 Collet P., Eckmann J.−P. and Lanford O. E. Universal properties of maps of an interval. *Commun. Math. Phys.* **76** (1980) 211.

3.13 Crawford J. D. and Knobloch E. Classification and unfolding of degenerate Hopf bifurcations with O(2) symmetry : No distinguished parameter. *Physica* **31D** (1988) 1.

3.14 Crutchfield J. P. and Packard N. H. Symbolic dynamics of noisy chaos. *Physica* **7D** (1983) 201.

3.15 Crutchfield J. P., Farmer J. D. and Huberman B. A. Fluctuations and simple chaotic dynamics. *Phys. Reports* **92** (1982) 45.

3.16 Crutchfield J. P., Nauenburg M. and Rudnick J. Scaling for external noise at the onset of chaos. *Phys. Rev. Lett.* **46** (1981) 933.

3.17 Cvitanovic' P., Jensen M. H., Kadanoff L. P. and Procaccia I. Renormalization, unstable manifolds, and the fractal structure of mode locking. *Phys. Rev. Lett.* **55** (1985) 343.

3.18 Derrida B., Gervois A. and Pomeau Y. Iteration of endomorphisms on the real axis and representation of numbers. *Ann. Inst. H. Poincaré* **29A** (1978) 305.

3.19 Ding E. J. and Hemmer P. C. Winding numbers for the supercritical sine circle map. *Physica* **32D** (1988) 153.

3.20 Eckmann J.－P., Thomas L. and Wittwer P. Intermittency in the presence of noise. *J. Phys.* **A14** (1981) 3153.

3.21 Elphick C., Tirapeugi E., Brachet M. E., Coullet P. and Iooss G. A simple global characterization for normal forms of singular vector fields. *Physica* **29D** (1987) 95.

3.22 Farmer J. D. and Satija I. I. Renormalization of the quasiperiodic transition to chaos for arbitrary winding numbers. *Phys. Rev.* **A31** (1985) 3520.

3.23 Feigenbaum M. J. Quantitative universality for a class of nonlinear transformations. *J. Stat. Phys.* **19** (1978) 25; The universal metric properties of nonlinear transformations. *J. Stat. Phys.* **21** (1979) 669.

3.24 Feigenbaum M. J. The onset spectrum of turbulence. *Phys. Lett.* **74A** (1980) 375.

3.25 Feigenbaum M. J. The transition to aperiodic behavior in turbulent systems. *Commun. Math. Phys.* **77** (1980) 65.

3.26 Feigenbaum M. J., Kadanoff L. P. and Shenker S. J. Quasiperiodicity in dissipative systems: A renormalization group analysis. *Physica* **5D** (1982) 370.

3.27 Franceschini W. Bifurcations of tori and phase locking in a dissipative system of differential equations. *Physica* **6D** (1983) 285.

3.28 Gaspard P. and Nicolis G. What can we learn from homoclinic orbits in chaotic dynamics? *J. Stat. Phys.* **31** (1983) 499.

3.29 Geisel T. and Nierwetberg J. Onset of diffusion and universal scaling in chaotic systems. *Phys. Rev. Lett.* **48** (1982) 7.

3.30 Geisel T. and Nierwetberg J. Intermittent diffusion: A chaotic scenario in unbounded systems. *Phys. Rev.* **A29** (1984) 2305.

3.31 Glass L. and Perez R. Fine structure of phase locking. *Phys. Rev. Lett.* **48** (1982) 1772.

3.32 Golubitsky M. and Schaeffer D. *Singularities and Groups in Bifurcation Theory*. New York, Springer, 1985.

3.33 Grassberger P. On the Hausdorff dimension of fractal attractors. *J. Stat. Phys.* **26** (1981) 697.

3.34 Grassberger P. Information flow and maximum entropy measures for $1 - D$ maps. *Physica* **14D** (1985).

3.35 Grebogi C., Ott E. and Yorke J. A. Crises, sudden changes in chaotic attractors, and transient chaos. *Physica* **7D** (1983) 181.

3.36 Grebogi C., Ott E. and Yorke J. A. Attractors on an $N-$torus: Quasiperiodicity versus chaos. *Physica* **15D** (1985) 354.

3.37 Grossmann S. and Thomae S. Invariant distributions and stationary correlation functions of the one－dimensional discrete processes. *Z. Naturforsch.* **32a** (1977) 1353.

3.38 Guckenheimer J. Bifurcations of maps of the interval. *Inventiones Math.* **39** (1977) 165.

3.39 Guckenheimer J. Sensitive dependence on initial conditions for one－dimensional maps. *Commun. Math. Phys.* **70** (1979) 133.

3.40 Hassard B. D., Kazarinoff N. D. and Wan Y.－H. *Theory and Applications of Hopf Bifurcation*. Cambridge, Cambridge University Press, 1981.

3.41 Herman M. R. Mesure de Lebesgue et nombre de rotation. In *Geometry and Topology*, ed. J. Palis and M. de Carmo, Lecture Notes in Mathematics **597**, New York, Springer, 1977, p. 271.

3.42 Hirsch J. E., Huberman B. A. and Scalapino D. J. A theory of intermittence. *Phys. Rev.* **A25** (1982) 519.

3.43 Hopf E. Abzweigung einer periodischen Lösung von einer stationaren Lösung eines Differentialsystems. *Ber. Verh. Sachs. Akad. Wiss. Leipzig Math.－Nat.* **94** (1942) 3.

3.44 Hopf E. A mathematical example displaying features of turbulence. *Commun. on Pure Appl. Math.* **1** (1948) 303.

3.45   Hu B. and Rudnick J. Exact solutions to the Feigenbaum renormalization group equations for intermittency. *Phys. Rev. Lett.* **48** (1982) 1645.

3.46   Iooss G. *Bifurcation of Maps and Applications.* Mathematical Studies **36**, Amsterdam, North Holland, 1979.

3.47   Jakobson M. K. Absolutely continuous invariant measure for one parameter families of one dimensional maps. *Commun. Math. Phys.* **81** (1981) 39.

3.48   Jensen M. H., Bak P. and Bohr T. Complete devil's staircase, fractal dimension, and universality of mode−locking structure in the circle map. *Phys. Rev. Lett.* **50** (1983) 1637.

3.49   Landau L. D. On the problem of turbulence. *C. R. Acad. Sci. URSS* **44** (1944) 311. In Collected Papers of Landau, ed. D. ter Haar, Oxford, Pergamon Press, 1965, p. 387.

3.50   Lanford O. E. A computer−assisted proof of the Feigenbaum conjectures. *Bull. Amer. Soc.* **6** (1982) 427.

3.51   Lorenz E. N. Noisy periodicity and reverse bifurcation. *Ann. N. Y. Acad. Sci.* **357** (1980) 282.

3.52   Lutzky M. Reverse multifurcations and universal constants. *Phys. Lett.* **128A** (1988) 332.

3.53   MacKay R. S. and Tresser C. Transition to topological chaos for circle maps. *Physica* **19D** (1986) 206.

3.54   Manneville P. Intermittency, self−similarity and 1/f spectrum in dissipative dynamical systems. *J. de Phys.* **41** (1980) 1235.

3.55   Manneville P. and Pomeau Y. Different ways to turbulence in dissipative systems. *Physica* **1D** (1980) 219.

3.56   Marsden J. E. and McCracken M. *The Hopf Bifurcation and Its Applications.* New York, Springer, 1976

3.57   Metropolis N., Stein M. L. and Stein P. R. On finite sets for transformations on the unit interval. *J. Comb. Theor.* **A15** (1973) 25.

3.58   Milnor J. and Thurston R. On iterated maps of the interval. Unpublished notes, Princeton, Univ. of Princeton, 1977.

3.59   Nauenberg M. and Rudnick J. Universality and the power spectrum at the onset of chaos. *Phys. Rev.* **B24** (1981) 493.

3.60   Newhouse S. The abundance of wild hyperbolic sets and non−smooth stable set of diffeo-morphisms. *Publ. Math. IHES* **50** (1980) 101.

3.61   Newhouse S., Palis J. and Takens F. Bifurcations and stability of families of diffeomorphisms. *Publ. Math. IHES* **57** (1983) 5.

3.62   Newhouse S., Ruelle D. and Takens F. Occurrence of strange axiom A attractors near quasiperiodic flows on T^m, $m \geq 3$. *Commun. Math. Phys.* **64** (1978) 35.

3.63   Oono Y. Period $\neq 2$ implies chaos. *Prog. Theor. Phys.* **59** (1978) 1029.

3.64   Ostlund S., Rand D., Sethna J. and Siggia E. Universal properties of the transition from quasiperiodicity to chaos in dissipative systems. *Physica* **8D** (1983) 303.

3.65   Pianigiani G. and Yorke J. A. Expanding maps on sets which are almost invariant: Decay and chaos. *Trans. Amer. Math. Soc.* **252** (1979) 351.

3.66   Pomeau Y. and Manneville P. Intermittent transition to turbulence in dissipative dynamical systems. *Commun. Math. Phys.* **74** (1980) 189.

3.67   Poston T. and Steward I. *Catastrophe Theory and Its Applications.* London, Pitman, 1978.

3.68   Shapiro A.P. and Luppov S. P. *Reccurent Equations in the Theory of Population Biology.* Moscow, Nauka, 1983 (in Russian).

3.69   Sharkovskii A. N. Coexistence of cycles of a continuous map of a line into itself. *Ukr. Math. Z.* **16** (1964) 61.

3.70   Shell M., Fraser S. and Kapral R. Subharmonic bifurcation in the sine map: An infinite hierarchy of cusp bistabilities. *Phys. Rev.* **A28** (1983) 373.

3.71   Shilnikov L. P. A case of the existence of a countable set of periodic motions. *Sov. Math. Dokl.* **6** (1965) 163.

3.72   Singer D. Stable orbits and bifurcations of maps of the interval. *SIAM J. Appl. Math.* **35** (1978) 260.

3.73   Štefan P. A theorem of Šarkovskii on the existence of periodic orbits of continuous endomorphisms of the real line. *Commun. Math. Phys.* **54** (1977) 237.

3.74   Thom R. *Structural Stability and Morphogenesis.* Reading, MA, Benjamin, 1975.

3.75   Umberger D. K., Farmer J. D. and Satija I. I. A universal strange attractor underlying the quasiperiodic transitions to chaos. *Phys. Lett.* **114A** (1986) 341.

3.76   Vul E.B., Sinai Ya. G. and Khanin K. M. Universality of Feigenbaum and thermodynamic formalism. *Usp. Mat. Nauk* **39** (1984) 3.

3.77   Yorke J. A., Grebogi C., Ott E. and Tedeschini—Lalli L. Scaling behavior of windows. *Phys. Rev. Lett.* **54** (1985) 1095.

3.78   Zeeman E. C. *Catastrophe Theory: Selected Papers 1972—1977.* Reading, MA, Addison—Wesley, 1977.

4

Numerical methods for studies of parametric dependences, bifurcations and chaos

Evolution problems are most often described by differential or difference equations in a finite-dimensional state space. However, both types of models can also possess infinite-dimensional state spaces, e.g. if they are in the form of partial differential equations, equations with time delay, integro-differential equations or functional iterations used in renormalization group theory. Numerical techniques used in computations of the time evolution of orbits differ according to the studied case. They can be relatively simple, as in the case of a low-dimensional system of ordinary differential equations, but often supercomputers have to be used[7.54], as in the case of larger systems of partial differential equations. The problem of finding special types of orbits and the determination of their stability is usually more difficult than direct integration.

In this chapter we shall limit our discussion to finite-dimensional systems in R^n where n is an acceptably small integer. Thus we shall work either with difference equations,

$$\mathbf{x}_{k+1} = f(\mathbf{x}_k, \alpha), \tag{4.1}$$

or with a system of ordinary differential equations (ODEs)

$$\frac{d\mathbf{x}}{dt} = v(\mathbf{x}, \alpha), \tag{4.2}$$

where $\mathbf{x} \in R^n$ is a state space variable and $\alpha \in R^d$ is a vector valued parameter. To determine an evolution of $\mathbf{x}$ with time for given initial conditions and a constant α is a trivial problem for Eq. (4.1). A large number of integration routines can be used to solve the same problem for Eq. (4.2)[4.45]. Usually single step integration routines with automatic control of step size are preferred, as the local error of the solution can be estimated and control of the step size will help to achieve an optimal computation time. Stiff systems form a special and often occurring type of ODEs. In such systems processes on considerably different time scales take place, which lead to fast changes of some components of the vector **x**. Special integration routines, such as Gear's algorithm[4.25], have to be used in such cases.

We always have to consider that the computer is a finite machine. Thus algorithms for finding a solution of ODEs not only replace the continuous time by a discrete one, but the entire state space consists of a network of discrete points. Floating point arithmetic enables us to work with relatively large or small numbers, but a loss of significant digits in the mantissa may occur. These errors can accumulate and invalidate numerical results[2.8, 4.67]. Such a situation may arise, for example, in computing Lyapunov exponents (LEs) and therefore the steplength in integration or other routines has to be carefully chosen.

Characterization of experimental data (time series) is another problem; LEs, dimensions of the attractors or entropies can be computed. The quality of estimates of these quantities depends both on the quality of the measured data (length of the record, sampling frequency, the level of noise, etc.) and on the computational algorithm used.

4.1 Continuation of periodic orbits and their bifurcations

The birth of chaos is often accompanied by local bifurcations of periodic orbits and the periodic behaviour often reemerges in the chaotic region. Knowledge of the parameter dependence of such orbits, even in cases when they are unstable, can facilitate studies of crises, intermittency, window structures, etc.

Let us consider a system of equations

$$F_i(x_1, ..., x_n, \alpha) = 0 , \qquad i = 1, ..., n \tag{4.3}$$

or equivalently

$$F(\mathbf{x}, \alpha) = \mathbf{0} , \tag{4.4}$$

where $\mathbf{F} = (F_1, ..., F_n)$, $\mathbf{x} = (x_1, ..., x_n)$ and α is a real valued parameter. If Eq. (4.4) has a solution $\mathbf{x}_0$ for certain α_0 and $\det[\mathbf{D_x} \mathbf{F}(\mathbf{x}, \alpha)] \neq 0$ at the point $(\mathbf{x}_0, \alpha_0)$, then according to the implicit function theorem, the solution of (4.4) can be expressed in the neighbourhood of this point as a parametric dependence of $\mathbf{x}$ on α. More generally, the set of all pairs $(\mathbf{x}, \alpha)$ satisfying (4.4) consists of curves in the product space $\mathbb{R}^n \times \mathbb{R}$. We want to determine these curves numerically, to associate them with families of periodic orbits and to indicate bifurcation points on them.

Direct use of the implicit function theorem is unsuccessful in the neighbourhood of points where the matrix $\mathbf{D_x} \mathbf{F}(\mathbf{x}, \alpha)$ becomes singular. A suitable method has to generate an entire curve including singular points. We shall take the arc-length of the solution curve in the product space as a new parameter for the continuation[4.40]. Let us denote $x_{n+1} = \alpha$, then the length c of the arc is determined by normalizing the length of the tangent vector $(dx_1/dc, ..., dx_{n+1}/dc)$,

$$\left(\frac{dx_1}{dc}\right)^2 + ... + \left(\frac{dx_n}{dc}\right)^2 + \left(\frac{dx_{n+1}}{dc}\right)^2 = 1 . \tag{4.5}$$

Upon differentiating (4.3) with respect to c we obtain

$$\frac{dF_i}{dc} = \sum_{j=1}^{n+1} \frac{\partial F_i}{\partial x_j} \frac{dx_j}{dc} = 0, \quad i = 1, ..., n . \tag{4.6}$$

Initial conditions for (4.5), (4.6) can be written in the form

$$c = 0: (\mathbf{x}, \alpha) = (\mathbf{x}_0, \alpha_0) . \tag{4.7}$$

Eqs (4.6) form a system of n linear equations in $n + 1$ unknowns dx_i/dc, $i = 1, ..., n + 1$. Let us assume that the n by n matrix which does not contain derivatives with respect to x_k,

$$J_k = \begin{bmatrix} \dfrac{\partial F_1}{\partial x_1}, & ..., & \dfrac{\partial F_1}{\partial x_{k-1}}, & \dfrac{\partial F_1}{\partial x_{k+1}}, & ..., & \dfrac{\partial F_1}{\partial x_{n+1}} \\[2ex] \dfrac{\partial F_2}{\partial x_1}, & ... & & & & \\[2ex] \vdots & & & & & \\[2ex] \dfrac{\partial F_n}{\partial x_1}, & ..., & \dfrac{\partial F_n}{\partial x_{k-1}}, & \dfrac{\partial F_n}{\partial x_{k+1}}, & ..., & \dfrac{\partial F_n}{\partial x_{n+1}} \end{bmatrix} \tag{4.8}$$

is regular for some k, $1 \le k \le n + 1$ at every point along the continued curve. Fixing some k, Eqs. (4.6) can be solved for n unknowns δ_i, where

$$\frac{dx_i}{dc} = \delta_i \frac{dx_k}{dc} , \quad i = 1, ..., k - 1, k + 1, ..., n + 1 , \tag{4.9}$$

provided we know all derivatives $\dfrac{\partial F_i}{\partial x_j}$, $j = 1, ..., n + 1$; $j \neq k$; $i = 1, ..., n$.

The remaining unknown derivative dx_k/dc is determined from (4.5) as

$$\frac{dx_k}{dc} = \left(1 + \sum_{\substack{i=1 \\ i \neq k}}^{n+1} \delta_i^2 \right)^{-1/2} . \tag{4.10}$$

Eqs. (4.9) and (4.10) form a set of $n + 1$ coupled differential equations; the coefficients δ_i are determined by solving the linear system (4.6) as described above. The Gauss elimination method with a pivoting can be used to find δ_i and a suitable index k, so that the matrix J_k will be regular. The system of differential equations (4.9) and (4.10) with the initial condition (4.7) can be solved by any numerical technique for the integration of initial-value problems. However, every integration step requires the solving of a system of linear equations. Hence, it is advantageous to use a numerical method which requires relatively few evaluations of the right-hand sides. For example, explicit multistep integration routines can be used[4.40].

Nevertheless, approximation errors might accumulate in the course of the integration. In such a case, Newton's method applied to (4.3) for a fixed x_k can be used at every integration step. This method can also be used to find the starting solution $(\mathbf{x}_0, \alpha_0)$. The entire algorithm together with the FORTRAN program is described in Refs 4.40 and 4.41. For other approaches see Refs 4.12 to 4.14, 4.31, 4.35 and 4.58.

Hence to find a solution curve we need to know the derivatives of $\mathbf{F}$ with respect to $\mathbf{x}$ and α and the starting point $(\mathbf{x}_0, \alpha_0)$. The method is directly applicable to the construction of a parameter dependence of stationary points of the system of ODEs with a scalar parameter α

$$\frac{dx_i}{dt} = v_i(x_1, ..., x_n, \alpha), \qquad i = 1, ..., n . \tag{4.11}$$

Here $F_i \equiv v_i$ and the required derivatives are obtained by differentiating v_i with respect to x_i and α.

To find the dependence of a periodic orbit on a parameter is not so straightforward. The case of systems with discrete time is relatively simple. A q-periodic orbit $\{\mathbf{x}_1^*, ..., \mathbf{x}_q^*\}$ of (4.1) for a given $\alpha \in \mathsf{R}$ can be recovered from any one of the q periodic points $\mathbf{x}_i^*$, $i = 1, ..., q$. These points are fixed under f^q hence it is sufficient to take only one of them for the continuation. The equation of the form (4.4) is a simple boundary value problem

$$F(\mathbf{x}, \alpha) = f^q(\mathbf{x}, \alpha) - \mathbf{x} = 0 . \tag{4.12}$$

The derivatives of $f^q(\mathbf{x}, \alpha)$ with respect to $\mathbf{x}$ and α are obtained from those of $f(\mathbf{x}, \alpha)$ according to the rules of differentiation of composed functions. This is an analogy of shooting technique for continuous time systems described below. The stability of an actual orbit along the continued family of orbits is given by the eigenvalues of the matrix $D_{\mathbf{x}} f^q$ evaluated at the chosen periodic point.

The situation is more complicated in continuous time systems. To obtain an equation of the form (4.4) we need to solve a boundary value problem. In addition, the period T of the continued orbit depends on α. We can use shooting technique[4.26, 4.41] to find the derivatives of F with respect to its arguments. Let

$$\mathbf{g}^t(\mathbf{x}, \alpha) = \left(g_1^t(\mathbf{x}, \alpha), ..., g_n^t(\mathbf{x}, \alpha)\right)$$

be the flow associated with Eqs. (4.11), i.e. $\mathbf{x}(t, \alpha) = \mathbf{g}^t(\mathbf{x}, \alpha)$. Then we can write down (4.4) in the form

$$F_i(\mathbf{x}, \alpha) = g_i^T(\mathbf{x}, \alpha) - x_i = 0 , \qquad i = 1, ..., n . \tag{4.13}$$

We seek a solution of (4.13) such that $T > 0$ and $g^t(\mathbf{x}, \alpha) \neq \mathbf{x}$ for $0 < t < T$. However, we have n equations for $n + 2$ unknowns $(x_1, ..., x_n, \alpha, T)$ hence an additional condition must be included to fix the point $\mathbf{x}$ on the orbit. This can be accomplished in several ways. For example, we can define a Poincaré section Σ and choose $\mathbf{x}$ in Σ; the simplest choice of Σ is to fix some x_m, $1 \leq m \leq n$. Now we have a standard situation for the application of the continuation algorithm described above, i.e. a system with n nonlinear equations and $n + 1$ unknowns $(x_1, ..., x_{m-1}, x_{m+1}, ..., x_n, \alpha, T)$. One has to change Σ adaptively in the process of continuation otherwise the intersection of the periodic orbit with Σ may be lost. For other ways of fixing conditions see, e.g. Refs 4.1, 4.14 and 4.33. After choosing initial values of the unknowns (and fixing x_m) we obtain the values of F_i in (4.13) by direct integration of Eqs. (4.11). The values of the derivatives of F_i with respect to x_i, α and T can be determined by solving the following set of variational equations

$$\frac{du_{ij}}{dt} = \sum_{s=1}^{n} \frac{\partial v_i}{\partial x_s} u_{sj}, \qquad i, j = 1, ..., n \tag{4.14}$$

with initial conditions

$$u_{ij}(0) = \delta_{ij} , \tag{4.15}$$

where δ_{ij} is the Kronecker δ and additional variational equations

$$\frac{dq_i}{dt} = \sum_{s=1}^{n} \frac{\partial v_i}{\partial x_s} \cdot q_s + \frac{\partial v_i}{\partial \alpha} \tag{4.16}$$

with initial conditions

$$q_i(0) = 0 . \tag{4.17}$$

In fact, the solutions of (4.14)–(4.17) at time T are

$$u_{ij}(T) = \frac{\partial g_i^T(\mathbf{x}, \alpha)}{\partial x_j} , \quad q_i(T) = \frac{\partial g_i^T(\mathbf{x}, \alpha)}{\partial \alpha} . \tag{4.18}$$

From the definition of the vector field we have

$$\frac{\partial g^T(\mathbf{x}, \alpha)}{\partial T} = v_i(g^T(\mathbf{x}, \alpha), \alpha) .$$

Hence the required derivatives of F_i in (4.13) are

$$\frac{\partial F_i}{\partial x_j} = u_{ij}(T) - \delta_{ij}, \quad \frac{\partial F_i}{\partial T} = v_i(\mathbf{x}(T, \alpha), \alpha), \quad \frac{\partial F_i}{\partial \alpha} = q_i(T) .$$

Now we have the necessary information for solving Eqs. (4.6), (4.9) and (4.10), as well as for the application of Newton's method.

The fixed value of x_m can be changed adaptively in the process of continuation, if necessary. The eigenvalues of the monodromy matrix $\{u_{ij}(T)\}$, $i, j = 1$, ..., n, evaluated at each point of the solution family are multipliers determining the stability of the periodic orbit.

Finally, the periodically forced nonautonomous systems of the form

$$\frac{d\mathbf{x}}{dt} = v(\mathbf{x}, t, \alpha), \quad v(\mathbf{x}, t + T, \alpha) = v(\mathbf{x}, t) \tag{4.19}$$

can be handled similarly to discrete time systems by considering the Poincaré mapping $P(\mathbf{x}, \alpha)$ as defined in Section 2.3. However, the derivatives of $P(\mathbf{x}, \alpha)$ with respect to $\mathbf{x}$ and α have to be computed via variational equations similar to those for the autonomous system (see Eqs (4.14)–(4.17)).

Eventual numerical instabilities of the boundary value problems both for discrete (if $q > 1$) and continuous time systems can be effectively removed by using multiple shooting technique. This method consists in partitioning the period of the orbit into intervals and computing the derivatives on each interval separately at the expense of increased number of equations.

For example, the boundary value problem for a q-periodic orbit of difference equations (4.1) solved by the multiple shooting method is

$$f(\mathbf{x}_i, \alpha) - \mathbf{x}_{i+1} = 0 , \quad i = 1, ..., q - 1; \quad \mathbf{x}_{q+1} = \mathbf{x}_1 . \tag{4.20}$$

The dimension of the problem is qn but we avoid computation of derivatives of f^q which may not be correct due to numerical round-off errors. This method is even more important for continuous time systems. For other approaches to computation of periodic orbits see Refs 4.35, 4.46, 4.50, 4.60, 4.62, 4.64.

The above described algorithm produces a solution diagram, i.e. a parameter dependence of stationary or periodic orbits. It can easily be extended to produce a bifurcation diagram, i.e. curves in a two-dimensional parameter space along which the bifurcations occur (see Chapter 3 and Appendix A). Using the computed linearization matrix of the continued orbit and its eigenvalues we can impose a constraint on this matrix such that an eigenvalue (or a pair of complex conjugate eigenvalues) takes on a prescribed value. This amounts to the addition of one (or two in the complex conjugate case) equation to the system (4.3). The solution of this extended system is a codimension one bifurcation point $(\mathbf{x}, \alpha_c)$ of a prescribed type[4.27] (i.e. period-doubling point, limit point or point of the torus bifurcation). A new parameter β can now be introduced and the continuation of the bifurcation point in dependence on β yields the desired curve in the (α, β) parameter plane. This procedure can be formally extended to cover bifurcations of a higher codimension.

A FORTRAN program for finding solution and bifurcation diagrams of stationary points of Eq. (4.2) and periodic orbits of Eqs (4.1), (4.2) and (4.19) according to methods described here is found in Appendix B. Recently, a number of authors, see Refs 4.1, 4.6, 4.13, 4.36, 4.61, have developed a software for continuation and bifurcation analysis of discrete and continuous time systems. The program presented in this book is simple enough to be handled and adapted by students. A more sophisticated version conceived as an open expert system[4.6, 4.51] enabling a more comfortable work is available in our research group.

Other special solutions of ODEs than stationary and periodic ones can also be studied numerically using Newton's method combined with continuation. In particular, orbits homoclinic to a stationary point (saddle-loops)[4.32, 4.44], invariant tori[4.29, 4.30, 4.34] and other manifolds[4.42, 4.43] can be found iteratively. Let us mention here that the Newton's iteration procedure itself may have non-attracting chaotic limit sets and/or attractors in infinity[4.4]. This may cause either divergence or transient chaotic dynamics before approaching a fixed point unless the initial estimate is sufficiently close to it.

4.2 Lyapunov exponents from mathematical models

Numerical computation of Lyapunov exponents (LEs) makes use of the fact that in the tangent space vectors expand (or contract) by the action of the linearized flow along an orbit. We shall describe the original algorithm for the computation of LEs proposed by Benettin *et al.*[4.3] and by Shimada and Nagashima[4.63] other for continuous time systems and make some remarks on other methods of computation.

Let us consider a continuous time system defined by the differential equation

$$\frac{dx}{dt} = v(\mathbf{x}), \qquad \mathbf{x} \in \mathbf{R}^n \tag{4.21}$$

(we do not consider an explicit dependence on a parameter here). All trajectories of (4.21) are given by the associated flow

$$\mathbf{x}(t) = g^t(\mathbf{x}) . \tag{4.22}$$

Let us fix an initial point $\mathbf{x}(0) = \mathbf{x}$ at time $t = 0$ in the state space and choose at random n linearly independent vectors $\mathbf{w}_1, ..., \mathbf{w}_n$ spanning the tangent space at $\mathbf{x}$. Consider next i-dimensional paralellepipeds formed by the vectors $\mathbf{w}_1, ..., ..., \mathbf{w}_i$, $i = 1, ..., n$. The linear flow $U^t_{\mathbf{x}(0)} = D_{\mathbf{x}} g^t(\mathbf{x})$ along the followed orbit

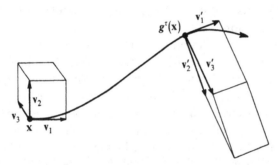

Fig. 4.1. Deformation of a three-dimensional parallelepiped by the flow $g^t(\mathbf{x})$; the parallelepiped generated by the vectors v_1, v_2 and v_3 at $\mathbf{x}$ is deformed to that generated by v'_1, v'_2 and v'_3 after time τ. The vector v_1 is tangent to the orbit and does not change substantially, the vectors v_2, v_3 grow with time and rotate but the volume of the parallelepiped is contracted by the (dissipative) flow.

will deform the i-dimensional paralellepiped to that formed by the vectors $U^t_{\mathbf{x}(0)} \mathbf{w}_1, ..., U^t_{\mathbf{x}(0)} \mathbf{w}_i$ after time t. Individual vectors in the tangent space at $\mathbf{x}$ may be expanded or contracted exponentially fast by the flow but the vectors tangent to the followed orbit do not have this property, see Fig. 4.1. We denote the volume of the i-dimensional paralellepiped (in the space $\mathbf{R}^i$) at time t by $V^i(t)$.

The following definition of LEs is convenient for numerical computations,

$$\lambda^{(i)} = \lim_{t \to \infty} t^{-1} \log \left(V^{(i)}(t)/V^{(i)}(0) \right) \tag{4.23}$$

here $\lambda^{(i)}$ are called multidimensional LEs. Almost every choice of the vectors $\mathbf{w}_1, ..., \mathbf{w}_n$ leads to a maximal possible (average) rate of an exponential growth of

$V^{(i)}(t)$ and (4.23) will yield a maximal $\lambda^{(i)}$, denoted by $\lambda^{(i)}_{max}$. Consider the (normal) LEs arranged in the decreasing order, $\lambda_1 \geq \lambda_2 \geq ... \geq \lambda_n$. A relation between λ_i and $\lambda^{(i)}_{max}$ is given by

$$\lambda^{(i)}_{max} = \sum_{j=1}^{i} \lambda_j, \qquad i = 1, ..., n .\tag{4.24}$$

Choosing a sufficiently large time T we can numerically compute an approximation of the limit in (4.23) for $i = 1, ..., n$, which together with (4.24) gives the estimates of LEs.

Similarly as in the computations of stability of periodic orbits, we need to integrate variational equations to obtain $U^t_{x(0)}$. We denote

$$U^t_{x(0)} = \{u_{ij}\}, \qquad v = (v_1, ..., v_n), \qquad x = (x_1, ..., x_n)$$

and write (see Eq. (4.14))

$$\frac{du_{ij}}{dt} = \sum_{s=1}^{n} \frac{\partial v_i}{\partial x_s} u_{sj}, \qquad i, j = 1, ..., n \tag{4.25}$$

or in a compact form as a matrix differential equation

$$\frac{dU^t_{x(0)}}{dt} = [D_x\, v(x(t))]\, U^t_{x(0)} \tag{4.26}$$

with $U^0_{x(0)}$ equal to the unit matrix. However, the accumulation of numerical errors and the loss of numerically obtainable information (because of the exponential expansion or contraction of the initial vectors) does not allow for a direct use of the definition (4.23).

The problem can be solved in the following way. We choose a suitable unit of time τ. According to the chain rule

$$U^{k\tau}_{x(0)} = U^{\tau}_{x((k-1)\tau)} \cdot \ ... \ \cdot U^{\tau}_{x(\tau)} \cdot U^{\tau}_{x(0)} \tag{4.27}$$

the definition (4.23) can be within a large but finite time interval $T = K\tau$ rewritten as

$$\lambda^{(i)} \approx \frac{1}{K} \sum_{k=1}^{K} \log\, (V^{(i)}(k\tau)/V^{(i)}(k-1)\,\tau)). \tag{4.28}$$

At each step of the procedure defined by the time unit of the length τ we can orthonormalize the vectors forming the deformed paralellepiped. This renormalization does not change the average growth rate of $V^{(i)}(k\tau)$ but the loss of numerical accuracy in the course of the integration of variational equations (4.26) is avoided.

Numerical errors occur also in the orthonormalization procedure and may accumulate if τ is too small. Thus τ has to be chosen carefully.

Starting with an orthonormal set of vectors $\bar{\mathbf{w}}_1, ..., \bar{\mathbf{w}}_n$ and using the Gramm–Schmidt orthogonalization procedure we have for $k = 1, 2, ...,$ recurrently

$$\mathbf{w}_i^k = U^\tau_{\mathbf{x}((k-1)\tau)} \bar{\mathbf{w}}^{k-1}, \qquad i = 1, ..., n,$$

$$\bar{\mathbf{w}}_1^k = \mathbf{w}_1^k / \|\mathbf{w}_1^k\|, \qquad d_1^k = \|\mathbf{w}_1^k\|,$$

and

$$\tilde{\mathbf{w}}_i^k = \mathbf{w}_i^k - \sum_{m=1}^{i-1} \langle \mathbf{w}_i^k, \bar{\mathbf{w}}_m^k \rangle \bar{\mathbf{w}}_m^k, \tag{4.29}$$

$$\bar{\mathbf{w}}_i^k = \tilde{\mathbf{w}}_i^k / \|\tilde{\mathbf{w}}_i^k\|, \qquad d_i^k = \|\tilde{\mathbf{w}}_i^k\|, \qquad i = 2, ..., n.$$

The volume ratio $V^{(i)}(k\tau)/V^{(i)}((k-1)\tau)$ is at each step equal to the product $d_1^k d_2^k ... d_i^k$; on introducing it to (4.28) and using (4.24) we have

$$\lambda_i \approx \frac{1}{K} \sum_{k=1}^{K} \log d_i^k \tag{4.30}$$

and λ_is are arranged in decreasing order.

Another method of computation of LEs proposed by Eckmann and Ruelle[2.16] uses a decomposition of U into the product of an orthogonal and an upper triangular matrix. The triangularization method yields more precisely orthogonal matrices than the Gramm–Schmidt orthogonalization. Moreover, this method is more convenient if we are interested only in the i largest LEs. For an algebraic interpretation of the geometric method described above see Greene and Kim[4.20] and Lorenz[4.47].

The LEs for a periodically driven system with driving period T are computed in the same way as above but the reorthonormalization time τ has to be either an integer multiple of T or equal to $\tau = T/k$, where k is an integer. The computations of LEs for discrete-time systems defined by a map $f(\mathbf{x})$ do not require the integration of variational equations, the derivatives are directly accessible from the map $f(\mathbf{x})$.

The knowledge of LEs allows us to evaluate the Lyapunov dimension D_L and the metric entropy h_μ of an attractor (see Section 2.5).

4.3 Computation of Poincaré sections for continuous orbits

An effective algorithm for the determination of intersections of a trajectory of the system of differential equations (4.21) with a codimension one cross-section given by

$$S(x_1, ..., x_n) = 0 \qquad (4.31)$$

was described by Hénon[4.23]. If we use a single-step method for the numerical integration of (4.21), we can apply the following procedure. At each integration step a point on the trajectory is obtained and the sign of $S(x_1, ..., x_n)$ is evaluated. The passage through the section is indicated by a sign change. The intersection could then be determined by an interpolation between the points on the trajectory with opposite signs of S. However, this procedure often leads to an error larger than the error of the integration method. This can be avoided by introducing a new variable

$$x_{n+1} = S(x_1, ..., x_n) \qquad (4.32)$$

and by adding the equation

$$\frac{dx_{n+1}}{dt} = v_{n+1}(x_1, ..., x_n) \qquad (4.33)$$

to the system (4.21); here

$$v_{n+1} = \sum_{i=1}^{n} v_i \frac{\partial S}{\partial x_i} .$$

A cross-section in the new system

$$\frac{dx_i}{dt} = v_i(x_1, ..., x_n), \qquad i = 1, ..., n + 1 \qquad (4.34)$$

is defined by the condition

$$x_{n+1} = 0 .$$

Now we can integrate the system (4.34) until the sign change of the variable x_{n+1} occurs, than we switch to the system

$$\frac{dx_i}{dx_{n+1}} = \frac{v_i(x_1, ..., x_n)}{v_{n+1}(x_1, ..., x_n)}, \qquad i = 1, ..., n,$$

$$\frac{dt}{dx_{n+1}} = \frac{1}{v_{n+1}(x_1, ..., x_n)}, \qquad\qquad (4.35)$$

where x_{n+1} is an independent variable; on using the last computed point as an initial condition we perform one integration step in Eq. (4.35) with the stepsize $\Delta x_{n+1} = -S$. Thus we obtain a point in the cross-section within the error of the used integration method.

Poincaré maps of multidimensional systems simplify the geometric interpretation of a chaotic attractor by reducing its dimension by one. If the dimension of the attractor in the cross-section of dimension $n - 1$ is still large (i.e. in the case of k positive LEs, $k > 1$) one can proceed further and look for transverse intersection of the attractor with a $(n - 1 + k)$ – dimensional hyperplane as described by Kostelich and Yorke[4.38].

The computations between two successive points of a Poincaré map requires numerical integration of ODEs. The computation time can be substantially decreased by using a simplicial approximation of the Poincaré map[4.68]. This consists in partitioning the cross-section into simplices and finding a linear map on each simplex corresponding to the given differential system. The vertices of each simplex are mapped exactly and other points are obtained through linear interpolation. Similar in spirit is the interpolated cell mapping approach[4.66].

4.4 Reconstruction of attractors from experimental time series

Before we describe numerical methods of computation of dimensions and entropies we shall discuss the method which enables us to realize such computations from time series of experimental variables: namely, the well-known 'reconstruction of the phase space' proposed by Takens[4.65] and by Packard *et al.*[4.53].

An experimental system can often be considered as a system with an infinite-dimensional state space; however, an asymptotic set where the dynamics of the system takes place is often of relatively low dimension. In the measurements we can sometimes obtain only a scalar signal which is recorded digitally, at constant time intervals Δt. Let us assume that we have an infinite time series $u_0, u_1, u_2, ...$ measured with infinite accuracy. From this series we can obtain a sequence of vectors in R^n, x_0, x_1, x_2, ..., defined by $x_i = (u_i, u_{i+k}, ..., u_{i+k(n-1)})$ where k, n are some chosen positive integers.

If the recorded data are stationary (without transients), we can assume that the sequence $\{x_i\}$ defines its ω-limit set A which can be considered to be an attractor. Also we can assume that $\{x_i\}$ defines an asymptotic measure on A as defined in Section 2.5, i.e. the measure of a subset K of A is given by a relative occurrence of points x_i in K. If n is sufficiently large then the attractor in the original state space is embedded in R^n, hence the trajectories on A do not intersect. The lowest value $n_{min} = m$ for which this occurs is called an *embedding dimension*, see Section 2.5.

Actual experimental time series are finite and of finite accuracy. If we assume that the level of noise is small and the length of the time series is sufficiently large, then $\{x_i\}$ is an acceptable approximation of the attractor. The time delay $T = k \, \Delta t$ by means of which the sequence $\{x_i\}$ is reconstructed has to be carefully chosen so that the individual terms of the sequence are as far as possible independent. Fraser and Swinney[4.17] have provided a criterion for the choice of T based on a minimum of a quantity which they have called a *mutual information*.

The method used to find embedding dimension m is to increase n systematically, until trajectories no longer appear to intersect. This rather subjective criterion has been removed by Broomhead and King[4.7] who have used a singular value decomposition method. This method yields directly a consistent estimate of m and in addition, a basis in R^n such that the attractor generated by $\{x_i\}$ expressed in this basis has an invariant geometry confined to a subspace of fixed dimension for $n > m$. The amount of noise is also clearly indicated.

The measured signal need not necessarily be scalar; it is certainly desirable to have several independent signals. If the number of measured signals is s, then u_i is an s-vector and $\{x_i\}$ is located in R^{ns}. The rest of the procedure is analogous to that described above, see Eckmann and Ruelle[2.16]. Crutchfield and McNamara[4.11] have developed a computer program which, given a time series, yields equations of motion (in a polynomial approximation) through an optimization procedure.

4.5 Lyapunov exponents from time series

Experimental data usually allow us to determine only trajectories but not the matrix of derivatives along the trajectory, analogous to the linearized flow $D_x g^t$ of Eq. (4.21). If we succeed in obtaining an estimate of $U_x^t(0) = D_x g^t$ from the experimental data, then the next procedure in evaluating LEs is similar to that used in numerical computations. However, only positive exponents are expected to be determined with confidence. The method described below was proposed by Eckmann and Ruelle[2.16] and by Sano and Sawada[4.59].

The reorthonormalization time τ will be an integer multiple of the time interval Δt between two successive measurements, $\tau = p \, \Delta t$. As in Section 4.2, τ must be chosen carefully. Let us suppose that we have the reconstructed sequence $\{\mathbf{x}_i\}$ in R^m m being determined as described in Section 4.4. Then $\mathbf{x}_i$ is mapped to $\mathbf{x}_{i+p}$ by the map $\mathbf{g}^\tau$ and the matrix of the derivatives $U_{\mathbf{x}_i}^\tau$ will be obtained by a best multilinear fit of the map which for $\mathbf{x}_j$ close to $\mathbf{x}_i$ maps $\mathbf{x}_j - \mathbf{x}_i$ to $\mathbf{x}_{j+p} - \mathbf{x}_{i+p}$. For example, we can consider all $\mathbf{x}_j$s contained in the ε-ball centred at $\mathbf{x}_i$ such that $\mathrm{d}(\mathbf{x}_i, \mathbf{x}_j) \leqq \varepsilon$ and $\mathrm{d}(\mathbf{x}_{i+p}, \mathbf{x}_{j+p}) \leqq \varepsilon$.

The value of ε must be determined empirically and the length of the sequence $\{\mathbf{x}_i\}$ must be sufficiently large. In principle, m points are enough to determine a linear mapping from R^m to itself but a higher number of points enables us to use linear regression analysis (the least squares fit). Once the matrices $U_{\mathbf{x}_{k+p}}^\tau$ $k = 0, 1, \ldots$ are given, we can choose an orthonormal basis $\mathbf{w}_1, \ldots, \mathbf{w}_m$ in R^m and proceed as in Section 4.2.

The matrix $U_{\mathbf{x}_i}^\tau$ is well defined only in directions with many points $\mathbf{x}_j$; this is expected to occur for unstable directions (which are contained within the attractor). The elements of $U_{\mathbf{x}_i}^\tau$ corresponding to contracting directions will be determined inaccurately because we have poor information about orbits which approach the attractor but do not belong to it. Thus only positive LEs can be determined with confidence.

A different method of the computation of LEs working with small vectors directly in the state space instead of the tangent space was proposed by Wolf *et al.*[4.71]. One starts with the first point $\mathbf{x}_0$ and its nearest neighbour $\mathbf{y}_0$. The distance L_0 between $\mathbf{x}_0$ and $\mathbf{y}_0$ is allowed to evolve in time until it exceeds a value ε at time k_1. Let $\mathbf{x}_{k_1}$, $\mathbf{z}_{k_1}$ denote the evolved points $\mathbf{x}_0$, $\mathbf{y}_0$, respectively, and let L_0' be the evolved distance L_0. A new neighbour $\mathbf{y}_{k_1}$ of $\mathbf{x}_{k_1}$ is sought such that the distance L_1 between $\mathbf{x}_{k_1}$ and $\mathbf{y}_{k_1}$ is less then ε and $\mathbf{y}_{k_1}$ lies as nearly as possible in the same direction from $\mathbf{x}_{k_1}$ as $\mathbf{z}_{k_1}$. The procedure is continued until the end of the sequence $\{\mathbf{x}_i\}_0^N$ is reached. The largest Lyapunov exponent is estimated as

$$\lambda_1 = \frac{1}{N \, \Delta t} \sum_{j=0}^{M-1} \log \frac{L_{k_j}'}{L_{k_j}} \tag{4.36}$$

where $k_0 = 0$, M is the number of replacement steps and N is the total number of points of the sequence $\{\mathbf{x}_i\}$. This method can in principle be extended to compute all nonnegative LEs but it becomes too complicated. Nevertheless the method seems to be more efficient in the computation of the largest LE than the previous one[4.69].

4.6 Dimensions and entropies of attractors

Numerical methods for the computation of dimensions and entropies are in principle the same both for chaos generated in model equations and in experiments. They require a knowledge of a sequence of points sampled equidistantly in time from a chaotic orbit. This sequence can be obtained from a numerical integration of a model as well as from an experimental time series, using the reconstruction procedure described in Section 4.4. In addition, knowledge of the asymptotic measure is required except for the computation of the capacity dimension. This measure can be determined numerically from its definition (see Section 2.5). However, we need not compute the measure explicitly when using an efficient numerical method. If we use an orbit reconstructed from a time series, we do not know a priori the embedding dimension m. We can use an estimate of m, see Refs 4.7, 4.8 or compute the required quantity subsequently for increasing values of n and take as a final result the asymptotically constant value of the sequence of estimates.

First attempts to compute the capacity dimension D_C and the information dimension D_I of the attractor (for definitions see Section 2.5, Eqs (2.41) and (2.42), respectively) used the box counting algorithms. For example, computation of the capacity proceeds as follows: the part of the state space containing the attractor is partitioned into boxes of size ε. Then the number $M(\varepsilon)$ of boxes containing at least one point is determined. It follows from (2.41) that for small ε

$$M(\varepsilon) \sim \varepsilon^{-D_C}.$$

The plot of $\log M(\varepsilon)$ against $\log \varepsilon$ is assumed to possess a linear part with the slope D_C. The value of $M(\varepsilon)$ is usually overestimated unless the value of ε is very small. High requirements of computer time and memory makes the algorithm impractical though an improved efficiency can sometimes be achieved[4.49]. The necessary number of boxes is very large even in a low-dimensional state space $(n \leq 3)$ and thus the computer memory required for larger n exceeds the capacity of available computers.

The computation of the information dimension D_I proceeds in a similar way, but, in addition, we have to compute explicitly the asymptotic measure of the partition. This measure often has a nonhomogeneous distribution; many boxes may have a very small measure and need not be counted in computations. Nevertheless, the algorithm is not very efficient.

The information dimension can be computed more efficiently using Young's result[4.72] which, when applied to the current problem, says that if A is an attractor in R^m, μ is the asymptotic measure and $B_x(\varepsilon)$ is the ball with radius ε centred at $x \in A$, then for almost all x (with respect to μ)

$$D_I = \lim_{\varepsilon \to 0} \frac{\log \mu(B_x(\varepsilon))}{\log \varepsilon}, \tag{4.37}$$

provided the limit in (4.37) exists. We can use the reconstructed finite orbit $x_1, ..., x_N$ in R^m to estimate $\mu(B_{x_i}(\varepsilon))$ on defining

$$C_i(\varepsilon) = N^{-1}\{\text{number of } x_j\text{s such that } d(x_i, x_j) \leq \varepsilon\}. \tag{4.38}$$

We may use the Euclidean norm or any equivalent norm; for example,

$$d(x_i, x_j) = \max_n |u_{i+nk} - u_{j+nk}|, \tag{4.39}$$

here u_{i+nk}, u_{j+nk}, $n = 0, ..., m-1$, are the components of x_i, x_j, respectively, and k is the time delay. An estimate of the information dimension D_I is then obtained as the slope of the linear part of the plot of $\log C_i(\varepsilon)$ against $\log \varepsilon$ for ε sufficiently small[4.21].

As in all other methods discussed here, the excessive level of noise in the data does not allow us to obtain reliable results. One drawback of the described method is the requirement of a large time series to get good estimates of $\mu(B_{x_i}(\varepsilon))$ for small values of ε. On the other hand, a small part of the data contained in the ε-ball centred at a particular x_i is used. Hence the averaging of $\log C_i(\varepsilon)$ over all (or at least several) centre points x_i will improve the statistics.

A method which takes all points x_i into consideration was proposed by Grassberger and Procaccia[2.25]. It is based on the definition of a generalized dimension D_q, see Section 2.5, Eq. (2.48). In particular, the correlation dimension D_2 can be obtained from a 'discrete correlation integral'

$$C(\varepsilon) = N^{-2}\{\text{number of pairs } (x_i, x_j) \text{ such that } d(x_i, x_j) \leq \varepsilon\}. \tag{4.40}$$

The correlation dimension D_2 can then be estimated from the relation

$$C(\varepsilon) \sim \varepsilon^{-D_2(n)} \tag{4.41}$$

using the plot of $\log C(\varepsilon)$ versus $\log \varepsilon$ for different values of n and taking that value of the slope $D_2(n)$ which becomes independent of n (this occurs for $n \geq m$). The correlation dimension forms a lower bound for the information dimension, $D_2 \leq D_I$.

This method enables us also to indicate a low level noise. Under the effects of a bounded noise chaotic attractor loses the fractal structure only on small scales, the larger ones being almost unaffected. A typical $\log C^m(\varepsilon)$ versus $\log \varepsilon$ plot is shown in Fig. 4.2 for the Hénon attractor (see Eq. (1.5)) in the presence of a bounded noise on three levels[4.2]. The curves shown in Fig. 4.2 in all three cases consist of approximately linear parts connected via a bend. This bend separates the deterministic and the noisy scales of the attractor. Larger scales are essenti-

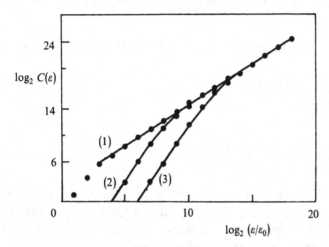

Fig. 4.2. The plot of $\log_2 C^m(\varepsilon)$ versus $\log_2 \varepsilon$ for the noisy Hénon attractor with $m = 3$. Curve (1) corresponds to the noiseless attractor and its slope gives $D_2 \approx 1.25$. Curves (2) and (3) correspond to noisy attractors. The bends on the curves determine the noise level below which the slope gives $D_2 \approx 3$.

ally unaffected by the noise and the slope of the corresponding linear part defines the correlation dimension. Smaller scales are completely noisy, the attractor fills sets of a nonzero volume in the state space on these scales and the slope of the corresponding linear part is close to the dimension of the state space. The generalized dimension D_q (Eq. (2.47)) is defined for any real number q, hence D_q can be seen as a function of q called a *dimension function*. This function can also be computed numerically[4.54, 4.55] both from models and from experiments.

The metric entropy h_μ and all generalized entropies h_q (see Section 2.5) can in principle be computed in a similar way. In particular, the metric entropy h_μ and the generalized second order entropy h_2 are of interest[2.16, 4.10, 4.19, 4.56]. To estimate the metric entropy we have to find the measure $C_i^l(\varepsilon)$ of the ε-ball $B_{x_i}(\varepsilon)$ centred at x_i such that the points contained in $B_{x_i}(\varepsilon)$ do not leave the ε-ball $B_{x_{i+k}}(\varepsilon)$ centred at x_{i+k}, $k = 1, 2, ..., l - 1$, i.e.

$$C_i^l(\varepsilon) = \lim_{N \to \infty} N^{-1} \{\text{number of } x_j \text{ such that } d(x_{j+k}, x_{i+k}) \leqq \varepsilon$$

$$\text{for all } k = 0, ..., l - 1\}. \tag{4.42}$$

The metric entropy h_μ can then be determined as[2.16]

$$h_\mu = \frac{1}{\Delta t} \lim_{\varepsilon \to 0} \lim_{l \to \infty} \lim_{N \to \infty} \frac{1}{N} \sum_i \log \left(C_i^l(\varepsilon)/C_i^{l+1}(\varepsilon) \right) \tag{4.43}$$

(Δt is the time interval between two consecutive measurements). Let us define the function

$$C^l(\varepsilon) = \lim_{N \to \infty} N^{-2} \{ \text{number of pairs } (x_i, x_j) \text{ such that } d(x_{i+k}, x_{j+k}) \leqq \varepsilon$$

$$\text{for all } k = 0, 1, ..., l - 1 \}. \tag{4.44}$$

The entropy h_2 is then[2.16]

$$h_2 = \frac{1}{\Delta t} \lim_{\varepsilon \to 0} \lim_{l \to \infty} \log \left(C^l(\varepsilon)/C^{l+1}(\varepsilon) \right). \tag{4.45}$$

Eqs. (4.43) and (4.45) can be used for computing the estimates of h_μ and h_2 as with the computation of D_I and D_2. Generalized entropy h_q for any real q can also be computed from a time series, see Refs 1.30, 4.15, 4.18, 4.54, 4.55. The methods of computation of dimensions and entropies can be numerically realized in a number of variants. Because of the finite number of data points and the properties of the computational algorithms themselves, systematic errors arise which can sometimes invalidate the results. Hence we have to proceed with caution in the computations. Details of numerical computations are discussed in Ref. 4.48. Dimensions and entropies can be used to predict the future from chaotic time series[4.16] and to reduce experimental noise[4.16, 4.39].

4.7 Power spectra

The power spectrum serves as a standard tool for the analysis of a measured signal $u(t)$. It can be either computed by standard software available on most computers (based on the efficient Fast Fourier Transform method[4.5, 4.52]) or evaluated on line on different types of spectrum analysers. The power spectra enable us to differentiate between periodic, quasiperiodic and chaotic signals considering both characteristic frequencies ('peaks') and the level of broadband noise; however, this distinction is often somewhat subjective.

The power spectrum of a scalar signal $u(t)$ can be defined in two different ways[4.28, 4.52]. The first definition uses the square of the Fourier transform of $u(t)$ averaged over time,

$$S(f) = \lim_{T \to \infty} T^{-1} \left| \int_0^T u(t) \, e^{-2\pi i f t} \, dt \right|^2. \tag{4.46}$$

Alternatively we can define $S(f)$ as a Fourier transform of the time correlation function $b(\tau)$ (see Section 2.5, Eq. (2.33))

$$S(f) = \int_{-\infty}^{\infty} b(\tau) \, e^{-2\pi i f \tau} \, d\tau .$$ (4.47)

The time integrals are replaced by sums if the time is discrete. The Wiener–Khinchin theorem[4.52] states that these two definitions coincide under rather general assumptions. In applications to dissipative dynamical systems we assume that $u(t)$ reflects the dynamics of the dynamical system $(\mathbf{X}, g^t(\mathbf{x}))$ on an ergodic attractor, i.e. $u(t) = \varphi(g^t(\mathbf{x}))$, where $\varphi: \mathbf{X} \to \mathsf{R}$ is an arbitrary continuous function called the *phase function*. The power spectrum can then be used to identify attractors which are mixing (i.e. chaotic). Such attractors have continuous power spectra, while periodic or quasiperiodic attractors exhibit discrete spectra. More specifically, the power spectrum of a periodic motion with frequency f_1 has Dirac δ-peaks at f_1 and its harmonics $2f_1, 3f_1, \ldots$. A quasiperiodic motion with m rationally independent frequencies $f_1, \ldots, f_m$ has Dirac δ-peaks at all linear combinations of the basic frequencies with integer coefficients.

In practice we have a finite discrete time series $u_0, \ldots, u_N$ of numbers with a limited precision and therefore the criterion of continuity cannot be used in a strict sense. All δ-peaks become finite in both height and width and there is a basic noise level at all frequencies. High level of this broadband noise is the main practical criterion for the determination of chaos. Spectral density calculations[4.57] can be used to characterize the amount of deterministic chaos in more detail.

For the time series $u_0, \ldots, u_N$ sampled with time increment Δt the Fourier transform is in the form

$$U(f_m) = \Delta t \sum_{k=0}^{N} p_k u_k \, e^{-2\pi i k f_m} , \qquad f_m = m/N \, \Delta t , \quad m = 0, 1, \ldots, N/2 .$$ (4.48)

To avoid undesiderable effects, the function p_k *(time window)* must be suitably chosen. The well-known *Fast Fourier Transform algorithm*[4.5] can be used to compute $U(f_m)$. The power spectrum is easily obtained from $U(f_m)$ but some kind of averaging is desirable[4.52]. The length N of the time series and the sampling period Δt must be chosen with a view to the characteristic frequencies of the recorded process.

Power spectra indicate universal features of the transition to chaos[3.24–3.26, 3.54, 3.64, 4.9, 4.37] as well as the presence of chaotic regimes but they do not give detailed information on the type of chaotic motion in the state space as is obtained from such characteristics as dimension, LEs or entropy. Hence, spectra are a useful tool to study the onset of chaos rather than the evolution of subsequent chaotic regimes.

Let us remark that power spectra can give a good indication of local bifurcations of periodic orbits in the presence of small random noise. The period-doubling, symmetry breaking, transcritical and other bifurcations can be indicated by the presence of various 'noisy precursors' in the power spectra, i.e. by a specific broadening of certain spectral lines[4.70].

REFERENCES

4.1   Aluko M. and Chang H.−C. PEFLOQ: An algorithm for the bifurcational analysis of periodic solutions of autonomous systems. *Comp. & Chem. Engng.* **8** (1984) 355.

4.2   Ben−Mizrachi A., Procaccia I. and Grassberger P. The characterization of experimental (noisy) strange attractor. *Phys. Rev.* **29A** (1984) 975.

4.3   Benettin G., Froeschlé C. and Scheidecker H. P. Kolmogorov entropy of a dynamical system with an increasing number of degrees of freedom. *Phys. Rev.* **19A** (1979) 454.

4.4   Benzinger H. E., Burns S. A. and Palmore J. I. Chaotic complex dynamics and Newton's method. *Phys. Lett.* **119A** (1987) 441.

4.5   Brigham E. O. *The Fast Fourier Transform.* Englewood Cliff, NJ, Prentice−Hall, 1974.

4.6   Brindley J., Kaas-Petersen C. and Spence A. Pathfollowing methods in bifurcation problems. Physica 34D (1989) 456.

4.7   Broomhead D. S. and King G. P. Extracting qualitative dynamics from experimental data. *Physica* **20D** (1986) 217.

4.8   Broomhead D. S., Jones R. and King G. P. Topological dimension and local coordinates from time series data. *J. Phys. A: Math. Gen.* **20** (1987) L563.

4.9   Brunsden V. and Holmes P. Power spectra of strange attractors near homoclinic orbits. *Phys. Rev. Lett.* **58** (1987) 1699.

4.10 Cohen A. and Procaccia I. On computing the Kolmogorov entropy from the time signals of dissipative and conservative dynamical systems. *Phys. Rev.* **31A** (1984) 1872.

4.11 Crutchfield J. P. and McNamara B. S. Equations of motion from a data series. *J. Compl. Syst.* **1** (1987) 417.

4.12 Deuflhard P., Fiedler B. and Kunkel P. Efficient numerical path following beyond critical points. Preprint, Heidelberg, Univ. of Heidelberg, 1984.

4.13 Doedel E. J. AUTO: A program for the automatic bifurcation analysis of autonomous systems. *Cong. Num.* **30** (1981) 265.

4.14 Doedel E. J. and Kernevez J. P. Software for continuation problems in ordinary differential equations with applications. Preprint, Pasadena, Calif. Inst. of Technology, 1985.

4.15 Fahner G. and Grassberger P. Entropy estimates for dynamical systems. *Complex Systems* **1** (1987) 1093.

4.16 Farmer J. D. and Sidorowich J. J. Exploiting chaos to predict the future and reduce noise. Submitted to *Rev. Modern Physics.*

4.17 Fraser A. M. and Swinney H. L. Independent coordinates for strange attractors from mutual information. *Phys. Rev.* **33A** (1986) 1134.

4.18 Grassberger P. Finite sample corrections to entropy and dimension estimates. *Phys. Lett.* **128A** (1988) 369.

4.19 Grassberger P. and Procaccia I. Estimating the Kolmogorov entropy from a chaotic signal. *Phys. Rev.* **28A** (1983) 2591.

4.20 Greene J. M. and Kim J.−S. The calculation of Lyapunov spectra. *Physica* **24D** (1987) 213.

4.21 Guckenheimer J. and Buzyna G. Dimension measurements for geostrophic turbulence. *Phys. Rev. Lett.* **51** (1983) 1438.

4.22 Hao Bai−lin Numerical methods to study chaos in ordinary differential equations. In *Directions in Chaos* Vol.2, ed. Hao Bai−lin, Singapore, World Scientific, 1988, p. 294.

4.23 Hénon M. On the numerical computation of Poincaré maps. *Physica* **5D** (1982) 412.

4.24 Higuchi T. Approach to an irregular time series on the basis of the fractal theory. *Physica* **31D** (1988) 277.

4.25 Hindmarsh A. C. Gear: Ordinary differential equation solver. Technical Report, Livermore, CA, Lawrence Livermore Laboratory, 1972.

4.26 Holodniok M. and Kubíček M. DERPER − An algorithm for continuation of periodic solutions in ordinary differential equations. *J. Comp. Phys.* **55** (1984) 254.

4.27 Holodniok M. and Kubíček M. Computation of period doubling bifurcation points in ordinary differential equations. TUM Report, München, Technische Universität, 1984.

4.28 Jenkins G. M. and Watts D. G. *Spectral Analysis and Its Applications.* San Francisco, Holden−Day, 1969.

4.29 Kaas−Petersen C. Computation of quasi−periodic solutions of forced dissipative systems I and II. *J. Comp. Phys.* **58** (1985) 395 and *J. Comp. Phys.* **64** (1986) 433.

4.30 Kaas−Petersen C. Computation, continuation, and bifurcation of torus solutions for dissipative maps and ordinary differential equations. *Physica* **25D** (1987) 288.

4.31 Keller H. B. Numerical solution of bifurcation and nonlinear eigenvalue problems. In *Applications of Bifurcation Theory*, ed. P. H. Rabinowitz, New York, Academic Press, 1977, p. 359.

4.32 Kevrekidis I. G. A numerical study of global bifurcations in chemical dynamics. *AICHE J.* **33** (1987) 1850.

4.33 Kevrekidis I. G., Aris R. and Schmidt L. D. The CSTR forced. *Chem. Engng. Sci.* **41** (1986) 1549.

4.34 Kevrekidis I. G., Aris R., Schmidt L. D. and Pelikan S. The numerical computation of invariant circles of maps. *Physica* **16D** (1985) 243.

4.35 Khibnik A. I. Periodic solutions of a system of n differential equations. Algorithms and programs in FORTRAN. Pushchino, USSR Academy of Sciences, 1979 (in Russian).

4.36 Khibnik A. I. and Shnol E. E. Programs for qualitative analysis of differential equations. Pushchino, USSR Academy of Sciences, 1982 (in Russian).

4.37 Kim S., Ostlund S. and Yu G. Fourier analysis of multiple−frequency dynamical systems. *Physica* **31D** (1988) 117.

4.38 Kostelich E. J. and Yorke J. A. Lorenz cross sections of the chaotic attractor of the double rotor. *Physica* **24D** (1987) 263.

4.39 Kostelich E. J. and Yorke J. A. Noise reduction in dynamical systems. Submitted to *Phys. Rev. Lett.* (1988).

4.40 Kubíček M. Algorithm 502. Dependence of solutions of nonlinear systems on a parameter. *ACM Trans. Math. Software* **2** (1976) 98.

4.41 Kubíček M. and Marek M. *Computational Methods in Bifurcation Theory and Dissipative Structures.* New York, Springer, 1983.

4.42 Küpper T., Mittelman H. D. and Weber H. (Eds.) *Numerical Methods for Bifurcation Problems.* Basel, Birkhauser, 1984.

4.43 Küpper T., Seydel R. and Troger H. (Eds.) *Bifurcation: Analysis, Algorithms, Applications.* Basel, Birkhauser, 1987.

4.44 Kuznetsov Yu. A. One−dimensional separatrices of differential equations depending on parameters. Algorithms and programs in FORTRAN. Pushchino, USSR Academy of Sciences, 1983 (in Russian).

4.45 Lambert J. D. *Computational Methods in Ordinary Differential Equations.* New York, Wiley, 1973.

4.46 Ling F. H. A numerical method for determining bifurcation curves of mappings. *Phys. Lett.* **110A** (1985) 116.

4.47 Lorenz E. N. Lyapunov numbers and the local structure of attractors. *Physica* **17D** (1985) 279.

4.48 Mayer—Kress G. (Ed.) *Dimensions and Entropies in Chaotic Systems. Quantification of Complex Behaviour.* Berlin, Springer, 1986.

4.49 McGuinness M. J. A computation of the limit capacity of the Lorenz attractor. *Physica* **16D** (1985) 265.

4.50 Mestel B. and Percival I. Newton method for highly unstable orbits. *Physica* **24D** (1987) 172.

4.51 Orság J., Rosendorf P., Schreiber I., Marek M., Kubíček M. and Holodniok M. Expert system for use of continuation techniques in nonlinear dynamics studies. In *Computer Application in the Chemical Industry.* Dechema-Monography 116, Weinheim/VCH Publishers, 1989.

4.52 Otnes R. K. and Enochson L. *Applied Time Series Analysis.* New York, Wiley, 1978.

4.53 Packard N. H., Crutchfield J. P., Farmer J. D. and Shaw R. S. Geometry from a time series. *Phys. Rev. Lett.* **45** (1980) 712.

4.54 Pawelzik K. and Schuster H. G. Generalized dimensions and entropies from a measured time series. *Phys. Rev.* **35A** (1987) 481.

4.55 Procaccia I. The characterization of fractal measures as interwoven sets of singularities: Global universality of the transition to chaos. In *Dimensions and Entropies in Chaotic Systems*, ed. G. Mayer—Kress, Berlin, Springer, 1986, p. 8.

4.56 Procaccia I. The static and dynamic invariants that characterize chaos and the relations between them in theory and experiments. *Physica Scripta* **T9** (1985) 40.

4.57 Rabinovitch A. and Thieberger R. Time series analysis of chaotic signals. *Physica* **28D** (1987) 409.

4.58 Rheinboldt W. C. *Numerical Analysis of Parametrized Nonlinear Equations.* New York, Wiley, 1986.

4.59 Sano M. and Sawada Y. Measurement of the Lyapunov spectrum from a chaotic time series. *Phys. Rev. Lett.* **55** (1985) 1082.

4.60 Schwartz I. B. Estimating regions of existence of unstable periodic orbits using computer based techniques. *SIAM J. Num. Anal.* **20** (1983) 106.

4.61 Seydel R. *From Equilibrium to Chaos: Practical Bifurcation and Stability Analysis.* New York, Elsevier, 1988.

4.62 Seydel R. Numerical computation of periodic orbits that bifurcate from stationary solutions of ordinary differential equations. *Appl. Math. Comput.* **9** (1981) 257.

4.63 Shimada I. and Nagashima T. A numerical approach to ergodic problem of dissipative systems. *Prog. Theor. Phys.* **61** (1979) 1605.

4.64 Siemens D. and Bucher M. New graphical method for the iteration of one—dimensional maps. *Physica* **20D** (1986) 363.

4.65 Takens F. Detecting strange attractors in turbulence. In *Dynamical Systems and Turbulence.* Lecture Notes in Mathematics **898**, Berlin, Springer, 1981, p. 366.

4.66 Tongue B. H. On obtaining global nonlinear system characteristics through interpolated cell mapping. *Physica* **28D** (1987) 401.

4.67 Tongue B. H. Characteristics of numerical simulations of chaotic systems. *ASME J. App. Mech.* **54** (1987) 1.

4.68 Varosi F., Grebogi C. and Yorke J. A. Simplicial approximation of Poincare maps of differential equations. *Phys. Lett.* **124A** (1987) 59.

4.69 Vastano J. A. and Kostelich E. J. Comparison of algorithms for determinig Lyapunov exponents from experimental data. In *Dimensions and Entropies in Chaotic systems*, ed. G. Mayer—Kress, Berlin, Springer, 1986, p. 100.

4.70 Wiesenfeld K. Noisy precursors of nonlinear instabilities. *J. Stat. Phys.* **38** (1985) 1071.

4.71 Wolf A., Swift J. B., Swinney H. L. and Vastano J. A. Determining Lyapunov exponents from a time series. *Physica* **16D** (1985) 285.

4.72 Young L.—S. Dimension, entropy and Lyapunov exponents. *Ergod. Theory & Dynam. Syst.* **2** (1982) 109.

5

Chaotic dynamics in experiments

5.1 Introduction

Chaos has been observed in a large number of both natural and artificial nonlinear dissipative dynamical systems. For example, it has been shown in observations and computations of the evolution of meteorological patterns in the Earth's atmosphere that the patterns of evolution are very sensitive to initial conditions and predictions necessarily diverge after several days. These observations can be interpreted as an indication of the presence of a chaotic attractor in the corresponding state space. Also in the studies of populations dynamics it is often observed that the sizes of populations fluctuate widely from one season to the next; this can again be interpreted on the basis of the existence of chaotic trajectories in the population state space.

Recent developments in the theory of nonlinear dynamical systems coupled with the use of computers in the acquisition and analysis of long time series of experimental data have supported an exponential increase of detailed studies of various nonlinear dissipative systems displaying aperiodic behaviour and lead to both the confirmation of the presence and a detailed characterization of chaotic attractors.

Several model systems have been studied mostly with the aim of verifying theoretical predictions. Among them are classical hydrodynamical systems for the study of the development of turbulence – the Taylor–Couette flow and the Rayleigh–Bénard convection, nonlinear electronic oscillatory circuits, oscillatory mechanical systems, lasers, chemical systems (well stirred reactors) and various oscillating or excitatory structures in biology (heart cells, neurones).

Experimental observations on nonlinear circuits give results which are closest to theoretical predictions. On the other hand, hydrodynamical systems are often the most complex and observations of the transition to turbulence require a careful choice of experimental conditions to eliminate effects of more than just a few characteristic modes on the dynamics of the system studied.

In experiments we very often observe, record and analyse a scalar variable (current, voltages, flow velocity, temperature, concentration of chemical component, etc.), let us denote it $x(t)$. The data are usually evaluated in the form of correlation functions[5.207] see also Section 2.5, Eqs (2.31–2.32)

$$C(\tau) = \lim_{T \to \infty} T^{-1} \int_0^T [x(t) - \langle x \rangle][x(t + \tau) - \langle x \rangle] \, dt , \qquad (5.1)$$

where $\langle x \rangle$ denotes mean value, or as power spectra, see also Section 4.7, Eqs (4.46–4.48)

$$P(\omega) = \lim_{T \to \infty} T^{-1} |\int_0^T x(t) \, e^{-i\omega t} \, dt|^2 . \qquad (5.2)$$

In practice, the overall time of an experiment is finite and data are discrete in time, so that the integrals are replaced by finite sums. The power spectrum can be computed via a Fast Fourier Transform (FFT) algorithm with appropriate windowing to suppress side lobes generated by the finite measurement time. When the correlation function $C(\tau)$ vanishes as τ is large or when the power spectrum contains broadband noise, i.e. it does not contain only instrumentally sharp peaks (of a width of order $2\pi/T$) above a certain level of noise, then we assume that the time series represents random data. This randomness can have two sources; a broadband power spectrum can be generated internally, due to the presence of a chaotic attractor and/or it can be caused by an external noise. A deterministic chaos with a low number of degrees of freedom and a stochastic noise (which may be also viewed as a deterministic chaos with a high number of degrees of freedom) can be distinguished by analysing state space portraits, Poincaré sections (constructed, e.g. by the time delay method from a single variable data), by the spectrum of Lyapunov exponents (LEs) and by the dimension of the attractors. Box-Jenkins method was also proposed for an analysis of time series of chaotic signals[5.221]. As we have learned in the previous chapters, the emergence of chaotic attractors is connected with characteristic scenarios – sequences of different dynamical regimes that can occur as a characteristic experimental parameter ('bifurcation parameter') is varied. When these scenarios (the period doubling, the intermittency, transitions via torus) are identified in the studied time series, we have again a positive verification of the presence of deterministic chaos.

In very precise experiments we can further study quantitative properties of asymptotic behaviour close to the transition to chaos (e.g. confirm the value of the Feigenbaum's δ or the ratios of the powers of various bands in the power spectra). We shall illustrate the above approaches to the characterization of an aperiodic behaviour on those model systems that have been studied most. First, we shall discuss observations reported on electric circuits where both the values of parameters and the levels of external noise can be best controlled. Then the most characteristic experimental results on low-dimensional chaos in physical, hydrodynamical, chemical and biological systems will be presented. Finally, we shall review typical observations of chaotic attractors in various other nonlinear dissipative systems.

5.2 Nonlinear circuits

Nonlinear circuits can also be seen as nonlinear analog computers. Experimental conditions can here be controlled with high accuracy and the circuits can be well described by systems of nonlinear ODEs. Period-doubling bifurcations, intermittency and transitions via a torus were identified and Feigenbaum's constants δ and α, the noise scaling exponent $\varkappa$ and the ratio of the successive period-doubling peaks in the power spectra were evaluated from the experiments and often confirmed the validity of values predicted for one-dimensional unimodal mappings.

We shall describe first the formation of chaotic oscillations in a simple electronic circuit and their approximate description by means of a one-dimensional mapping and then discuss in more detail reported experimental results on two nonlinear driven (nonautonomous) oscillators. The first, a two-dimensional LRC circuit with a varactor as a nonlinear element, was first used in this connection by Linsay[5.169] and also studied in detail by Jeffries and coworkers[5.134, 5.135] and by other authors, see references in[5.41, 5.45, 5.133, 5.156, 1.191]. The second, a nonautonomous generator with an inertial nonlinearity, was investigated by Anishchenko and coworkers[5.7–5.13] and by Landa and coworkers[5.33]. In the first case a detailed evaluation of quantitative relations valid in the period-doubling bifurcation route to chaos will be illustrated and in the second one a more or less qualitative picture of the transition to chaos via the two-dimensional torus destruction will be discussed.

5.2.1 Noise generator

Pikovsky and Rabinovich[5.147, 5.215, 5.216] discussed an electronic circuit schematically shown in Fig. 5.1a, where the chaotic behaviour (noise generation) results from the switching mechanism (similar to the Van der Pol generator). The circuit differs in having a tunnel diode used in series with an inductor. The vacuum tube (or transistor) works in a linear regime as a source of energy. The tunnel diode with the characteristics shown in Fig. 5.1 (*b*) serves as a nonlinear-

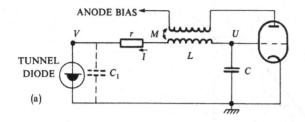

(a)

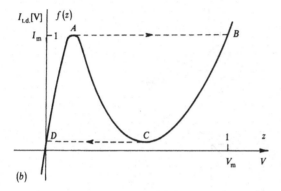

(b)

(c)

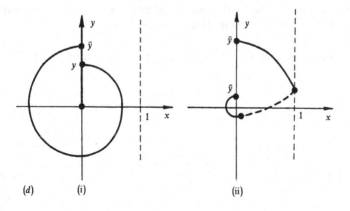

(d) (i) (ii)

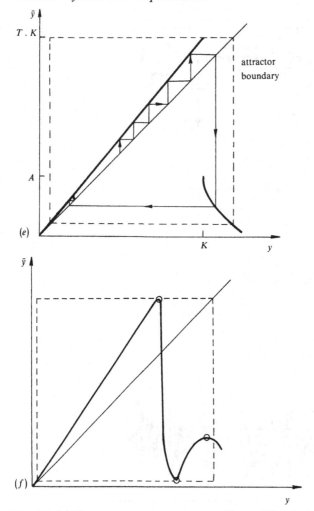

Fig. 5.1. (*a*) Electronic noise generator according to Pikovsky and Rabinovich; (*b*) tunnel diode voltage-current characteristics; (*c*) recording of $U(f)$; (*d*) construction of the Poincaré mapping (*i*) trajectory lies on one slow surface, (*ii*) trajectory crosses over to another slow surface); (*e*) Poincaré map; (*f*) more realistic Poincaré map.

ity. Considering a small capacity C_1 of the tunnel diode we obtain from Kirchhoff's laws

$$LC \frac{dI}{dt} = (MS - rC)\, I + C(U - V)\,, \tag{5.3a}$$

$$C \frac{dU}{dt} = -I\,, \tag{5.3b}$$

$$C_1 \frac{dV}{dt} = I - I_{td}(V) ; \qquad (5.3c)$$

here S is the tube transconductance, L denotes inductance, C capacity, r resistance, U, V voltages and I_{td} tunnel diode current. Let us describe the function of the generator qualitatively. When the current I and the voltage are low, the tunnel diode does not affect oscillations in the circuit $(V \approx 0)$ and (5.3a, b) are decoupled from (5.3c) and they increase in amplitude according to the tube energy. When I exceeds the threshold value I_m, the diode switches along the dashed line A, B $\big($Fig. 5.1$(b)\big)$ and the diode voltage becomes V_m. Then the current decreases again and the diode returns along the line CD to the state $V \approx 0$. This switching process consumes part of the circuit energy; the oscillations start to increase again until I exceeds I_m and the process is repeated. The output signal $U(t)$ is in the form of a sequence of pulses with an exponentially increasing amplitude, see Fig. 5.1(c).

Let us try to determine whether the sequence of pulses will be periodic or chaotic. We introduce dimensionless variables $x = I/I_m$, $y = UC^{1/2}/(I_m L^{1/2})$, $z = V/V_m$ into (5.3a–c) and obtain

$$\dot{x} = 2hx + y - gz , \qquad (5.4a)$$

$$\dot{y} = -x , \qquad (5.4b)$$

$$\varepsilon \dot{z} = x - f(z), \qquad (5.4c)$$

where $h = (MS - rC)(LC)^{1/2}/2$, $g = V_m C^{1/2}/(I_m L^{1/2})$, $\varepsilon = gC_1/C \ll 1$, $f(z) = I_{td}(V_m z)/I_m$ are dimensionless diode characteristics and $\tau = t(LC)^{-1/2}$ is a dimensionless time. The small parameter ε suggests a division of motions in the state space into slow and fast ones[5.216]. The slow motions are restricted to the stable branches $(f'(z) > 0)$ of the slow manifold $x = f(z)$ and the fast ones are straight lines $x = \text{const}$, $y = \text{const}$, and z goes over from one stable branch of the manifold to another. Oscillations in the circuit correspond to slow motions and the switching to fast ones. We can now construct a Poincaré mapping in the following way: we choose a line on the slow surface $\Sigma \equiv \{Y = y|_{x=0}, \ y > 0\}$. Next, we choose on this line a point where certain trajectory passes and denote the point where this trajectory returns to Σ as $\bar{y}$; $\bar{y} = F(y)$ is then the Poincaré mapping. Every trajectory leaving Σ may behave in two different ways. When y is sufficiently small, the trajectory revolves around the origin once and returns on to Σ; for these trajectories $\bar{y} = F_1(y)$. If y exceeds a certain critical value, then the trajectory reaches the boundary of the slow surface without finishing the revolution, then crosses over to the other slow surface, moves there for some time, jumps back and completes the revolution and returns to Σ; for such trajectories $\bar{y} = F_2(y)$. Both forms of the Poincaré mappings are schematically shown in Fig. 5.1(d). If we assume that $z \approx 0$, we obtain from (5.4)

$$\bar{y} = F_1(y) = \exp(2\pi h) y = Ty, \qquad T > 1. \qquad (5.5)$$

The function F_2 can be approximated as

$$\bar{y} = F_2(y) = A - \sqrt{y - K}, \qquad (5.6)$$

where K is the threshold value of y. When $y < K$, (5.5) holds and when $y > K$ (5.6) is used. The constant A describes the shift of the trajectories while moving on the second slow surface. On combining (5.5) and (5.6) we obtain an approximate description of the Poincaré mapping shown in Fig. 5.1(e). From the figure we can infer that for the values of y from the interval

$$\langle A - (TK - K)^{1/2} < y < TK \rangle$$

the trajectories cannot escape from the above interval, hence it contains an attractor. Further, for

$$0 < T - 1 < (4K)^{-1}, \quad |d\bar{y}/dy| > 1,$$

hence the attractor is expanding and, moreover, the mapping is ergodic and mixing[5.215, 5.216] The experimental recording of $U(t)$ and iterations of the Poincaré mapping (schematically shown in Fig. 5.1(e)) give qualitatively the same results.

The above discussion of the approximate Poincaré mapping was based on several idealizations. A more realistic form of the Poincaré mapping is shown in Fig. 5.1(f). The mapping contains three points where $d\bar{y}/dy = 0$. This mapping is not expanding for all values of the parameters. Hence, chaotic regimes may coexist with periodic ones as was observed in experiments.

Detailed studies of evolution of chaos in a distributed noise generator with a time delay were reported by Katz and coworkers[5.140, 5.141] and by Landa and coworkers[5.159]. Geometrical and computer analysis of a simple autonomous electrical circuit with a piecewise-linear resistor as nonlinear element, producing a chaotic attractor with a 'double-scroll' structure was reported by Matsumoto, Chua and Komuro[5.186]. Chaotic transitions in the forced negative-resistance oscillators were described by Ueda and Akamatsu[5.280].

5.2.2 Experiments on period-doubling

The experimental system studied by Linsay[5.169] and Jeffries with coworkers[5.269] is a series LRC circuit driven by a controlled oscillator, shown schematically in Fig. 5.2. A nonlinear capacitance is represented by the silicone varactor diode IN953. The capacitance C across the varactor varies with the applied voltage V as

$$C(V) = \frac{C_0}{(1 + V/b)^\beta} \qquad (5.7)$$

where in the concrete experiment $C_0 = 300$ pF, $b = 0.6$ and $\beta \approx 0.5$.

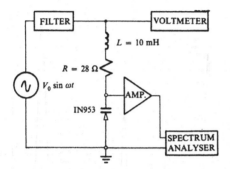

Fig. 5.2. LRC circuit used by Jeffries, schematically.

The circuit can be described by the equation

$$L\ddot{q} + R\dot{q} + V_c = V_d(t) = V_0 \sin(2\pi ft) \qquad (5.8)$$

where V_c is the voltage across a varactor. Under a reversed voltage $V_c = q/C$ and under a forward voltage the varactor behaves like a normal conducting diode. At low values of V_0, the circuit behaves like a high-Q resonant circuit at $f_{res} = 93$ kHz. In the experiment the value of f is fixed near f_{res}, the driving voltage

Table 5.1 *Measured sequence of periodic regimes*[5.269]

Period	V_{0n} (rms volts)	Period	V_{0n} (rms volts)
2	0.639	5	2.353
4	1.567	10	2.363
8	1.785	7	2.693
16	1.836	14	2.696
32	1.853	3	3.081
chaos	1.856	6	3.081
12	1.901	12	3.711
24	1.902	24	3.821
6	2.073	9	4.145
12	2.074	18	4.154

V_0 is varied and the varactor voltage $V_c(t)$ is measured. Hence V_0 corresponds here to the characteristic (bifurcation) parameter and $V_c(t)$ is the followed state variable. The display of $V_c(t)$ revealed the existence of periodic solutions on frequencies $f/2^n$ up to $f/32$ bifurcating at the values V_{0n} and then the existence of chaos. The display is very similar to the diagram obtained for the logistic

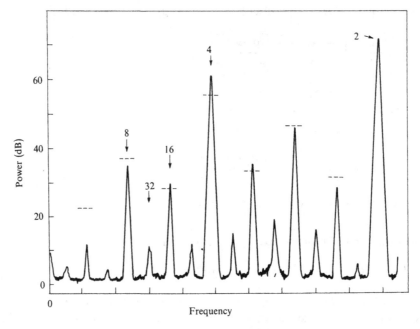

Fig. 5.3. Power spectral density versus frequency for $f = 98$ kHz, dynamic range 70 dB, with subharmonics to $f/32$; dashed bars – theoretical predictions.

mapping, discussed in Chapter 3 and shown in Fig. 3.14. Then a number of windows of periodic solutions appeared in the sequence shown in Table 5.1. The display $V_c(t) \approx V_0(t)$ allowed a direct measurement of the number α, characterizing the universal metric scaling of period doubling bifurcations, see Ref. 3.23. For example, data for period 16 yielded the values $\alpha = 2.35$ and 2.61. An average value determined from the series of ten measurements was $\alpha = 2.41 \pm 0.1$. The entire sequence of periodic windows shown in Table 5.1 is consistent with the universal U-sequence, see Refs 3.18, 3.57. The dependence of the power spectral density of V_c, exhibiting subharmonics to $f/32$ for $f = 98$ kHz is shown in Fig. 5.3. The dashed bars correspond to the theoretical predictions, i.e. that the average heights of the peaks for a period is 13.21 dB below the height of the previous period. The data can be considered to be in agreement with the theory.

From values of V_{0n} for the first four period-doubling bifurcation windows Feigenbaum's convergence rate parameter can be computed

$$\delta_1 = \frac{V_{02} - V_{01}}{V_{03} - V_{02}} = 4.257 \pm 0.1 \,,$$

$$\delta_2 = \frac{V_{03} - V_{02}}{V_{04} - V_{03}} = 4.275 \pm 0.1 \,.$$

Hence, the computed values are relatively close to the theoretical asymptotic value $\delta = 4.669...$. The authors also observed the effects of added random noise voltage $V_n(t)$ to $V_d(t)$. When the magnitudes of the added noise were increased from $V_n = 10$ to 62, 400 and 2 500 mV_{rms}, respectively, the windows for periods 16, 8, 4 and 2 were successively obliterated. From these data the noise factor $\varkappa$ was computed as $\varkappa = 6.3$, which is again in principal agreement with the theoretical value $\varkappa = 6.65 ...$.

5.2.3 Other routes to chaos

An intermittent route to chaos was also observed in this system[5.134]. The period 5 window was chosen to demonstrate that the scaling behaviour[3.66], relating the average length $\langle l \rangle$ of the 'laminar' period to the distance from the attracting fixed point in the state space ε,

$$\langle l \rangle = \varepsilon^{-1/2}$$

holds. Moreover, the scaling behaviour of the standard deviation g of a white noise source added to the system

$$\langle l \rangle = g^{-2/3}$$

was also tested. The averaged length $\langle l \rangle$ defined as

$$\langle l \rangle = \frac{\text{frequency of 'laminar' events}}{\text{frequency of 'turbulent' events}}$$

was measured as a function of $\varepsilon \approx V_{05} - V_0$, where V_{05} was the measured driving oscillator voltage for the bifurcation of the period 5 window and V_0 was the voltage just below the bifurcation. The evaluation of data gave

$$\langle l \rangle \approx \varepsilon^{-(0.45 \pm 0.05)}$$

and

$$\langle l \rangle \approx g^{-(0.65\pm0.05)} \, ,$$

hence an excellent agreement with the theoretical predictions. However, a window with no hysteresis and without higher dimensional effects had to be carefully chosen for the testing of the intermittency relations.

Jeffries and Perez[5.134] also described experimental observations of three cases of an interior crisis of the chaotic attractor, i.e. of sudden and discontinuous changes in the attractor, predicted by Grebogi, Ott and Yorke[3.35], when the chaotic attractor is intersected by a coexisting unstable periodic orbit.

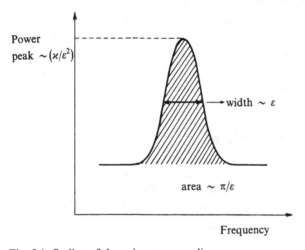

Fig. 5.4. Scaling of the noisy precursor lines.

Measurements of power spectra of the same periodically forced circuit were also used to test the ideas introduced by Jeffries and Wiesenfeld[5.136] on the effects of external noise on the power spectra close to the bifurcation points. The addition of external noise introduces new lines in the power spectra which become more important as the distance from the bifurcation point decreases. If we denote ε as a natural bifurcation parameter which is zero at the bifurcation point and the power of the noise as $\varkappa$, then (see Fig. 5.4) the power contained in the noise added (precursor) spectral line grows proportionally to ε^{-1}, the width of the line increases as approximately ε, hence the area under the peak grows as approximately π/ε and the peak height as π/ε^2. Jeffries and Wiesenfeld[5.136] tested these scalings by following experimental power spectra near the period-doubling and Hopf bifurcation points and found excellent agreement with theoretical predictions. Van-Buskirk and Jeffries also reported observations of chaotic dynamics of coupled oscillators[5.281]

Finally, Linsay[5.170] used a similar oscillatory circuit to test experimentally the theoretical prediction by Yorke and coworkers[5.299]. They predicted global scaling for the periodic windows which exhibit period-doubling. Their predictions are based on the properties of a canonical quadratic mapping. The authors claim that the distance from the initial saddle-node bifurcation (where the window of periodic solutions is created) – corresponding to voltage V_s in the experiment – to the point of the crisis of the attractor (experimental value V_c) is 9/4 times the distance from the saddle node to the first period-doubling bifurcation (V_d), i.e.

$$9/4 \approx m_c = (V_c - V_s)/(V_d - V_s) .$$

Linsay[5.170] used experimental results for period 5, 4 and 3 windows and found that the experimental values of m_c approach the value 2. However, in all cases the measured values of m_c did not deviate from 9/4 by more than 30 %. Linsay[5.170] also determined the correlation dimension D_2 of the state space to be 1.6 ± 0.1. Hence, the actual experimental system is not one-dimensional and it is remarkable that the theoretical predictions hold so well.

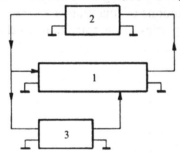

Fig. 5.5. Structural scheme of the autogenerator; 1 – two-cascade linear amplifier; 2–selective chain with a positive feedback (symmetric Wien's bridge); 3–one-half-period quadratic detector with an RC filter.

The use of RCL circuits in the studies of chaotic attractors was discussed in several other papers, see for example Refs 5.41, 5.265. It was pointed out that a critical property of the varactor diode necessary for frequency division (period-doubling) and bifurcation to chaos is a charge storage. Jeffries and Usher[5.133] suggested and demonstrated experimentally that a diode with a charge storage could find an application in technical elements (prescalers) for a frequency division in the range 1–100 GHz where standard techniques become difficult.

Transition to chaos via quasiperiodic oscillations with two incommensurate frequencies (two-dimensional torus) were studied both on a computer and on a corresponding nonlinear circuit by Anishchenko and coworkers[5.7–5.13]. They used a nonautonomous version of a generator with an inertial nonlinearity, schematically shown in Fig. 5.5. Here 1 denotes a two-cascade linear amplifier,

2 a selective chain with positive feedback in the form of a symmetric Wien's bridge, 3 a half-period quadratic detector with an RC filter. The mathematical model of the nonautonomous system is in the form

$$\dot{x} = mx + y - xz + B_0 \sin p\tau ,$$

$$\dot{y} = -x ,$$

$$\dot{z} = -gz + gf(x), \qquad f(x) = \begin{cases} x^2 & x > 0 \\ 0 & x \leq 0 \end{cases}. \qquad (5.9)$$

Here g describes the inertiality and B_0 and p are the amplitude and frequency of the external forcing, respectively. Power spectra, Poincaré mappings and projections of the trajectories in the state space were used to study various forms of the transitions to chaos. Anishchenko and coworkers combined numerical studies with studies of the nonlinear circuit tuned in such a way that its properties closely approximated the model equations. Experiments on the nonlinear circuit were used to determine qualitatively the behaviour of the system in the parametric plane and a detailed mathematical modelling with the use of continuation techniques then served for computations of quantitative characteristics, for example of LEs and asymptotic properties of the transitions to chaos.

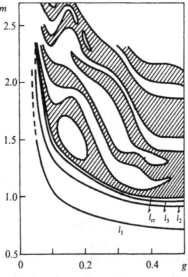

Fig. 5.6. Experimental oscillatory regimes in the parameter plane m–g: l_1, l_2, l_3 – first three period-doubling bifurcation curves; l_{cr} – formation of chaotic attractor.

For example, an autonomous system (5.9) (i.e. when $B_0 = 0$) exhibits a transition to chaos via the period-doubling mechanism, as follows from the experimentally determined bifurcation diagram in the parameter plane $m - g$ shown in Fig. 5.6. Here the lines l_1, l_2, l_3 correspond to subsequent bifurcations of regimes of periodic oscillations of period 2, 4 and 8, respectively and l_{cr} denotes the formation of chaotic oscillations (chaotic regimes are shown as cross-hatched regions). Numerical simulations were then used to verify the validity of asymptotic relations for δ in one-dimensional maps with a quadratic maximum. Similarly, in the nonautonomous system (5.9) bifurcation diagrams in the parameter plane $m - p$ at constant values of g and B_0 were constructed both as results of experiments on the circuit and by numerical simulations. The computed bifurcation diagram is presented in Fig. 5.7(a), the experimental one in Fig. 5.7(b). Let us denote l_0, l_1 and l_2 the lines in the parameter plane where multipliers of the corresponding orbit are purely imaginary, equal to $+1$ and -1, respectively. The range of existence of stable oscillations on the frequency of the external forcing (region 1 in Fig. 5.7) is bounded by the bifurcation lines l_0, l_1, l_2^1. When crossing the line l_0, a torus bifurcates from the limit cycle. The rotation number monotonically increases along l_0. It successively passes the points of strong resonances. At the points D in addition to the pair of purely imaginary multipliers the third multiplier also crosses the unit circle at $+1$, the line l_0 ends and is substituted by l_1 (the points A, B, D have a codimension 2). When crossing l_1, stable and unstable limit cycles come together and disappear. When leaving the region 1 via crossing l_2^1, a period-doubling of the original limit cycle occurs. Increase of m causes a sequence of period-doubling bifurcations to occur (lines l_2^k, $k = 1, 2, 3$) and the birth of chaos CA_1 (l_{cr1}, region 6). Points E correspond to intersections of l_1 and l_2. In the regions $E - L$ the line l_2^1 coincides with l_1^1. Resonance tongues start at A and B and they are bounded by the lines l_1 where a corresponding limit cycle on the torus disappears. The rotation numbers close to l_0, $\varrho(p, m)$, form a devil's staircase. When we move in the tongue B in the direction of increasing m, then at l_0 a stable two-dimensional torus supercritically bifurcates. At points K the resonance $\varrho = 1:2$ occurs. Above l_0 the destruction of the torus occurs on the bifurcation line l_{cr2} and the chaos CA_2 (region 5) is formed.

In the experimental bifurcation diagram (Fig. 5.7b) R_m corresponds to m and the external forcing frequency f_1 to p (the range $5.5 < f_1 < 15$ corresponds to $0.8 < p < 2.2$, $p = f_1/f_0$), respectively.

The character of the experimentally observed bifurcations in the neighbourhood of strong resonance corresponds to that of the computer modelling. The characteristic regimes qualitatively repeat at forcing frequencies f_0 and $2f_0$. The diagram contains more than 30 regions formed by intersections of the bifurcation lines l_i, $i = 0, 1, 2, 3$ and $l_{cr1, 2}$. The bifurcation curves l_3 and l_{cr2} correspond to a doubling of the torus and its destruction when a chaos arises, respectively. The tongue starting at A corresponds to the rotation number $\varrho = 1/4$ and that starting at B to $\varrho = 1/3$. The small tongues for weak resonances, even if observed in the experiments, are not shown in Fig. 5.7b.

Seven regions with topologically different types of oscillations (regions 1 – 7 in Fig. 5.7(b)) were found. Region 1 shows oscillations with forcing frequency f_1. When $R_m > 33$ kΩ, then a synchronization at the basic frequency pf_0 occurs. A synchronization on the second harmonics, $f_1 \approx 2f_0$ is observed in region 2.

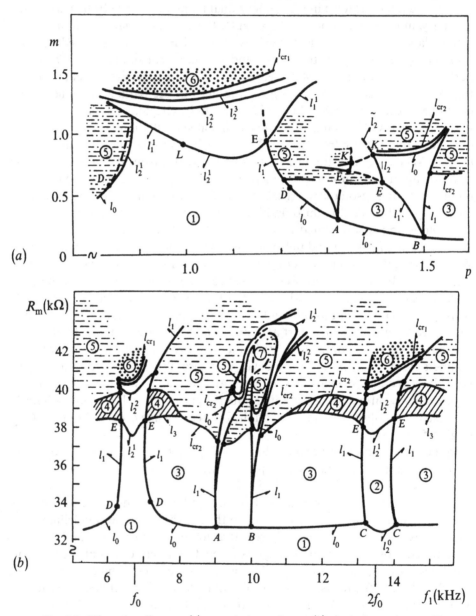

Fig. 5.7. Bifurcation diagram: (a) numerical experiment; (b) physical experiment, $U_0 = 0.5$B. For notation see text.

Stable beats with different rotation numbers occur in region 3. The oscillations in the state space are located on the surface of a 2-torus in the region 4 (a torus-doubling occurred on the original 1-torus). Narrow ranges of weak resonances are contained both in region 3 and region 4 (in region 4 they occur on 2-tori). Regions 5 contain chaotic trajectories which arise via several mechanisms of the destruction of two-dimensional tori when crossing l_{cr2}. In regions 6, located within the resonances 1:1 and 1:2 is observed chaos, formed via the period-doubling mechanism. Region 7 where the tongues A and B overlap contains a number of various oscillatory regimes, for example, an intermittency of the type limit cycle-chaos.

Four different types of routes to chaos were observed.

(1) Period-doubling of the resonance cycles inside the synchronization regions 1 and 2; this is realized when we are crossing the lines l_2^k transversaly in the parameter plane;

(2) soft (supercritical) destruction of two-periodic trajectories on a 1-torus (or in a 2-torus in the region 4) when crossing l_{cr2};

(3) intermittency between CA_2 and the synchronized cycles arising when moving inside the regions of synchronization in the direction of regions 5 across l_1. When returning inside the synchronization region with initial conditions located on CA_2 a metastable chaos is also observed;

(4) interactions of the type chaos-chaos (a competition between CA_1 and CA_2 in time), observed when moving between the regions 5 and 6 or in the region 7.

The doubling of the torus observed in experiments was also confirmed by numerical simulations of the model (5.9). Results of the computations for $p = = 0.111$, $B_0 = 1.2$ are shown for three values of m in the form of Poincaré maps and power spectra in Fig. 5.8. The bifurcation occurring for values of m between 0.7745 and 0.7750 can be seen as a formation of two closed invariant curves which move away with the increase of m. From experiments it follows that the cascade of the torus-doublings is finite and depends on the value of B, as is shown in Fig. 5.9. For small values of B the transition to chaos occurs via the destruction of the 4-torus and for large values of B no torus-doubling occurs and the chaos is formed directly from the 1-torus. Experimental results are supported by computations presented in the form of Poincaré maps and power spectra in Fig. 5.10. The figure illustrates the evolution of one of the two invariant intersections of the 2-torus by a surface $x = 0$ depending on the value of the parameter m for $B = 0.3$ and $p = 0.111$. The 2-torus is shown in Fig. 5.10(a), the 4-torus resulting after the torus-doubling is in Figs. 5.10(b) and (c); here we can observe increasing wrinkles on the torus; the doubling sequence ends because of the loss of smoothness of the torus, the chaotic attractor is formed in its neighbourhood (see Fig. 5.10(d)) and evolves with increasing m (see Fig. 5.10(e)). The Lyapunov dimension of the chaotic attractor increases with increasing m. The doubling of the torus and breakage of the torus can be indicated in the power spectra. We can observe harmonics $f_0/2$ (formation of a 2-torus) and $f_0/4$ (formation of a 4-torus). The computed values of the integral power $P_I = \int P(f)\, df$ confirm a soft ('supercritical') transition to chaos and its development.

A detailed experimental and simulation study of a forced magnetic oscillator, illustrating bifurcations to quasiperiodicity, entrainment (Arnold) horns and chaos was reported by Bryant and Jeffries[5.43]. They also studied transient behaviour of the system using various initial conditions ('an initialization technique') and were thus able to observe coexisting attractors. All three examples

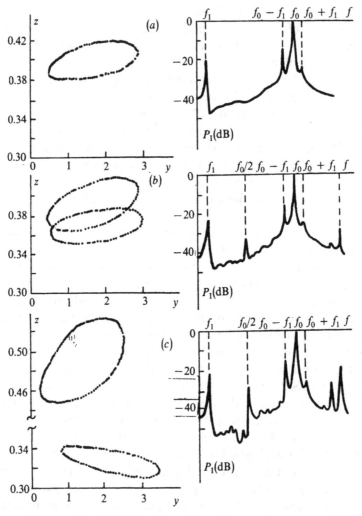

Fig. 5.8. Torus-doubling Poincaré sections and power spectra: $p = 0.111$, $B_0 = 1.2$; (a) $m = 0.774\,5$; (b) $m = 0.775\,0$; (c) $m = 0.8$.

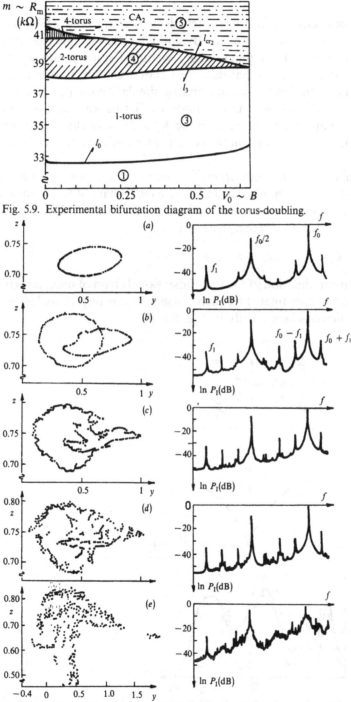

Fig. 5.9. Experimental bifurcation diagram of the torus-doubling.

Fig. 5.10. Formation of chaotic attractor CA_2 via the breakage of two-dimensional 4-torus. Numerical simulation, $B = 0.3$, $p = 0.111$, $g = 0.3$; Poincaré sections and power spectra.

included electronic circuits with just a few degrees of freedom which could be described by low-dimensional models. By coupling such circuits together into networks, we can form multidimensional dynamical systems and study their behaviour in a similar way to that described above. Gollub, Romer and Soco-lar[5.86] studied both on a model circuit and by simulation of the corresponding model equations two coupled tunnel diode oscillators and demonstrated by means of return maps and power spectra the dependence of chaotic regimes and synchronized periodic solutions on the coupling strength and frequency ratio of the oscillators.

Ezerskii, Rabinovich and coworkers[5.63] investigated experimentally spatial and temporal spectra in a nonlinear one-dimensional chain of 60 parametrically excited *LC* line elements, shown schematically in Fig. 5.11(*a*). The dependence of the charge Q_n on the voltage u_n was of the form

$$Q_n(u_n) = C_c(u_n + \alpha_1 u_n^2 + \alpha_2 u_n^3 + \ldots)$$

and the linear approximation of the one-dimensional chain of oscillators may be characterized in the investigated frequency region by the dispersion law $\omega(k) = \omega_{cr}|\sin k/2|$ depicted schematically in Fig. 5.11(*b*).

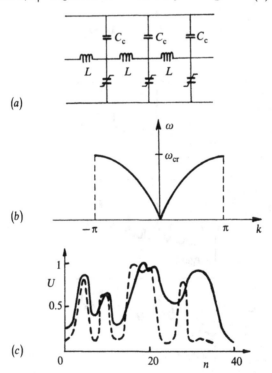

Fig. 5.11. (*a*) One-dimensional LC circuit; (*b*) dispersion characteristics; (*c*) spatial structure of the field; —— $U_P < U_{P3}$; ---- $U_P > U_{P5}$.

The properties of the chain excited by a spatially homogeneous noise source were monitored by determining the structure of the temporal power spectra along the chain. The electrical field in the network for an intense spatially homogeneous noise source could be described by the equation

$$\partial^2 Q/\partial t^2 - (u_{n-1} - 2u_n + u_{n+1}) = C_{cB} L(\tilde{U}_P), \tag{5.10}$$

where C_{cB} is a coefficient characterizing the coupling of the external and internal fields, L is a linear differential operator and $\tilde{U}_P = U_P \sin \omega_P t$, $\omega_P/2 \leqq \omega_{cr} = (2\pi)^{-1} \sqrt{LC}$.

When the forcing amplitude was increased, the following sequence of transitions was observed. First, for $U_P > U_{P1}(\omega)$ a pair of opposing waves was excited with wave numbers k_1 and $-k_1$. This corresponds to a spatially homogeneous regime at a frequency $\omega_P/2$. Then at $U_P = U_{P2}(\omega_P)$ several more modes with wave numbers $k_1 \pm \Delta k(k_1 \gg \Delta k)$ were excited but the power spectrum contained only a single frequency (all modes were synchronized). The width of the existence of the spatially homogeneous regime was dependent on the forcing frequency; at some frequencies a spatially inhomogeneous regime was established directly. Further increase in the forcing amplitude leads to the destruction of the synchronized regime and at $U_P \geqq U_{P3}(\omega_P)$ a three-frequency regime was established with satellite frequencies located symmetrically with respect to the frequency $\omega_P/2$. The structure of the field on each spectral component was spatially inhomogeneous. The next increase of the forcing amplitude caused a time modulation of the spectral components at $U_P = U_{P4}(\omega_P)$. Finally, at $U_P = U_{P5}(\omega_P)$ the nonstationary regime with a discrete spectrum changed jumpwise into a stochastic regime characterized by a broadband power spectrum. The spatial distribution of the field intensity in the prechaotic regime $(U_P < U_{P5})$ and in the chaotic regime is shown in Fig. 5.11(c).

The authors[5.63] also found that the introduction of noise did not influence qualitatively the character of the transition or the power spectrum, but lowered the value of U_{P5}. The fact that the spatial structures of the field in the prechaotic and chaotic regimes were quite similar (see Fig. 5.11(c)) also confirmed that the chaotic regime arose due to a nonlinear interaction of a small number of modes and was not connected with an enhancement of fluctuations. An analogous sequence of transitions was also observed in the network of 30 coupled LC elements.

Gaponov–Grekhov, Rabinovich and Starobinets[5.69–5.71] investigated routes to chaos and the subsequent increase in the dimension of chaotic attractors for a sequence of coupled oscillators of the van der Pol type described by a set of differential-difference equations

$$\frac{da_j}{dt} = a_j - (1 + i\beta) |a_j|^2 + e(1 - ic) (a_{j+1} + a_{j-1} - 2a_j),$$

$$a_j(t) = a_{j+N}(t), \quad j = 1, 2, ..., N, \tag{5.11}$$

which in the long-wave limit coincides with the one-dimensional Ginzburg–Landau equation

$$\frac{da}{dt} = a - (1 + i\beta) |a|^2 + D\nabla^2 a .$$ (5.11a)

Here $a_j(t)$ denotes a complex amplitude of the j-th oscillator and $\beta|a_j|^2$ its frequency. The parameter e describes a dissipative and c a reactive coupling. The authors demonstrated both for $N = 9$ and 50 that chaos with the Lyapunov dimension D_L of the order of 10 can develop both evolutionarily ('supercritically') via an intermittency[5.71] and jumpwise[5.70] ('subcritically') when the parameter e is varied. The dependence of the Lyapunov dimension of the chaotic attractor on e is shown in Fig. 5.12. Here $\beta = c = 1.71$, $N = 9$. For $\varepsilon \in (0, e^*)$ a limit cycle exists and for $e \in (0.75, 9.93)$ a two-dimensional torus is present. We can observe that a chaotic attractor appears abruptly both for $e \approx 0.1$ (here it bifurcates from the limit cycle) and at $e \approx 1$ (bifurcates from the 2-torus). Intermittency routes to chaos, involving two and three-dimensional tori were also observed. The authors[5.70] showed that an important role in the understanding of chaotic behaviour is played by elementary excitation – stationary waves of the type $a_j(t) = A_0 \exp\left[i(\omega t) - k_j + \varphi j)\right]$ with an amplitude depending on the wave number

$$A_0^2 = 1 - 4e \sin^2 k/2$$

with the dispersion relation

$$\omega(k) = -\beta + 4e(\beta + c) \sin^2 k/2 .$$ (5.12)

Spatially homogeneous oscillations $(k = 0)$ and dissipative structures

$$\left(\omega = 0, \sin^2 \frac{k}{2} = \frac{\beta}{4e(\beta + c)}\right)$$

are particularly important.

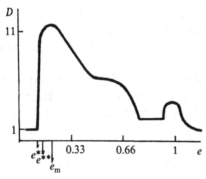

Fig. 5.12. Dependence of the dimension of chaotic attractor D on the intensity of dissipative coupling e.

The abrupt appearance of a high-dimensional chaotic attractor is connected with the coexistence of a stable attractor (a limit cycle or a torus) with an unstable (repelling) high-dimensional chaotic attractor. When the regular solution (periodic, quasiperiodic) disappears, a repelling chaotic set becomes a part of an attracting high-dimensional chaotic attractor.

For low values of e $(e < 0.25$ at $c = 1.71)$ a new type of transition to chaos was observed. The frequency of periodic oscillations decreased to zero when the boundary of chaotic behaviour in the (e, β) parameter plane was approached. This boundary coincided with the surface $\omega(k_m, \beta, e, c) = 0$, with ω given by (5.12) and k_m being a maximal possible wave number for given $N(k_m = \pi$ for even $N)$. The behaviour can be understood when following the development of solutions with increasing e. When $0 < e < e_1(\beta)$ $(e_1 = 10^{-4}$ for $c = \beta = 1.71)$, a single frequency regime $\omega = \beta$ is established in the system. Increase of e (i.e. of the intensity of the coupling) leads to the suppression of longwave modes and to the synchronization of shortwave modes on the frequency $\omega = \omega(k_m)$. When $e \to e^*$, where e^* is determined from $\omega(k_m, \beta, c, e^*) = 0$, the frequency of the regime of mutual synchronization monotonically decreases and at $e = e^*$ equals zero and the developed chaos appears without hysteresis. The authors propose the following picture in the state space: the stable limit cycle for $e < e^*$ approaches the steady states of (5.11), $a_j = A_0 \exp\left[-i(k_j + \varphi_0)\right]$, $\varphi_0 \in [0, 2\pi]$, located on the circle $A_0^2 = 1 - 4e \sin^2(k_m/2)$, passes through this circle and either disappears or becomes unstable. The high-dimensional chaotic set at the same time becomes attracting at $e = e^*$. The dimension of the chaotic attractor is already high at $e = e^*$ (the Lyapunov dimension $D_L \approx 8$ for $N = 9$) and increases until $e = e_m$. For $e > e_m$ the dimension falls almost monotonically (see Fig. 5.12).

The evolution of chaotic trajectories depended on the chosen initial conditions. When initial conditions formed from the zero $(k_0 = 0)$ and third $(k_3 = \pm 2\pi/(N/3))$ modes were used, the chaotic regime arose at a higher value of $e = e^{**}$, determined from the relation $\omega(e^{**}, k_3) = 0$, see Fig. 5.12. The chaotic set corresponding to modes which are multiples of k_4 already exists in the state space; however, because the excitations which are multiples of k_4 are absent from initial conditions and the nonlinearity $|a|^2 a$ does not create them, an independent chaotic set composed from the k_3 modes arises at $e = e^{**}$ with the dimension $D \approx 10$. This chaotic set is for general boundary conditions unified with other nonattracting sets and forms a high-dimensional strange attractor, corresponding to interactions of several chaotic sets reflecting oscillations in various groups of spatial modes.

Hence, in a distributed system we can observe both a continuous and an abrupt development of chaotic attractors with dimensions either dependent or practically independent on characteristic parameters. The constancy of the dimension of the attractor when increasing the number of interacting modes reflects the fact that the newly introduced modes become synchronized with the earlier ones. The increase of the dimension is connected with the destruction of such structures. Several aspects of chaotic solutions of the Ginzburg–Landau equation will be discussed later. Most of the above described routes of the transition to chaos were also observed in experiments in a distributed system with delayed feedback, formed by a system consisting of the chain of a nonlinear active element – an electron beam – a travelling electromagnetic wave, a non-linear filter and a holding circuit[5.141]. Similarly as in simple dynamic systems, a sequence of period-doubling bifurcations with the rate of convergence described by Feigenbaum's δ, a destruction of the three-dimensional torus, a 'hard' onset of chaos, an intermittent transition and a sequence of torus doublings were observed as typical routes to chaos. The observed routes differed when starting from different initial conditions for the same values of parameters. An intermittency of the type 'chaos–chaos' was also observed, hence the coexistence of different chaotic attractors was indicated. It was proposed that the developed chaotic set far from an equilibrium state consisted of several chaotic subsets formed in different parts of the state space.

5.3 Mechanical and electromechanical systems

Mechanical and electromechanical systems can often be both precisely controlled and well described by model equations (similar to electric circuits) and thus they may serve for demonstrations and more quantitative studies of chaotic behaviour. From a large number of available examples we shall discuss in more detail experimental results obtained by Holmes and Moon[5.118–5.119] for forced vibrations of a buckled beam and briefly review several other interesting experimental observations of chaos in typical mechanical systems.

The apparatus used by Holmes and Moon[5.118–119] is shown schematically in Fig. 5.13(a). The apparatus consists of a steel (ferromagnetic) cantilevered beam suspended vertically. The clamped end is attached to a vibration shaker (vibrated with a steady sinusoidal motion). Below the free end of the beam were placed permanent magnets which rested on a steel base. Strain gauges were attached to the beam near the clamped end and a linear variable differential transformer was attached to the shaker platform to measure the forced vibration amplitude of the beam base. The nonperiodic motion was characterized by time histories, Fourier analysis and Poincaré maps. An example of the time course of the forced chaotic vibrations is shown in Fig. 5.13(b). The determined Poincaré maps depended on the phase angle of the driving motion θ. The oscillations were studied analytically and numerically. The mathematical model using a one-degree-of-freedom approximation is in the form

$$\ddot{A} + \gamma' A - \tfrac{1}{2}(1 - A^2)\, A = f \cos \omega t , \tag{5.13}$$

where A is the modal amplitude of the first vibration mode. Studies of this system have demonstrated the existence of chaotic solutions for certain sets of parameters (γ, f, ω) and chosen initial conditions ('Duffing–Holmes strange attractor'). Holmes[5.118] has derived a necessary condition for chaos

$$f \geq (\gamma \sqrt{2/3\pi\omega}) \cos h(\pi\omega/\sqrt{2}) \tag{5.14}$$

and Moon[5.194] has presented a heuristic condition for chaos existence

$$f \geq \frac{\alpha}{2\omega} \left[\left(1 - \omega^2 - \frac{3}{8} \frac{\alpha^2}{\omega^2} \right)^2 + \gamma^2 \omega^2 \right]^{1/2} \tag{5.15}$$

where the parameter α is close to 1. Experimental measurements of the critical forcing amplitude between a chaotic motion and periodic motions have shown

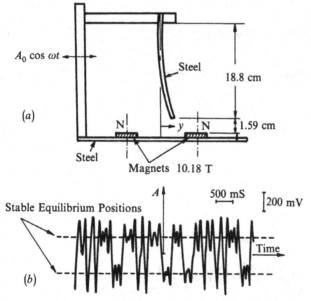

Fig. 5.13. (*a*) Experimental apparatus of a buckled beam and two magnets according to Moon and Holmes; (*b*) bending strain vs time for forced chaotic vibrations of the buckled beam.

a nonsmooth boundary depending on the driving frequency. Moon[5.194] has proposed that the boundary between chaotic and periodic oscillations is fractal and that this behaviour follows from the presence of higher vibration modes in the beam, not represented in the model (5.13). A typical comparison of experimental and theoretical boundaries (Eq. 5.15) is shown in Fig. 5.14. The erratic nature of the boundary between the chaotic and the periodic motion was much larger than the experimental errors in the measurements. To measure the fractal dimension of the boundary the curve between two points was approximated by a connected set of N straight lines of length l. The length of the approximate curve is then $L = N \cdot l$. For a smooth curve $N \to \lambda l^{-1}$ where λ is its length. For a fractal curve $N \to \lambda l^{-D}$ and $L = \lambda l^{1-D}$; when D is not an integer, the curve is fractal. For the data shown in Fig. 5.14 the value of $D = 1.28$ was determined. Hence the data support the conjecture of Grebogi, Ott and Yorke[3.35] on the existence of fractal boundaries between periodic and chaotic motions. Moon and Li[5.195] have computed capacity dimensions of Poincaré maps generated both in experiments and from numerical solutions of Eq. (5.13) using the algorithm proposed by Grassberger and Procaccia[2.25, 4.19]. The vector $x_i = \{A(t_i), \dot{A}(t_i)\}$ in the state space was sampled on the trajectory at times t_i; for a set of N samples $(N = 10^3$–10^4 points) a ball of a radius r centred at each x_j was constructed, the number of points within this ball was counted, and the 'discrete correlation integral' (see also Eq. 4.40)

$$C(r) = \frac{1}{N^2} \sum_i \sum_j H(r - |x_i - x_j|) \tag{5.16}$$

was computed. Here $H(s) = 1$, $s \geqq 0$; $H(s) = 0$, $s < 0$. The correlation dimension is then

$$d = \lim_{r \to 0} \frac{\ln C(r)}{\ln r}. \tag{5.17}$$

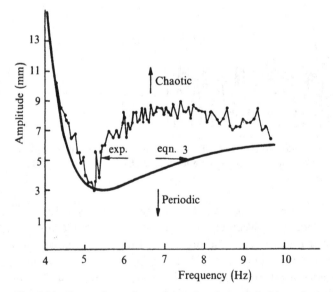

Fig. 5.14. Comparison of experimental and theoretical boundaries between periodic and chaotic vibrations.

From numerical computations it was found that $2 < d(\gamma) < 2.8$ for $0.23 >$
$> \gamma > 0.05$. The dimension was independent of the phase of the forcing
function. The dependence of the logarithm of the 'correlation integral' on $\ln r$
for an experimental Poincaré map (forcing frequency 8.5 Hz, nondimensional
damping $\gamma = 0.013$) is shown in Fig. 5.15. A dimension ranging between 1.63
and 1.75 was evaluated from these data, again independent of the forcing phase.
However, only the dimension of Poincaré maps and not of the attractor itself was

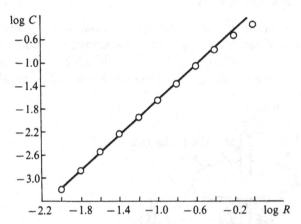

Fig. 5.15. Dependence of the correlation function C on the radius of the ball R;
$d \in (1.63; 1.75)$.

evaluated in this paper. Moon and Holmes[5.196] discussed magnetoelastic strange
attractor and Moon, Cusumano and Holmes[5.197] presented evidence for existence of homoclinic orbits as a precursor to chaos in a magnetic pendulum.

A very interesting magnetomechanical system is a rotating disc in a magnetic
field, studied by Robbins[5.229]. Chaotic current oscillations in the coil under
constant torque are observed, see Fig. 5.16. The rotation Ω and currents I_1, I_2
(see Fig. 5.16) can be described by the set of equations

$$J\dot{\Omega} = -k\Omega - \mu_2 I_1(I_1 + I_2) + T,$$

$$L_1\dot{I}_1 = -RI_1 - R_3 I_2 + \mu_1 \Omega I_1,$$

$$L_2\dot{I}_2 = -R_2 I_2 + \mu_2 \Omega I_1. \tag{5.18}$$

Robbins[5.229] has demonstrated the existence of chaotic solutions with current
reversals in time, which are qualitatively similar to those observed in experiments, see Fig. 5.16. This system is of interest in geophysics, where similar

models are used for the interpretation of reversals of the Earth's magnetic field. It also suggests that chaotic vibrations might be possible in turbine generator systems.

A forced nonlinear torsion pendulum is another model mechanical system used for detailed studies of chaotic behaviour[5.144]. The system can be described by the following equation of motion

$$\theta\ddot{\phi} + R\dot{\phi} + A\phi + B\sin\phi = F_0\cos\omega t \tag{5.19}$$

where ϕ is the angular position, θ the moment of inertia, R the damping parameter and $F_0\cos\omega t$ describes the driving torque. Experimental studies have demonstrated the existence of a period-doubling cascade which occurs both when the driving frequency and the damping parameter are varied[5.26] or when

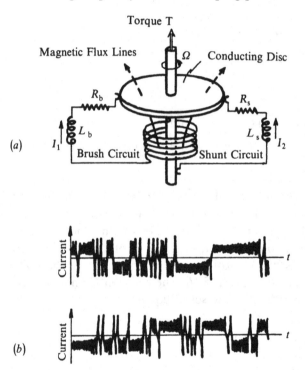

(a)

(b)

Fig. 5.16. (a) Rotating dynamo and electrical circuits; (b) chaotic current oscillations in the coil under constant torque.

the driving amplitude is increased and the damping force decreased[5.164]. Maurer and Libchaber[5.187] have also illustrated coexistence of various periodic and chaotic solutions and the presence of transient chaotic behaviour. Libchaber and coworkers[5.52] also discussed routes to chaos in the forced pendulum. Shaw and Wiggins[5.251] described chaotic dynamics of a whirling pendulum. Croquette and

Poitou[5.51] described period-doubling bifurcations and chaos in the motions of a compass. Bifurcation sequence leading to chaos for compass needle in an oscillating magnetic field was studied by Schmidt[5.241].

Chaotic dynamics plays a role of increasing importance in dynamical problems of machines and mechanisms. Thus clattering vibrations in gear-boxes may be described as a sequence of impulsive processes[5.155]. The motion between two impacts during teeth-meshing can be represented by iterated invertible mappings. Chaotic behaviour of a one-stage and two-stage gearing is indicated by positive LEs[5.212, 5.213]. The discrete mappings are used to estimate the boundaries of the phase space where chaotic motions occur[5.121]. Hiller and Schnelle[5.113] have discussed the period-doubling route to chaos in the four-bar mechanism with elastic coupler, which appears both in the mechanism of a motion picture camera and in the cloth advancer of a sewing machine. Stelter and Popp[5.259] have described occurrence of chaos in slip-stick phenomena, arising as an undesired noise in bowed music instruments, in curving tram wheels and in machine-tool clattering. Isomäki, Boehm and Räty[5.131] demonstrated both Feigenbaum and intermittent scenarios of transitions to chaos in a driven elastic impact oscillator. Chaotic dynamics of an impact oscillator was also discussed by Thompson and Ghaffari[5.271].

Instabilities and chaos often occur in manufacturing machines. Grabec[5.89, 5.90] has described the generation of chaos in the plastic flow in the cutting process. He considers that the energy consumed by the cutting process is supplied by a machine mechanism pushing the workpiece with a certain velocity against the tool. There the cutting zone is formed in which the input material is transformed by plastic flow into the output chip. The resulting force consists of cutting and friction components, F_x, F_y, normal and parallel to the tool surface, respectively. The transmission of this force over the tool and the machine (total mass m) causes elastic deformations, which can be represented by the tool displacement (x, y). If the rigidity in the x and y directions are denoted by k and q, respectively, then the dynamics can be described by the equations

$$m\ddot{x} + kx = F_x,$$

$$m\ddot{y} + qy = F_y.$$

Properties of the cutting forces (F_x, F_y) are described by empirical relations[5.89]. When the empirical parameters corresponding to the cutting of low carbon content steel have been chosen and the slow variations of the cut depth have been followed, the formation of a chaotic attractor in the phase space has been observed. The fractal dimension of the attractor has changed from 1 to 2.8 in the transition from periodic to chaotic oscillations. Such behaviour might be expected as the model system of equations corresponds to coupled nonlinear two-dimensional oscillators.

Most of the above described chaotic behaviours in machines and mechanisms have resulted from the studies of standard mathematical models. This has been caused by the recent trend to a more theoretical foundation and by the use of computer modelling in design. Thus Thompson and Virgin[5.270] discussed connections between spatial chaos and localization phenomena in nonlinear elasticity problems. An analysis of chaotic oscillations in actual mechanisms combining modelling with analysis of experimental observations will certainly not only bring additional examples of chaotic attractors but also will help in better design.

5.4 Solid state systems

Coupled and forced nonlinear oscillators form the basis of the description of many solid state systems. Hence, we may expect the existence of chaotic motions in appropriate ranges of parameters. Here we shall briefly discuss several experimental observations which illustrate characteristic features of chaotic behaviour in a solid state.

5.4.1 Semiconductors

The first example of solid state turbulence in a semiconductor was described by Aoki and coworkers[5.16]. They observed both periodic oscillations and chaos in the firing density wave of filaments in n-GaAs at 4.2 K. A schematic

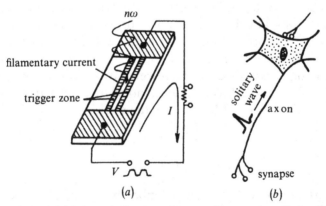

Fig. 5.17. (*a*) Schematic drawing of current filaments in n-GaAs; (*b*) solitary wave propagation in a nerve axon.

drawing of current filaments is shown in Fig. 5.17(*a*) and period-doubling bifurcations of the oscillations observed at 812 nm as a function of the applied electric field are shown in Fig. 5.18 (*a*) to (*d*); here the field intensity is (*a*) 3.8 V/cm, (*b*) 3.25 V/cm, (*c*) 3.1 V/cm, (*d*) 3.0 V/cm. The Poincaré return map of chaotic oscillations in Fig. 5.18 (d) obtained as the next amplitude plot of the peak amplitudes J_n and J_{n+1} is depicted in Fig. 5.18(e) and supports the deterministic character of the chaotic dynamics.

The temporary behaviour of the firing density wave showed stationary, periodic and chaotic states depending on the values of the control parameters. An

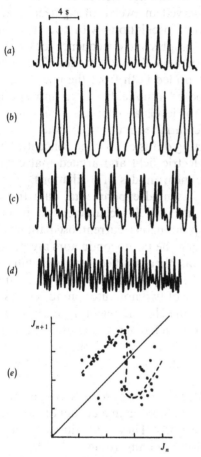

Fig. 5.18. Period-doubling bifurcations of oscillations in n-GaAs observed as a function of the applied electric field: (a) 3.8 V/cm; (b) 3.25 V/cm; (c) 3.1 V/cm; (d) 3.0 V/cm; (e) Poincaré return map of the time series in (d).

analogy with a nerve axon was suggested, see Fig. 5.17(b). The filamentary channel corresponds to the axon, the trigger zone in the filamentary channel to the synapse and the photon flux to the external stimuli. Numerical investigations based on the models of random nets of neurons[5.14][5.15] then revealed the character of the basin of the strange attractor, which had the structure of a folded torus.

Teitsworth, Westervelt and Haller[5.268], Teitsworth and Westervelt[5.266] and Westervelt and Teitsworth[5.290] found self-generated limit cycle oscillations and chaos in the conductance of liquid helium cooled far-infrared photoconductors. The authors studied eighteen devices from four different crystals of a p-type ultrapure germanium. The majority of germanium samples studied displayed complex behaviour, often similar to that of coupled nonlinear oscillations. A period-doubling route to chaos was observed in twelve of eighteen samples, taken from all four crystals. Quasiperiodic oscillations with two incommensurate frequencies were identified in four samples and a frequency locking in six samples. Finally, intermittent switchings between different modes of oscillations were also observed. The regimes were identified both from the phase portraits and the power spectra. The authors proposed a model based on impact ionization of a charge stored on shallow acceptor levels and space – charge injection written in the form of three partial differential equations (considering a one-dimensional sample geometry and neglecting diffusion) for the concentration of holes, ionized acceptors and the local electric field and argued that complex regimes resulted from interaction of nonlinear modes of oscillations.

Later Teitsworth and Westervelt[5.267] presented experimental data from extrinsic Ge photoconductors showing nonlinear dc I–V characteristics including a negative differential resistance with spontaneous current oscillations. This behaviour may cause chaotic response when the photoconductors are periodically driven. Period-doubling sequence was then studied experimentally in a systematic series of experiments with a driven photoconductor system and its presence has been confirmed by simulations of two nonautonomous ODEs with harmonic forcing term. The noise level in the chaotic region is more than four orders of magnitude larger than outside it which destroys the usefulness of this far-infrared detector in many applications. Detailed analysis of shapes of resonance tongues in the driving amplitude–driving frequency phase plane can help to keep noise levels down.

Held, Jeffries and Haller[5.108, 5.111] studied transitions from spontaneous current oscillations to chaos for an electron–hole plasma in a crystal of germanium at 77 K in parallel electric and magnetic fields. They have illustrated in state space portraits the existence of the period-doubling route to chaos and also transitions including quasiperiodic states. When increasing the applied electric field, regions of broadband noise with noise-free periodic windows were also observed. Held and Jeffries[5.107, 5.109] studied also spatial structure of chaotic instabilities in this system and found both spatially coherent and incoherent chaotic helical waves. Quasiperiodic transitions to chaos was reported by the same authors[5.110] as the result of ac perturbations. Gwinn and Westervelt[5.96] reported later experimental verification of the universality of $f(\alpha)$, the fractal dimension of the subset of the attractor with scaling index α, at the transition from the quasiperiodicity to chaos with the golden-mean winding number for the cooled extrinsic p-type Ge.

Martin, Leber and Martienssen studied oscillatory and chaotic states of the electrical conduction in barium sodium niobate single crystals[5.182]. The control parameters were crystal temperature, current density and flow rate of humidified oxygen. Depending on temperature they observed self-generated voltage oscilla-

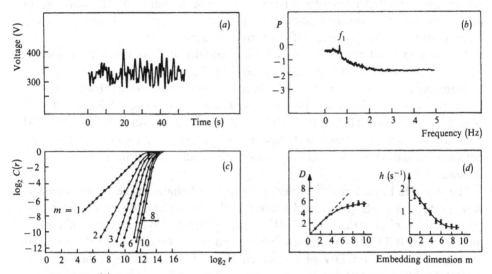

Fig 5.19. (a) Self-generated oscillations of the voltage across the barium sodium niobate crystal; (b) normalized power spectrum for the time series in (a); (c) the correlation integral $C(r)$ (computed from the measured voltage signal shown in (a)) plotted versus the hypercube edge length r for different values of the embedding dimension m; (d) correlation exponent D and the entropy parameter h_2 (computed from the data in (c)) as a function of the embedding dimension m.

tions which became unstable and a chaotic state developed via the Ruelle –Takens–Newhouse scenario. Three incommensurate frequencies subsequently appeared in the power spectrum and then chaotic state resulted. The time course of chaotic oscillations is shown in Fig. 5.19(a) and the corresponding power spectrum exhibiting broadband noise is depicted in Fig. 5.19(b). However, in most of the experiments only two fundamental frequencies appeared before the onset of chaos. The authors also computed a 'correlation integral' $C(r)$ for different values of the embedding dimension m according to the procedure proposed by Grassberger and Procaccia[2.25]. In Fig. 5.19(c) is shown as a function of $\log_2 r$ (r is a hypercube edge length in the Grassberger–Procaccia algorithm). We can see that $C(r)$ scales as r_2^D, i.e. *the correlation dimension D_2 becomes a constant when increasing m.* The dependence of the correlation dimension is plotted in Fig. 5.19(d) as a function of m and it was found that the approached asymptotic value is $D_2 = 5.3 \pm 0.2$. In case of a noise signal, the value of D_2 would increase proportionally to m, hence the constant value indicates the existence of a strange attractor. Finally, the dependence of the entropy h_2 on m

is also depicted in Fig. 5.19(d). Grassberger and Procaccia have shown that the correlation integral scales with the embedding dimension m like $C(r) \sim$ $\sim \exp(-m\tau h_2)$. As can be inferred from the figure h_2 decreases with m and reaches a value $h_2 \sim 0.4~s^{-1}$. The nonperiodic evolution of trajectories, the power spectrum with broadband noise, a constant value of the correlation dimension and the nonzero positive asymptotic value of the Kolmogorov entropy confirmed the existence of a strange attractor.

Martin and Martienssen have shown in more detailed studies with ac and dc fields[5.180, 5.181] that the dynamical behaviour of the system can be modelled by a one-dimensional circle map. Mode locking has been observed in the system prior to the transition to chaos and the mode-locked intervals have exhibited a devil's staircase behaviour. The authors suggested that chaotic state of the electrical conductivity in the presence of dc-field have resulted from a nonlinear coupling between oscillatory modes spatially distributed along the cathode surface of the crystal.

Sherwin, Hall and Zettl[5.252] studied the onset of chaos in the charge-density-wave state of a $(TaSe_4)_2I$ crystal when coupled to an external inductance and driven by an *rf* electric field. Zettl[5.301] has reported investigations of charge density wave (CDW) in materials $NbSe_3$ and TaS_3. Phase-slip centre-induced discontinuities in the CDW phase velocity lead to mode locked solutions with period-doubling route to chaos.

Gibson and Jeffries[5.73] observed period-doubling bifurcations, chaos and periodic windows in gallium doped yttrium iron garnet. Hoffnagle *et al.*[5.115] reported quasiperiodic order-chaos transition of two trapped Ba^+ ions.

Seiler, Littler, Justice and Milonni[5.249] have discussed chaotic behaviour in the electric properties of an n-InSb narrow gap semiconductor. The power spectra indicated the presence of the period-doubling scenario of the onset of chaos. Landsberg, Schöll and Shukla[5.160] presented a simple model for chaos in semiconductors based on a Chapman-Kolmogorov equation.

Peinke *et al.*[5.208] have described spontaneous oscillations and the transition to chaos in the post-breakdown regime of p-Ge samples in the temperature range between 1.7 and 4.2 K. they demonstrated on the phase portraits and the power spectra the presence of three mechanisms of the transition to chaos: the intermittent switching between two oscillatory states, the period-doubling scenario and the scenario involving two incommensurate frequencies. Transition from a chaotic attractor to a hyperchaotic one, reflecting a loss of spatial coherence was reported in another paper[5.209].

The use of many semiconductor devices is limited by the onset of a high level of noise or instabilities and breakdowns when working in nonlinear regimes or at high electric fields[5.151]. Detailed studies of the mechanisms of their generation in the frame of the theory of nonlinear dynamical systems will hopefully lead to improvements in the functions of practical semiconductor devices.

5.4.2 Josephson junctions

The Josephson junction is a weak link between two bulk superconductors. Since Huberman and coworkers[5.123] predicted an occurrence of chaotic behaviour in Josephson junctions, a large number of papers have been published on the subject. Most of them reported results of digital and analogue simulations[5.25, 5.190] but several experimental studies also appeared, see, for example, Refs 5.126 and 5.201 and references therein.

When the *rf* driven Josephson junctions are used as parametric amplifiers, high level noise appears and grows with the gain level. It is related to the intrinsic nonlinear dynamics of the junction[5.24].

The current conservation equation for the resistively shunted junction can be written in the form

$$C\dot{V} + V/R + I_c \sin \varphi = I_r \cos \omega t + I_0 \qquad (5.20)$$

where R, C, I_c, I_r and I_0 denote the resistance, capacity, critical current, *rf* current and dc bias, respectively. The phase difference across the junction is given by the Josephson equation

$$\dot{\varphi} = \frac{2eV}{h} . \qquad (5.21)$$

Eqs. (5.20, 5.21) can be written in the dimensionless form as

$$\ddot{\varphi} + \beta_c^{-1/2}\dot{\varphi} + \sin \varphi = i_0 + i_r \cos \omega t , \qquad (5.22)$$

which is also the equation of a forced damped pendulum[5.92, 5.93], hence a full repertoir of the ways of transition to chaos may be expected.

Yeh and Kao[5.297] have demonstrated on the basis of power spectra and state space trajectories for a sine-wave-driven Josephson junction analogue that the transition occurs by the period-doubling route and the universal power laws known for one-dimensional unimodal mappings are valid. The same authors later[5.298] demonstrated by means of numerical and analogue solutions of the differential equation

$$\ddot{\varphi} + \beta_c^{-1/2}\dot{\varphi} + \sin \varphi = A \sin \omega t \qquad (5.23)$$

that two types of intermittent transitions are also present and that the measured average number of pseudoperiodic oscillations between two successive turbulent bursts followed theoretically predicted power-law scaling. Intermittent chaos was also reported by Ben-Jacob and coworkers[5.31]

Seifert[5.247, 5.248] has also found both the period-doubling and intermittent scenarios in ac and dc driven Josephson junctions and demonstrated the existence of different chaotic regimes within a single locked state.

He, Yeh and Kao[5.106] have described experimental observations of the transition from quasiperiodicity to chaos by using a Josephson junction analogue (hence also a damped pendulum analogue), described by the model equation

$$\ddot{\varphi} + \beta_c^{-1/2}\dot{\varphi} + \sin \varphi = A_1 \sin \omega_1 t + A_2 \sin (\omega_2 t + \gamma) . \qquad (5.24)$$

The real-current driven Josephson junction is also described essentially by Eq. (5.24) but the analogue circuit simulator permits greater versatility. When the ratio of the driving frequencies ω_1/ω_2 is chosen to approximate the reciprocal of the golden mean $\frac{1}{2}(\sqrt{5} - 1)$ then it is found that the transition from the quasiperiodicity to chaos obeys the scaling behaviour predicted for circle maps by Feigenbaum, Kadanoff and Shenker[3.26]. The experimental convergence ratios δ and α also approximate well the theoretical values (the authors determined $\alpha = -2.75 \pm 0.2$, $\delta = 1.63 \pm 0.2$, compared to theoretical values -2.618 and -1.618, respectively). Transition to chaos by interaction of resonances was also studied by Bohr, Bak and Jensen[5.36].

Sakai and Yamaguchi[5.238] integrated numerically model equations, constructed one-dimensional maps and then analysed in detail the mechanism of period-doubling bifurcations and sudden transitions of chaotic orbits to periodic ones ('transfer crises') and interior crises of chaotic attractors. Noldeke and Seifert[5.202] have demonstrated by means of digital and analogue simulations of model equations that both Pomeau–Manneville type I and type III intermittency may occur in the system.

Iansiti and coworkers[5.126] have studied both by digital simulations and experiments a Josephson system in a highly nonlinear regime. They demonstrated that simulated and experimental current-voltage curves agreed satisfactorily and determined by detailed computations that high values of noise observed in experiments correspond in simulations to intrinsic chaotic motions in some parts of the state space and to noise-induced transitions between periodic orbits in other parts of the state space. The authors studied in detail the basin-boundary of periodic attractors and determined that it has a fractal structure with a correlation dimension $D_C = 1.75$. The power spectrum of the solutions in the fractal basin-boundary regime showed a noise-induced intermittency leading to a large low-frequency noise. The spectrum scaled approximately as $1/f$ for two frequency decades. Thus they confirmed that the presence of a fractal basin boundary in the dynamics of a nonlinear system may give rise both to chaos and a low-frequency noise when thermal and other types of noise are considered.

Noldeke[5.201] has found chaotic loss of the phase lock in the current–voltage characteristics of an $Sn/Sn_xO_y/Sn$ Josephson tunnel junction under 10 GHz microwave irradiation. The experimental results obtained at 3.7 K were consistent with the results of analogue simulations based on a model of the type (5.22) when an external noise corresponding to 17 K was included in the simulations. Gruner[5.91] has reviewed various nonlinear phenomena including transition to chaos in Josephson junctions and in CDW systems in the presence of dc and ac drives. A high level of noise seriously limits possible applications of Josephson devices. It is to be hoped that detailed studies of the origin of the deterministic noise will help to control it better.

5.5 Chaos in nonlinear optics

The laser is a light source with properties such as high intensity, spectral purity and high directionality which are utilized in a number of technical applications. It consists typically of a rod of an active medium where the laser-active atoms are pumped from the outside[5.34, 5.97, 5.239]. It represents a typical nonlinear oscillator system. The unimodal laser can be described by a system of Maxwell–Bloch equations. For the laser cavity tuned to the resonant frequency of the two-level atoms we can write[5.97]

$$\frac{d\hat{E}}{dt} + \left[c\frac{\partial \hat{E}}{\partial x} \right] = -\varkappa\hat{E} + \varkappa\hat{P} , \tag{5.25}$$

$$\frac{d\hat{P}}{dt} = \gamma\hat{E}\hat{D} - \gamma\hat{P} , \tag{5.26}$$

$$\frac{d\hat{D}}{dt} = \gamma_{\parallel}(\lambda + 1) - \gamma_{\parallel}\hat{D} - \gamma_{\parallel}\lambda\hat{E}\hat{P} \tag{5.27}$$

where $\hat{E}$ describes the electromagnetic field, $\varkappa$ is the cavity decay rate, γ the dephasing rate of the atoms, $\gamma_{\parallel}$ the decay of the population inversion, λ the pumping parameter, $\hat{D}$ the population inversion and $\hat{P}$ the atomic polarization. Eqs (5.26) and (5.27) follow from the Schroedinger equations for the two-level atom and Eq. (5.25) is an inhomogeneous wave equation (reduced form of Maxwell equations).

Haken[5.98] has demonstrated that the set of Eqs (5.25–5.27) is equivalent to the Lorenz model (the choice of parameters can describe random spiking in the superradiance regime). Then Ikeda[5.127] studied theoretically multiple stationary states (optical bistability) and the instability of transmitted light in a ring cavity device and predicted a sequence of bifurcations and chaos. Graham and coworkers[5.245, 5.295] have demonstrated theoretically that chaos can exist in a laser under a modulated external field and found intermittency in a system with modulation inversion. Mayer, Risken and Vollmer[5.188] have also discussed periodic and chaotic pulses in a ring laser and Velarde and Antoranz[5.282] discussed chaos–optical turbulence in a laser with a saturable absorber. A review of the early theoretical work on laser instabilities can be found in Casperson[5.46].

Abraham[5.1] discussed the significance of the research on lasers with unstable and/or chaotic output. Blow and Doran[5.35] have described a sequence of period-doubling bifurcations followed by an inverse sequence (band-merging sequence) in an optical ring cavity where the propagation of an envelope pulse was described by the nonlinear Schroedinger equation. They also described the bistability and the merging of two chaotic branches and have shown that the global and local properties of the pulse sequences have the same dynamical behaviour.

Moloney[5.192] discussed self-focusing induced optical turbulence in transverse outputs of an optical resonator. The dynamical behaviour of the electromagnetic field in the resonator was described by the nonlinear equation

$$2i \frac{\partial}{\partial \zeta} G_n + \frac{\partial^2}{\partial y^2} G_n - \frac{G_n}{1 + 1 \, |G_n|^2} = 0 \tag{5.28}$$

with the boundary conditions

$$G_n(y, 0) = a(y) + R \, e^{ikl} G_{n-1}(y, p) \,. \tag{5.28a}$$

Here $G_n(y, \zeta)$ is the normalized intracavity field amplitude where y and ζ refer, respectively, to the coordinates in the transverse and propagation directions. When $a(0)$ (the peak input Gaussian amplitude) and p (the effective nonlinear medium) are varied, various types of bifurcation sequences were observed. They included period-doubling bifurcations, transitions via torus (including the torus doubling) and an intermittent chaotic attractor. The author also observed the coexistence of various attractors over wide ranges of parameters. Chaos in all optical Fabry-Perot resonator was also studied by Harrison, Firth and Al-Saidi[5.99].

A special issue on 'Instabilities in active optical media'[5.303] gives a number of examples of studies of chaotic behaviour in nonlinear optics. We shall now review several experimental studies in a nearly chronological order, which confirmed the presence of chaotic behaviour in actual laser systems.

Gibbs *et al.*[5.72] used a modified hybrid bistable device (they introduced an electronic delay line with a delay t_R in the feedback) with a ring cavity and demonstrated both the presence of periodic signals with the doubled periods $2t_R$ and $4t_R$ when the input power was increased and then chaos. This device could be described by a single equation for the phase as shown by Ikeda[5.127]

$$\tau \dot{\varphi}(t) = -\varphi + A^2 \{1 + 2B \cos [\varphi(t - t_R) - \varphi_0]\}$$
$$= -\varphi + g(p, \varphi(t - t_R)) \,, \tag{5.29}$$

where τ is the relaxation time of the nonlinear medium and $g(p, \varphi)$ is a nonlinear function characterizing the cavity and the medium. When $\tau \ll t_R$ (the region where Gibbs *et al.* worked), Eq. (5.29) reduces to one-dimensional mapping

$$\varphi_n = g(p, \varphi_{n-1}),\tag{5.29a}$$

and this supports the presence of period-doubling and chaos. Ikeda and Kondo[5.128] and Hopf and coworkers[5.122] later demonstrated the presence of chaos in this system also for $\tau \gg t_R$. Hoffer, Chyba and Abraham found both period-doubling and quasiperiodicity in a unidirectional, single mode, inhomogeneously broadened Xe ring laser[5.114]. Chaos in a solid-state ring laser was also studied by Khandokhin and Khanin[5.145, 5.146]. Transitions to chaos in a homogeneously broadened one- and two-mode ring laser was investigated by Klische and Weiss[5.149].

Purely optical chaos was observed by Arecchi and coworkers[5.17] in a Q-switched CO_2 laser. The authors observed subharmonic bifurcations and the coexistence of two independent attractors. A jump between two chaotic attractors caused $1/f$ type divergence in the power spectrum. Harrison with coworkers[5.100] observed period-doublings in an all-optical resonator containing NH_3 gas, very similar to that discussed theoretically by Ikeda[5.127]. Ikeda *et al.* also reported observations of bifurcations to chaos in an all-optical Fabry–Perot geometry resonator[5.129]. In addition to the $t_R \rightarrow 2t_R \rightarrow 4t_R$ route they also found $2t_R/3$ oscillation which after a period-doubling gave $5t_R/3$ and hence chaos. Gioggia and Abraham[5.76] used a Fabry–Perot dc-excited xenon laser to demonstrate on the power spectra that the period-doubling, the two-frequency and the intermittency routes lead to chaos.

In another paper, Abraham and coworkers[5.2] reviewed experimental evidence for self-pulsing and chaos in CW excited lasers. They described various types of noise and chaos observed in single-mode, two-mode and three-mode lasers and illustrated increasing complexity arising from the interaction of modes.

Brun and coworkers[5.42] discussed both experimental observations and computer simulations of various periodic and chaotic solutions for a single-mode, solid state nuclear spin-flip, ruby NMR laser. They have derived a system of Bloch equations similar to Eqs (5.25)–(5.27) and demonstrated its correspondence with the experimental results obtained with the homogeneously broadened, single-mode ring laser. Then they reduced the system to two variables by eliminating one variable and studied transitions to chaos when one of the adjustable parameters of the model was modulated sinusoidally or when the field gradient was imposed on the system. They illustrated cascades of period-doubling bifurcations, noisy bands, periodic windows in the chaotic range, intermittency, jumps from one basin of attraction to another, hystereses and crises in dependence on the variation of parameters. Both experimental observations and computer simulations were very sensitive both to small variations of parameters and to the choice of initial conditions, hence the agreement between them has to be taken more as illustrative than as quantitative. Review of numerical investigations of instabilities and routes to turbulence in a phase-conjugate resonator was published by Reiner, Belić and Meystre[5.225]. Lauterborn and Steinhoff[5.162] used the concept of nonlinear resonances and torsion numbers to describe numerically determined bifurcation structure of a laser with pump modulation.

Wiesenfeld and Mc Namara[5.291] proposed that any dynamical system close to the onset of a period-doubling bifurcation could be used to amplify perturbations near half the fundamental frequency, hence as a small-signal amplifier. They supported their proposition by computer simulations of the driven Duffing oscillator. Derighetti *et al.*[5.53] tested the idea experimentally and found that near the onset of a period-doubling bifurcation a parametrically modulated NMR laser could be used as a detector of weak input signals with a strongly peaked response curve centred near half the modulation frequency. The sensitivity of the device increased with the approach to the bifurcation point.

Mukai and Otsuka[5.198] observed a successive subharmonic oscillation cascade up to the fifth order followed by chaos in an AlGaAs laser diode coupled to an external cavity. They detected power spectra corresponding to f_c/m ($m = = 2, 3, 4, 5$) subharmonic oscillations which then broke into chaos, called this new route to chaos a 'successive subharmonic modulation route' and attributed it to the nonlinear interaction between the lasing and amplified spontaneous emission modes. The authors verified the proposed physical mechanism by numerical simulations based on the generalized van der Pol equation.

Puccioni *et al.*[5.220] measured the fractal dimensions and the correlation entropies of periodic and chaotic attractors for a CO_2 laser. They used the method of Grassberger and Procaccia[2.25] and found that for the $f/8$ subharmonic the correlation dimension was $D_2 = 1.5$ which is in agreement with the theoretical prediction for the dimension of the attractor at the accumulation point (infinite periodicity) of the logistic map ($D_2 \approx 0.5378$). When the system entered the chaotic region, the dimension suddenly jumped to a higher value, $D_2 = 2.4$. The authors modelled the observed behaviour by means of the Maxwell–Bloch equations and found a quantitative correspondence between the experimental and theoretical results even in the chaotic region. Hong-jun-Zhang and coworkers[5.120] studied chaos in liquid crystal optical bistability.

A stable mode-locked operation with shorter and narrower pulses increases the number of both spectroscopic and communication applications. Studies of the nature and causes of transitions from stable mode-locked pulses to more complex and chaotic pulses can thus contribute to the removal of some obstacles to such applications.

5.6 Chaos in chemical systems

Chemical reactions taking place far from equilibrium can be described by a set of coupled nonlinear differential equations. In homogeneous isothermal systems the right-hand sides of these systems are polynomials. This led Ruelle[5.237] to suggest that such chemical systems might exhibit a deterministic chaos. In nonisothermal conditions an exponential dependence of the reaction rate constants on temperature has also to be considered. Even more complicated reaction rate functions arise in heterogeneous reactions. Limitations to mass and heat transport add other complications into the corresponding mass and enthalpy balances, leading to various forms of nonlinear reaction-diffusion-convection problems[5.300]. Studies of various nonlinear phenomena such as multiple stationary states, oscillations and waves in chemical reactors, heterogenous catalytic reactions and combustion have been a very active branch of research at least in the last thirty years[5.18, 5.152, 5.171, 5.177, 5.242].

However, most detailed studies of the development and properties of chaos in chemical systems have been performed in the continuous stirred tank reactor (CSTR) with the Belousov–Zhabotinski (BZ) reaction. The BZ reaction is a homogeneous oxidation of malonic acid by bromate in the presence of a catalyst ($Ce^{3+/4+}$, ferroin, etc.)[5.302, 5.67]. When conducted in a CSTR undamped oscillations of reaction intermediates (e.g. Br^- ions or redox potential) are often observed. The reviews of various aspects of the BZ reaction are contained in Ref. 5.67. Here we shall first review the development of observations of the chaotic behaviour in the BZ reaction and then briefly discuss examples of chaotic behaviour in other types of chemical system.

In the first experimental studies of the BZ reaction in the CSTR, published in 1974, irregular (aperiodic) oscillations were observed at the boundaries of the oscillatory regions[5.153]. In 1975 several other aperiodic regimes were described not only in the single CSTR[5.177] but also in two CSTRs coupled by a mutual mass exchange[5.176]. However, the aperiodic behaviour was discussed only in qualitative terms, referring to the aperiodic time course of the redox potential. In 1977 Olsen and Degn[5.204] reported experimental observations of an aperiodic behaviour in a peroxidase enzyme system. They concluded on the basis of the observations of 'period 3' regimes that the recorded aperiodic behaviour corresponded to chaos.

Qualitative observations of aperiodic behaviour in the BZ reaction were also reported by Schmitz, Graziani and Hudson[5.243], Rössler and Wegmann[5.230, 5.288], Yamazaki, Oono and Hirakawa[5.296], Roux and coworkers[5.231, 5.232, 5.236] and Hudson, Hart and Marinko[5.125]. The reported observations were largely restricted to the measurement of some dependent variable as a function of time and to the construction of approximate one-dimensional mappings for an explanation of chaotic behaviour. The power spectra and the autocorrelation function were then presented by Vidal *et al.*[5.283]; Tomita and Tsuda[5.275, 5.276] and Pikov-sky[5.214] constructed a next-amplitude map for some experimental results of Hudson, Hart and Marinko[5.125] as Olsen and Degn had done for the enzyme reaction[5.204]. Tomita and Tsuda also developed a model for chaotic behaviour based on their return map. The aperiodic behaviour of the CSTR forced by the concentration forcing was also reported[5.173], Roux *et al.*[5.233, 5.234] presented three dimensional state space plots for the BZ reaction. Hudson and Mankin[5.124] analysed in more detail experimental results illustrating the chaotic nature of oscillations. They compared power spectra and three-dimensional state space trajectories for chaotic and periodic flows. They also constructed one-dimensional next-amplitude maps which could reproduce chaotic behaviour. From the map they computed a positive value of the LE $\lambda \approx 0.6$ and thus confirmed the deterministic character of chaos.

However, the quantitative characterization of the aperiodic behaviour required both a better control of experiments to decrease the level of noise and a better characterization of trajectories in the state space. The N-dimensional state space portraits could be, in principle, constructed from measurements of the time dependence of the concentration of all N chemical components. This is connected with experimental difficulties, particularly in the case of the BZ reaction, where the concentrations of many oscillating reaction intermediates cannot be followed continuously. Hence the idea of the reconstruction of a multidimensional phase portrait from the measurements of a single variable, proposed among others by Takens[5.264] is particularly useful here. For the observable $x(t)$ and time delay T an m-dimensional state space is constructed from the vectors $\{x(t_k), x(t_k + T), ..., x(t_k + (m - 1)\, T)\}$ where $t_k = k\,\Delta t$, $k = 1, 2, ...$ and if $m > 2N + 1$ it is expected that this portrait will have the same properties (e.g. the same spectrum of LE) as one constructed from the measurements of N independent concentrations. An example of the characterization of the time series of the recorded time course of the concentration of Br^- ions is shown schematically in Fig. 5.20(*a*)–(*h*)[5.262]. The time course of periodic oscillations is shown in Fig. 5.20(*a*) together with the corresponding power spectra (Fig. 5.20(*c*)) and the state space portrait in Fig. 5.20(*e*) (here $T = 8.8$ s); we can observe that it is a single closed curve. A chaotic course of oscillations is depicted in Fig. 5.20(*b*), the corresponding power spectrum exhibiting the broadband noise in Fig. 5.20(*d*) and a two-dimensional projection of the state space portrait in Fig. 5.20(*f*) (again $T = 8.8$ s). A Poincaré section constructed by the intersection of positively directed trajectories with the plane (normal to the paper) passing through the dashed line in Fig. 5.20(*f*) is shown in 5.20(*g*). The points on the Poincaré section lie along a smooth curve (the actual dimension of the Poincaré section is slightly greater than one). If the distance along the curve is parametrized by a coordinate y, then the coordinate values give a sequence $\{y_n\}$ which defines a one-dimensional mapping $y_{n+1} = f(y_n)$, which is shown in

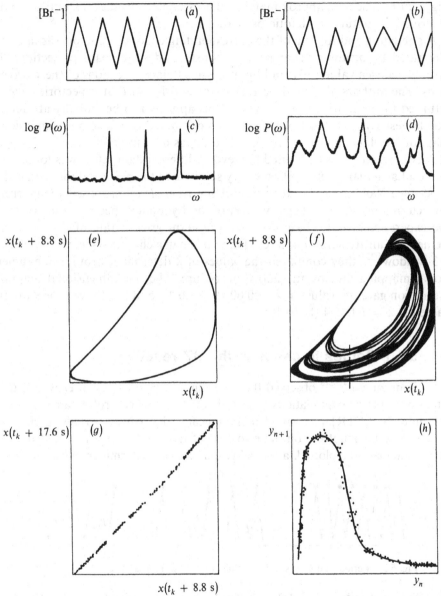

Fig. 5.20. Experimental characteristics of the BZ reaction: (*a*) periodic course of the potential of the bromide electrode; (*b*) aperiodic course of the potential; (*c*) power spectrum of the time series in (*a*); (*d*) power spectrum of the time series in (*b*); (*e*) a two-dimensional projection of the state space portrait for a periodic state, $T = 8.8$ s; (*f*) a two-dimensional projection of the state space portrait for an aperiodic state, $T = 8.8$ s; (*g*) a Poincaré section formed by the intersection of trajectories in a three-dimensional state space with the plane passing through the dashed line in (*f*); (*h*) a one-dimensional map constructed from the data in (*f*).

Fig. 5.20(h). The data approximately fall on a single-valued curve, indicating the deterministic nature of chaotic oscillations.

A stretching and folding of the corresponding chaotic attractor was later[5.235] illustrated by means of changes of reconstructed state space projections for properly chosen values of T and by successive Poincaré sections of these projections. The authors also used perturbations and found that trajectories always returned to the same attractor, which thus appeared to be globally attracting. The largest Lyapunov exponent was also computed from the one-dimensional map for the data shown in Fig. 5.20 and it was determined that $\lambda = 0.3 \pm 1$. However, when λ was computed for several different data sets it was found that the results (actual value of λ) were very sensitive to the noise and the data fitting procedure. Wolf and coworkers[5.293] and Simoyi *et al.*[5.254] proposed an algorithm for determining the non-negative part of the Lyapunov spectrum directly from the experimental data, based on the long-term growth rate of small volume elements in an attractor. For a set of data taken in a chaotic regime near a period – 3 window[5.254] they compared the values of λ determined from a one-dimensional mapping and by the above procedure. The one-dimensional mapping estimation gave the value of $\lambda = 0.004\,9 \pm 20\,\%$ and the above procedure the value of $\lambda = 0.005\,4 \pm 10\,\%$.

5.6.1 Routes to chaos in the BZ reaction

Pomeau *et al.*[5.218] observed the occurrence of type I intermittency as a transition from periodic oscillations to chaotic ones. As the control parameter (a flow rate into the CSTR) was varied monotonically, the regular periodic oscillations began to be interrupted from time to time at random by large peaks. If the next amplitude map was plotted, a typical graph of type I intermittency was obtained.

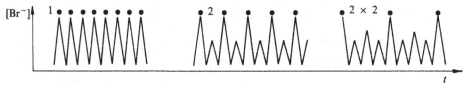

Fig. 5.21. Period-doubling sequence; time series with periods 1, 2, 2^2.

Simoyi et al.[5.254] observed the period-doubling sequence depicted in Fig. 5.21. Here the oscillations of bromide ions with periods T_0, $2T_0$ and $2^2 T_0$ are shown schematically. The authors claim that the period-doubling is fairly common in the BZ reaction but often the interval of the control parameter (e.g. flow rate) where the bifurcated periodic states are observed is very small (less than 2 % relative) and may be embedded in experimental noise. Hence the doubling is easy to miss unless a very fine variation of the control parameter is used.

The transition to chaos via quasiperiodicity (i.e. the destruction of a two torus, 'wrinkles on a torus') were described by Roux and coworkers[5.234, 5.235]. It was manifested by an abrupt transition from large amplitude periodic oscillations to modulated quasiperiodic oscillations where the two characteristic frequencies were 107 mHz and 3 mHz. Further evolution could be followed on Poincaré sections obtained from the reconstructed state space trajectories. Wrinkles developed on the side of the torus and they were then folded and stretched. The authors attempted to explain the observed behaviour by means of a circle mapping reconstructed from the experimental trajectories.

5.6.2 U-sequence in the BZ reaction

Simoyi *et al.*[5.254] reported observations of a sequence of periodic states which is similar to that which occurs in one-dimensional mappings with a single extreme. The observed elements of the U-sequence are given in Table 5.2. The

Table 5.2 *Parts of U-sequence observed by Simoyi, Wolf and Swinney*[5.254]

Period	Sequence	Pattern
1		0
2	R	0–1
2 × 2	RLR	2–0–3–1
2^2 × 2	RLR^3LR	2–6–0–4–3–7–5–1
10	RLR^3LRLR	2–8–6–0–4–3–9–5–7–1
6	RLR^3	2–0–4–3–5–1
5	RLR^2	2–0–4–3–1
3	RL	2–0–1
2 × 3	RL^2RL	2–5–3–0–4–1
9	RL^2RLR^2L	2–8–5–3–0–6–4–7–1
5	RL^2R	2–3–0–4–1
4	RL^2	2–3–0–1
2 × 4	RL^3RL^2	2–6–3–7–4–0–5–1

notation is similar to that discussed for the logistic mapping (symbolic dynamics theory) in Section 5.3. Unobserved U-sequence states might exist in small intervals of the control parameter but the order of observed states and the order of iteration pattern agreed with the theoretically predicted ones.

5.6.3 Periodic-chaotic sequence

The first observations of periodic behaviour in the BZ reaction showed that when the control parameter was varied a succession of periodic and aperiodic (chaotic) states was observed. The complex periodic and aperiodic oscillations usually consisted of combinations of large (L) and small amplitude oscillations (peaks), see schematic Fig. 5.22. Let us adopt the following convention to characterize an observed state: a letter P or C (P – periodic, C – chaotic) and two

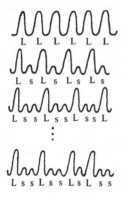

Fig. 5.22. A sequence of periodic and aperiodic regimes observed when the flow-rate is varied, schematically; L – large amplitude oscillation, s – small amplitude oscillation.

indices i and j standing for the number of large and small amplitude oscillations, respectively. Then, for example, P_1^2 means a periodic regime with one large and two small amplitude oscillations. The observed chaotic regimes could often be described as $C_i^{j, j+1}$, which means that such regimes looked like a random mixture of periodic regimes P_i^j and P_i^{j+1}. The observed periodic-chaotic sequence could be then described as

$$P_1^0, C_1^{0,1}, P_1^1, C_1^{1,2}, P_1^2, ..., P_1^n, C_1^{n,n+1}, P_1^{n+1}, ... \tag{5.30}$$

The sequence is sometimes called 'the period-adding sequence' and has been reported in a number of papers[5.60, 5.124, 5.206, 5.215, 5.227, 5.235, 5.278]. These observations may be explained in two qualitatively different ways. Either the observed chaotic states are manifestations of a truly deterministic chaos, or they just reflect random switching between two neighbouring periodic states P_1^n and P_1^{n+1} caused by the effects of noise.

Maselko and Swinney[5.183] observed a complex sequence of periodic states which looked very similar to the sequence of periodic states observed in the periodic-chaotic sequence. Each state was an admixture of small and large amplitude oscillations with the number of oscillations per period ranging up to 26 and the oscillation pattern even in the most complex states was a combination of the simplest oscillation patterns. The ratio of the number of small to the total number of oscillations per period was a function similar to a devil's staircase. No chaotic states were observed.

Bagley, Mayer-Kress and Farmer[5.23] demonstrated that these complex sequences of periodic states could be generated by simple one-dimensional mappings. They constructed a mapping which reproduced most of the devil's staircase observed by Maselko and Swinney. Maselko and Swinney later[5.184] have shown that the Farey triangle construction[5.148] can provide both description of the sequences and reveal their selfsimilar structure. With the Farey triangle one can predict that between any two states (p, q, r) and (p', q', r') there will be the daughter state (p + p', q + q', r + r') existing over a narrower parameter range than the patern states.

These observations supported scepticism about the deterministic character of the chaos in the BZ reaction. This scepticism also followed from the results of numerical simulations of mathematical models of the BZ reaction, which predicted only complex periodic regimes and no chaos[5.68]. Even if some simulations based on different versions of the most popular model 'Oregonator'[5.67] led to prediction of chaotic behaviour, the predictions did not agree with available experimental data[5.228, 5.279].

Richetti and Arneodo[5.226] constructed a seven-variable model and in numerical simulations observed a sequence of periodic and chaotic states similar to that observed in the above quoted experiments. Richetti *et al.*[5.227] reported a set of experiments, in which they located in the BZ system a situation where both a local instability (subcritical Hopf bifurcation) and a global instability (a homoclinic bifurcation) compete. When increasing the control parameter (flow rate) they observed a discontinuous transition to the chaotic regime, which they called 'homoclinic chaos'. They illustrated on state space portraits and Poincaré maps the mechanism of the homoclinic reinjection process. They also studied quantitatively the transition to quasiperiodicity in a codimension three experiment, i.e. close to the conditions where the Hopf bifurcation (codimension one) and the hysteresis bifurcation (codimension two) occur simultaneously. When shifting the experimental conditions away from the 'codimension three point' they observed breaking of the underlying torus into a fractal object. They detected wrinkles on the toroidal surface and demonstrated a mixing of the reconstructed trajectories. When further releasing the constraints from the bicriticality, they observed how a transition from quasiperiodicity to chaos was transformed into alternating periodic-chaotic sequences which looked very much like the sequences discussed above. They recorded a periodic-chaotic sequence

$$P_0^1,\ C_{n_1...n_i}^1(n_i \geq 4),\ C_{3,4}^1,\ P_3^1,\ C_3^{1,2},\ P_3^2,\ P_3^3,\ ...,$$

$$...,\ C_3^{4,5},\ P_2^1,\ P_2^{1,2},\ P_2^2,\ P_2^3,\ ...,\ C_2^n\ (n \sim 10),\ C_{1,2}^n,\ C_1^n. \tag{5.31}$$

Let us note that the periodic-chaotic sequence (5.30) contains just one characteristic large amplitude oscillation – fundamental frequency – (one frequency periodic-chaotic sequence); while the sequence (5.31) contains two characteristic frequencies (two frequency periodic-chaotic sequence) which indicates a route to chaos via the breakage of a two-torus.

The authors selected from (5.31) a typical chaotic state $C_3^{4,5}$, where the chaotic aspect of the recorded concentration signal was mainly contained in the small amplitude oscillations; their number was distributed at random between 4 and

Table 5.3 *Reaction scheme and mathematical model of the BZ reaction used by Ricchetti et al*[5.227]

The reaction scheme

$$BrO_3^- + Br^- + 2\,H^+ \xrightarrow{k_1} HBrO_2 + HOBr \tag{R1}$$

$$HBrO_2 + Br^- + H^+ \xrightarrow{k_2} 2\,HOBr \tag{R2}$$

$$HOBr + Br^- + H^+ \xrightarrow{k_3} Br_2 + H_2O \tag{R3}$$

$$BrO_3^- + HBrO_2 + H^+ \underset{k_{-4}}{\overset{k_4}{\rightleftharpoons}} 2\,BrO_2 + H_2O \tag{R4}$$

$$2\,HBrO_2 \xrightarrow{k_5} HOBr + BrO_3^- + H^+ \tag{R5}$$

$$BrO_2 + Ce^{3+} + H^+ \xrightarrow{k_6} Ce^{4+} + HBrO_2 \tag{R6}$$

$$HOBr + MA \xrightarrow{k_7} BrMA + H_2O \tag{R7}$$

$$BrMA + Ce^{4+} \xrightarrow{k_8} Br^- + R^\bullet + Ce^{3+} + H^+ \tag{R8}$$

$$R^\bullet + Ce^{4+} \xrightarrow{k_9} Ce^{3+} + P \tag{R9}$$

Here P denotes – inert organic product, BrMA – bromomalonic acid, $R^\bullet$ – oxidized derivative of malonic acid (MA)

7-variable model system of ODEs

$$\dot{x}_1 = -a_1x_1 - a_2x_1x_2 - a_3x_1x_3 + a_8x_5x_6 + k_0(x_1^0 - x_1)$$

$$\dot{x}_2 = a_1x_1 - a_2x_1x_2 - a_4x_2 + a_5x_4^4 - 2k_5x_2^2 + a_6x_4 - k_0x_2$$

$$\dot{x}_3 = a_1x_1 + 2a_2x_1x_2 - a_3x_1x_3 + k_5x_2^2 - a_7x_3 - k_0x_3$$

$$\dot{x}_4 = 2a_4x_2 - 2a_5x_4^2 - a_6x_4 - k_0x_4$$

$$\dot{x}_5 = a_6x_4 - k_8x_5x_6 - k_9x_5x_7 - k_0x_5$$

$$\dot{x}_6 = a_7x_3 - k_8x_5x_6 - k_0x_6$$

$$\dot{x}_7 = k_8x_5x_6 - k_9x_5x_7 - k_0x_7$$

Here the variables x_i denote concentrations $x_1 = [Br^-]$, $x_2 = [HBrO_2]$, $x_3 = [HOBr]$, $x_4 = [BrO_2]$, $x_5 = [Ce^{4+}]$, $x_6 = [BrMA]$, $x_7 = [R^\bullet]$. The parameters a_i include rate constants k_i and concentrations of the inlet species

$a_1 = k_1[BrO_3^-][H^+]^2$, $a_2 = k_2[H^+]$, $a_3 = k_3[H^+]$,

$a_4 = k_4[BrO_3^-][H^+]$, $a_5 = k_4[H_2O]$, $a_6 = k_6[Ce^{3+}][H^+]$,

$a_7 = k_7[MA]$, and x_1^0 denotes $[Br^-]_0$

5 and the authors demonstrated on the reconstructed state space portraits the presence of two underlying saddle-foci causing the two frequency dynamics.

A comparison of the experimental results with the results of numerical simulations helped in the interpretation of the periodic-chaotic sequences. The reaction scheme of the Oregonator type used by Richetti *et al.*[5.227] together with the mathematical model used in computations is shown in Table 5.3. The model belongs to the class of systems which give chaotic behaviour close to homoclinic situations, discussed by Shilnikov[3.71]. The parameter k_0 proportional to the flow-rate was taken as the control parameter. For the set of rate constants

$$k_1 = 2.1 \text{ M}^{-3} \text{ s}^{-1}; \qquad\qquad k_2 = 2 \times 10^9 \text{ M}^{-2} \text{ s}^{-1};$$

$$k_3 = 8 \times 10^9 \text{ M}^{-2} \text{ s}^{-1}; \qquad\qquad k_4 = 10^4 \text{ M}^{-2} \text{ s}^{-1};$$

$$k_{-4} = 111 \times 10^7 \text{ M}^{-2} \text{ s}^{-1}; \qquad k_5 = 4 \times 10^7 \text{ M}^{-1} \text{ s}^{-1};$$

$$k_6 = 10^3 \text{ M}^{-1} \text{ s}^{-1}; \qquad\qquad k_7 = 10^6 \text{ M}^{-1} \text{ s}^{-1};$$

$$k_8 = 5 \times 10^3 \text{ M}^{-1} \text{ s}^{-1}; \qquad\quad k_9 = 10^8 \text{ M}^{-1} \text{ s}^{-1}$$

(here M denotes mole) and the values of inlet concentrations $[\text{BrO}_3^-] = 5 \times \times 10^{-2}$ M, $[\text{MA}] = 10^{-3}$ M, $[\text{Ce}^{3+}] = 1.05 \times 10^{-4}$ M, $[\text{H}^+] = 1.5$ M, $[\text{Br}^-]_0 = 1.8 \times 10^{-7}$ M a periodic-chaotic sequence very similar to the sequence (5.30) could be reproduced when the value of k_6 was increased, starting from $k_6 = 4.69 \times 10^{-3}$ M^{-1} s^{-1}. The periodic-chaotic sequence involving two characteristic frequencies, (5.31), could be modelled for the set of inlet concentrations $[\text{BrO}_3^-] = 5 \times 10^{-3}$ M, $[\text{MA}] = 5 \times 10^{-3}$ M, $[\text{Ce}^{3+}] = 2 \times \times 10^{-3}$ M, $[\text{H}^+] = 1.5$ M, $[\text{Br}^-]_0 = 3.5 \times 10^{-7}$ M.

Richetti *at al.*[5.227] used in their simulations first a normal form approach to understand the evolution of the dynamics of the BZ system when it moved away from the local situation of the corresponding codimension and then studied the actual dynamics in the model given in Table 5.3. This approach combined with the analysis of both the state space portraits and the Poincaré sections helped in the detailed analysis of this complex sequence of periodic-chaotic regimes.

Chaotic responses were also observed in the forced BZ reaction by Nagashima[5.199] and by Dolník et al.[5.55]. A survey of chemical systems exhibiting complex dynamics was published by Epstein[5.61]. An example of forced oscillations will be discussed in detail in the next chapter.

5.6.4 Chaotic oscillations in other chemical systems

The BZ reaction is also used in studies of interactions of oscillations in systems of coupled cells[5.174]. When the intensity of mass exchange between two neighbouring oscillating cells is varied, various periodic and aperiodic regimes are observed. Four examples of chaotic regimes are presented in Fig. 5.23. The

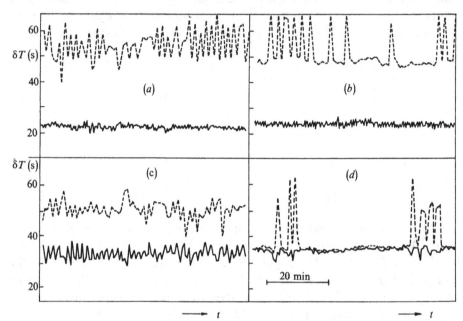

Fig. 5.23. Chaotic oscillations in two coupled oscillating cells with the BZ reaction; dependence of period T on time: (a) ratio of periods of non-interacting cells is 1:2; (b) intermittency; (c) ratio of periods of non-interacting cells is 5:7; (d) intermittency, stronger interaction; — left cell, - - - - right cell:

sequences of the time intervals between the succesive maxima of oscillation δT are shown in dependence on time. Fig. 5.23(a) illustrates the aperiodic course of oscillations in the cell with longer oscillations (the oscillations in the other cell are nearly periodic). Fig. 5.23(b) shows the intermittency in the slower oscillator for a higher intensity of mass exchange between the cells. Fig. 5.23(c) depicts chaotic oscillations in both cells and Fig. 5.23(d) again documents the presence of the intermittency.

If the BZ reaction takes place in a tubular reactor (i.e. in a distributed system), then various types of concentration waves travelling through the system can be observed. Examples of periodic, complex periodic and chaotic waves are presented in Fig. 5.24 (a)–(c)[5.223], in the form of the time course of a redox-potential. Chaotic concentration pulse waves were also observed in the modelling of a reaction–diffusion system with periodic perturbations at one end of the reactor[5.250]. Various periodic and nonperiodic concentration patterns are also observed in two and three-dimensional systems[5.67].

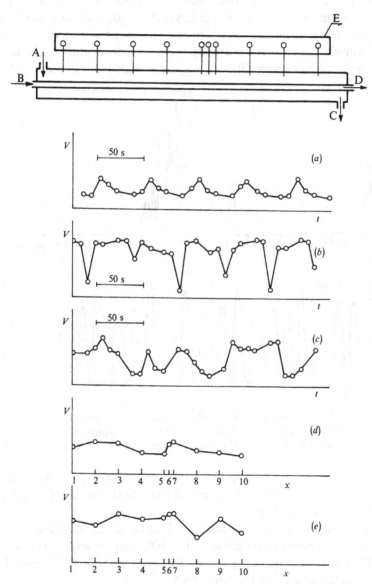

Fig. 5.24. Tubular reactor for measurement of reaction–diffusion concentration patterns in the BZ reaction; B, D – thermostating fluid; A, C – inlet and outlet of the reaction medium; E – measuring Pt electrodes; time course of oscillations at a fixed point in the reactor: (*a*) simply periodic; (*b*) complex periodic; (*c*) chaotic; (*d*) longitudinal profile for the case (*b*); (*e*) longitudinal profile for the case (*c*).

Oscillations occur also in chemical reactions which are catalysed on solid surfaces. Chaotic oscillations were observed in a number of catalytic oxidation reactions[5.205, 5.244]. However, the mechanisms of both periodic and chaotic oscillations are in most cases still not clear[5.224]. Oscillations also occur at the oxidation of carbon monoxide on the catalyst Pt/Al_2O_3 used in catalytic mufflers. Testing of catalytic mufflers occurs under conditions of periodic concentration forcing (simulating an average driving cycle). Experimental observations of concentration forcing in an oscillatory regime revealed the existence of both

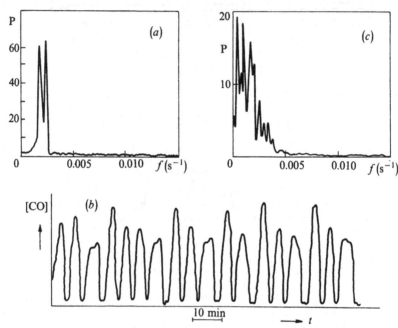

Fig. 5.25. Synchronized and aperiodic oscillatory regimes in the forced catalytic oxidation of CO: (*a*) power spectrum, synchronization at the ratio of frequencies 4:3; (*b*) chaotic oscillations of the outlet CO concentration; (*c*) power spectrum for the time series (*b*).

synchronized periodic regimes (see the power spectrum in Fig. 5.25(*a*)) and chaotic regimes (see the power spectrum and the time course of oscillations shown in Fig. 5.25 (*b*) and (*c*))[5.175].

Eiswirth, Krischer and Ertl[5.59] demonstrated for catalytic oxidation of CO on Pt(110) surface the presence of period-doubling bifurcations leading to aperiodic behaviour. They also determined positive LEs and identified the embedding dimension of the chaotic attractor equal to 5.

5.7 Chaotic oscillations in biological systems

Oscillations occur in biological systems on various levels of organization, starting from enzymatic reactions to cell membranes, organelles, excitatory cells (neurones, heart cells, etc.), excitatory tissues, organs, physiological control systems[5.137], up to ecological communities[5.47]. The spontaneous oscillations and adaptations of internal activities to cyclic changes in the environment both play important roles in the temporal organization of life. Hence the occurrence of chaos in biological systems should be ubiquitous, considering the fact that in driven limit cycle type oscillations its presence is generic, when both the driving frequency and the amplitude are varied.

Different nonperiodic (noisy) oscillations have been observed in all the above mentioned biological structures. An interpretation in terms of the deterministic chaos was first reported by Olsen and coworkers[5.204] for the oxidation of NADH catalysed by horseradish peroxidase. By varying enzyme concentration it was possible to observe reversible transitions between periodic and chaotic behaviour. A simple model in the form of four ordinary differential equations based on the reaction mechanism of a branched chain reaction with autocatalysis, where the enzyme participated in the branching reactions, was proposed by Olsen[5.203]. He demonstrated that the model was capable of reproducing both periodic and chaotic dynamic patterns. Markus, Kuschmitz and Hess[5.179] discussed the variety of periodic and chaotic regimes observed in yeast glycolysis pathways. When the input concentration of glucose was varied periodically, chaotic NADH fluorescence was observed. Period-doubling bifurcations and intermittency were predicted in the mathematical model of glycolysis[5.178, 5.179]. The effect of various levels of external fluctuations on various observed time patterns was investigated by means of numerical simulations of a two-enzyme model by Hess and Markus[5.112]. The chaotic behaviour of periodically forced glycolysis was earlier predicted by Tomita and Daido[5.274].

Chaotic cardiac dynamics has been studied in a number of theoretical papers[5.78, 5.94, 5.142, 5.255, 5.289]. Several experimental observations also exist. Guevara, Glass and Shrier[5.95] perturbed a spontaneous rhythmic activity of aggregates of embryonic chick heart cells by the injection of single current pulses and periodic trains of current pulses. They observed both periodic and aperiodic oscillations arising after a sequence of period-doubling bifurcations. A phase transition curve determined on the basis of single pulse experiments was used to predict the results of the continuous periodic forcing experiments and the connections with circle maps was discussed in detail in another paper[5.79].

Goldberger, Bhargava and West[5.82] identified in a record of heartbeats a sequence of periodic states followed by subharmonic bifurcations (periods 2, 4 and 8) and a return to an aperiodic steady state. They reported that the data were consistent with the hypothesis that under pathologic conditions the cardiac electrophysiological system behaves as if there were nonlinear coupling of multiple oscillatory pacemakers.

Mackey and Glass[5.77, 5.172] have proposed that clinical disorders often observed in the formation of various blood particles (hematopoiesis) which occur in the form of periodic or irregular variations of particle numbers contains both periodic and chaotic dynamics resulting from the properties of the governing feedback system.

Chaotic behaviour was predicted to occur on different organizational levels in neurophysiology. Carpenter[5.44] demonstrated that solutions of generalizations of the Hodgkin–Huxley equations could be chaotic. Chay[5.49] presented chaotic numerical solutions for a model of a single neuron. Sbitnev[5.240] described a generation of chaos in the interaction of inhibitory and excitatory populations of neurones.

Experimental evidence for the presence of chaos was obtained in periodically stimulated neurones. Holden and coworkers[5.116, 5.117] reported qualitative observations of chaotic activity in a molluscan neuron with a cyclic input and after treatment with high concentrations of chemicals. Matsumoto et al.[5.185] found nonperiodic responses of the membrane potential in squid giant axons.

A more quantitative demonstration of the presence of deterministic chaos was presented by Hayashi and coworkers[5.101–5.105]. In Ref. 5.103 the nonperiodic firing of the Onchidium verruculatum giant neuron under sinusoidal stimulation was studied. Stroboscopic portraits of the attractor were used to demonstrate that mixing occurs on the chaotic attractor. Nonperiodic behaviour was also observed in the internodal cell of Nitella flexitis under sinusoidal stimulation. In a later paper[5.105] the authors described in more detail by means of power spectra and stroboscopic mappings quasiperiodic and chaotic regimes present in this system. A transition to chaos via intermittency was also described in the Onchidium pacemaker neuron. Later, chaotic responses in this system were classified into three groups: chaos, intermittency and random alternation and it was shown that 1/2- and 1/1- harmonic responses bifurcate to chaos via intermittency and random alternation, respectively[5.102].

It was also suggested that chaos might occur in neurological disorders such as tremor, dyskinesia and epilepsy[5.222]. Babloyantz and Destexhe[5.20, 5.21] analysed an EEG recording of a human epileptic seizure and showed the existence of a chaotic attractor. They demonstrated by using the Grassberger – Procaccia algorithm that a sudden jump between the dimensionality of a brain attractor in deep sleep (4.05 ± 0.05) and of the epileptic state (2.05 ± 0.09) exists. A decaying autocorrelation function and the positive largest LE provided additional evidence for the presence of chaotic dynamics.

There is great interest in the possible implications of the presence of chaos in biological information processing and speculations about its role in the organization and development of living systems[5.200, 5.272, 5.273, 5.277].

5.8 Transition to turbulence in hydrodynamics

Experimental and theoretical studies of turbulence have attracted some of the greatest names in physics, mechanics and engineering in the last hundred years[5.48, 5.130, 5.139]. Even Leonardo da Vinci studied trajectories of vortices in the flow around solid bodies and represented them graphically[5.163]. However, most of the problems of importance in engineering are still handled by a mixture of empirical facts and phenomenological modelling which often leads to unexpected discrepancies between predictions and actual behaviour[5.168]. Flows with fully developed turbulence are often well characterized by statistical methods[5.193]. But the processes by which turbulence develops are relatively poorly understood. In the course of transition from laminar to turbulent flows the fluid usually undergoes a sequence of instabilities, which lead to a change in the spatial or temporal structure of the flow and, finally, turbulent motion appears.

The use of computers for data acquisition and laser Doppler techniques for the remote measurements of local velocities qualitatively advanced experimentation procedures and enabled detailed studies of the mechanism of the transition to turbulence and identification of various routes to chaos[5.58, 5.263].

The two model experimental systems used most frequently in experimental studies were the Rayleigh–Bénard convection problem and the Taylor–Couette problem. Both systems are continuum hydrodynamic systems which generally have an infinite number of degrees of freedom. However, we can expect that just beyond the onset of chaos there are only a few degrees of freedom that are excited.

Here we shall discuss several experimental observations of transitions to chaotic behaviour in the Rayleigh–Bénard and the Taylor–Couette problems and then briefly review examples of chaotic behaviour in other types of hydrodynamic systems. Comparison of routes to chaos in the Rayleigh-Bénard and Taylor-Couette problems with routes observed in the Belousov-Zhabotinskii reaction and in electrical oscillators was made by Swinney[5.261]

5.8.1 Taylor–Couette flow

In this system, schematically shown in Fig. 5.26, a fluid is contained between concentric cylinders that rotate independently with angular velocities ω_i (inner) and ω_o (outer). The characteristic parameters are the dimensionless Reynolds numbers $Re_i = (b - a)\, a\omega_i/v$ and $Re_o = (b - a)\, b\omega_o/v$, where a and b are the radii of the inner and outer cylinders, respectively, and v is the kinematic viscosity of the fluid. Most experiments were conducted with $Re_o = 0$. The radius ratio a/b, the aspect ratio $\Gamma = L/(b - a)$ and the boundary conditions at the ends are the other characteristic parameters of the system. Experiments[5.29, 5.56, 5.66, 5.85] have shown that a flow between the concentric cylinders with the inner cylinder rotating and the outer cylinder at rest exhibits the following sequence of well-defined dynamical regimes with increasing Reynolds number Re_i: a basic azimuthal flow, Taylor vortex flow, wavy vortex flow, modulated wavy vortex flow and nonperiodic flow.

Fig. 5.27(a) shows the velocity record and Fig. 5.27(b) the corresponding power spectrum for data obtained in the doubly periodic (quasiperiodic) modulated wavy vortex flow at $Re_i/Re_{ic} = 10.2$ (Re_{ic} corresponds to the onset of the Taylor vortex flow)[5.39, 5.262]. The frequencies ω_1 and ω_2 correspond to two sets of travelling azimuthal waves that have different wave speeds. When Re_i is increased, at $Re_i/Re_{ic} = 12.3$, a broadband noise appears in the power spectra (onset of aperiodic behaviour), see Fig. 5.28(b) (the corresponding time course of flow velocities is given in Fig. 5.28(a).

To confirm that the broadband noise arises from a deterministic dynamics and not as a manifestation of both an external and an internal noise, Brandstätter *et al.*[5.39] used a reconstruction of a multidimensional phase portrait from the measurements of a single velocity by using time delays. The Poincaré sections given by the intersections of orbits of three dimensional state space portraits with planes are given in Fig. 5.27(c) and Fig. 5.28(c). It is evident that the orbits for the doubly periodic flow lie on the surface of a torus and that the torus becomes fuzzy for the nonperiodic flow.

Brandstätter *et al.*[5.40] and Di Prima and Swinney[5.54] have computed the largest LE and the metric entropy and have shown that the system displays a sensitive dependence on initial conditions and that the motion was restricted to an attractor of dimension 5 for Reynolds numbers up to 30 % above the onset of chaos. The LE, the entropy and the dimension generally increased with the Reynolds number. When Re_i is increased the sharp frequency components disappear from the power spectrum leaving only broad components and a background noise. When it is increased still further, Walden and Donnely[5.285]

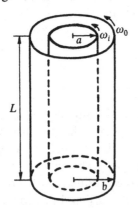

Fig. 5.26. Taylor–Couette system, schematically; a, b – radii of cylinders, L – length, ω_i, ω_o – frequencies of rotation of the inner and the outer cylinders, respectively.

observed sharp peaks to re-emerge on top of the background noise. The sharp peaks are observed for sufficiently large aspect ratios $\Gamma(\Gamma > 28)$. L'vov and Predtechensky[5.157, 5.158] have also observed re-emergence of sharp peaks. Moreover, they described a sequence of transitions, chaotic-quasiperiodic-chaotic, for an annulus with $\Gamma = 30$.

Belyaev *et al.*[5.29, 5.30] studied a transition to chaos in a rotating spherical layer of liquid and found from the power spectra that it was again close to the Ruelle – Takens picture.

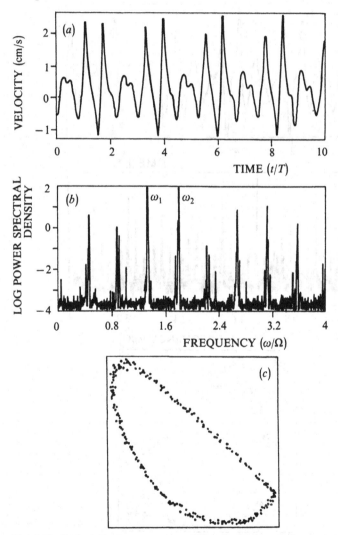

Fig. 5.27. Taylor–Couette system, modulated wavy vortex flow, $Re_i/Re_c = 10.2$: (*a*) velocity time series (the cylinder period $T = 1.845$ s); (*b*) the velocity power spectrum (frequencies in the units of cylinder frequency); (*c*) a Poincaré section of the attractor.

However, in the studies of the Taylor–Couette flow we have to bear in mind that at a fixed Re_i in a system with an aspect ratio of only 20, there can be more than a hundred different stable spatial states of the flow, corresponding to different numbers of vortices, waves and wave modulation patterns. In systems with counter-rotating cylinders it was found that intermittency, sudden transitions from the laminar to the turbulent flow and fully developed turbulence with no axial periodicity occurred.

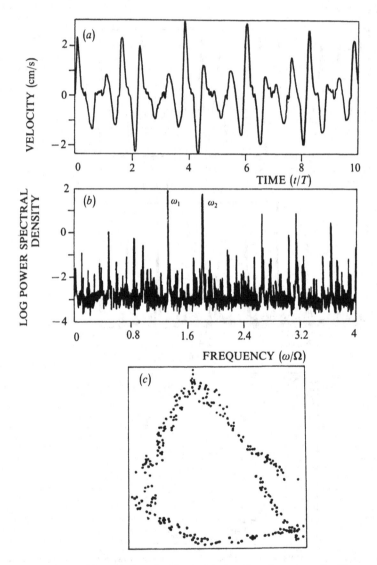

Fig. 5.28. Weakly turbulent flow in the Taylor–Couette system, $Re_i/Re_{ic} = 12.3$: (*a*) velocity time series; (*b*) the velocity power spectrum; (*c*) a Poincaré section.

5.8.2 Rayleigh–Bénard convection

In a Rayleigh–Bénard system a fluid is contained between parallel plates heated from below as shown schematically in Fig. 5.29. The behaviour is most often studied as a function of the dimensionless Rayleigh number $Ra = (g\alpha d^3/\varkappa v)\,\Delta T$, where g is the gravitational acceleration, α the thermal expan-

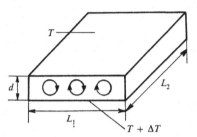

Fig. 5.29. The Rayleigh–Bénard experiment; the layer of liquid between parallel plates heated from below. For symbols see text.

sion coefficient, d the separation between the plates, $\varkappa$ the thermal diffusivity and v the kinematic viscosity. Other characteristic variables are the Prandtl number, $Pr = v/\varkappa$, the aspect ratios, $\Gamma_1 = L_1/d$ and $\Gamma_2 = L_2/d$ and the boundary conditions at the side walls. Cylindrical systems have a single aspect ratio defined as $\Gamma = R/d$, where R and d are the radius and the height of the cylinder, respectively. When the aspect ratios are small, experiments on convecting layers provide realizable systems for very precise studies of nonlinear phenomena in dissipative systems, including the onset not only of turbulence but also of a nonlinear pattern formation. Therefore it is not surprising that by far the highest number of experimental studies of deterministic chaos have been reported in the literature with this system. We shall first present several characteristic examples and then briefly comment on a number of them to give the reader a flavour of the intensive research in this area.

An example of the onset of turbulence via a quasiperiodic regime in a small aspect-ratio cylindrical layer of liquid after Ahlers and Behringer[5.3], is shown in Fig. 5.30. Each figure contains the measured power P versus frequency f for temperature fluctuations δT in the total temperature difference ΔT (expressed in a dimensionless form $\delta T/\Delta T_c$), for different values of Ra/Ra_c (indicated by the number in the upper right hand corner). The power spectrum for $Ra/Ra_c = 10.67$ consists of sharp peaks at two frequencies and their combinations. Over a narrow range of Rayleigh numbers the peaks broaden and then disappear into a continuous spectrum. Similar results were described by Gollub and Benson[5.83] for a low aspect ratio rectangular layer of water. They observed several routes to turbulence: quasiperiodicity and phase locking, period-doubling bifurcations, three frequencies route and intermittency.

Maurer and Libchaber[5.187] also described the frequency locking and the onset of turbulence.

Already in early quantitative studies of the transition to turbulence a period-doubling scenario has been identified. An example of such observation is shown

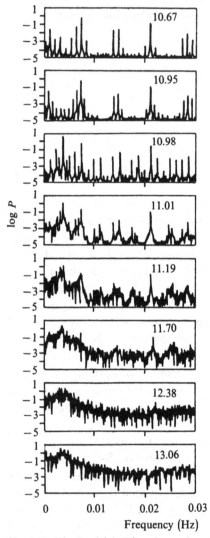

Fig. 5.30. The Rayleigh–Bénard experiment; the onset of turbulence in a small aspect-ratio cylindrical layer of liquid helium, $\Gamma = 2.08$, according to Ahlers and Behringer.

in Fig. 5.31. Here the amplitude versus frequency for temperature fluctuations illustrate sharp peaks at frequencies f, $f/2$ and $f/4$. A complete cascade is not seen in this case as it is embedded in the external noise. Because of the lack of resolution in Ra/Ra_c the Feigenbaum δ could not be determined. Also in the observations of Gollub and Benson[5.83] and Libchaber[5.165] the period-doubling bifurcations were reported with the power spectra showing a regular decrease in subharmonic amplitudes.

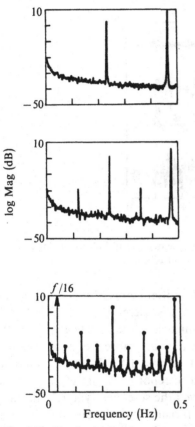

Fig. 5.31. The Rayleigh–Bénard experiment; the onset of turbulence via period-doubling in a layer of liquid helium, after Maurer and Libchaber.

Giglio, Musazzi and Perini[5.74] observed a reproducible sequence of period-doubling bifurcations up to $f/16$. They computed universal scaling numbers and found that $\delta \approx 4.3$ and $\alpha \approx 4.0$. The same authors[5.75] made an estimate of the fractal dimension by means of a correlation integral technique.

Libchaber, Laroche and Fauve[5.167] observed a period-doubling cascade in a Rayleigh–Bénard experiment in mercury with a dc magnetic field applied along the convective roll axis. They measured the value of $\delta \approx 4.4$ and found that the ratio of successive subharmonics was of the order of 14 dB for the lowest measured subharmonics.

Arneodo *et al.*[5.19] observed an uncompleted cascade in the Rayleigh–Bénard experiment, where the period-doubling sequence was interrupted by another subharmonic bifurcation (for example, by a period three oscillation).

The onset of turbulence via the intermittency observed by Maurer and Libchaber[5.166, 5.187] in a rectangular container using a liquid helium layer is shown

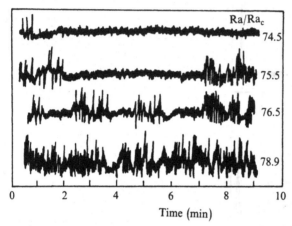

Fig. 5.32. The Rayleigh–Bénard experiment; the onset of turbulence via intermittency disrupting doubly periodic flows, after Maurer and Libchaber.

schematically in Fig. 5.32. Here the intermittent noise bursts disrupt doubly periodic (quasiperiodic) flow. For the lowest Rayleigh number $Ra/Ra_c = 74.5$ the signal is doubly periodic with two frequencies $f_1 \approx 1.0$ Hz and $f_2 \approx 0.3$ Hz. With increasing Ra/Ra_c the turbulent bursts are more frequent. The laminar regions vary as $(Ra - Ra_t)^{-\beta}$, where $1 < \beta < 1.5$ (Ra_t corresponds to the turbulent state). The intermittency was also reported by Bergé *et al.*[5.32] and by Dubois, Rubio and Bergé[5.57], who observed the so-called type III intermittency in the Pomeau–Manneville classification (arising with the destabilization of the limit cycle simultaneously with the period-doubling bifurcation). The distribution of the laminar period lengths determined from the constructed return map agreed with theoretical predictions. The effects of fluctuations on the transition to turbulence was examined by introducing a defined noise into the boundary temperature by Gollub and Steinman[5.84]. They found that the sequence of transitions and the statistical properties of the turbulent state were affected only in a limited way.

If the aspect-ratio is sufficiently large, a high number of modes may be excited and more complicated routes to turbulence may be observed, as described by Ahlers and Walden[5.4] and by Walden[5.284].

A quasiperiodic route to chaos including two or three frequencies was observed in systems with small aspect-ratios. More than three modes with incommensurate frequencies can reach a finite amplitude before chaos occurs in systems with larger aspect-ratios. Walden *et al.*[5.286] used a rectangular box with an intermediate aspect ratio 9.5 × 4.6. They observed quasiperiodic states with four and five incommensurate frequencies, predominantly localized in different parts of the cell. This indicates that, in contrast to systems with small aspect ratios, more than three modes with incommensurate frequencies can reach a finite amplitude before chaos occurs and points to the importance of the spatial dependence of the mode amplitudes in determining the route to chaos in such systems. Pocheau, Croquette and Le Gal[5.217] demonstrated through a pattern visualization that in a low Prandtl number experiment in a cylindrical container with $\Gamma = 7.66$ the weak turbulence corresponds to an erratic wandering of rolls. Fauve *et al.*[5.65] studied different types of transitions to chaos in a Rayleigh–Bénard convection in a cylindrical container of mercury subjected to a horizontal magnetic field. The magnetic field intensity was a control parameter for the number of degrees of freedom of the system and determined the nature of the transition to chaos as the Rayleigh number was varied. They demonstrated that structural changes of the spatial pattern were connected with different time-dependent behaviours and routes to chaos.

Stavans, Heslot and Libchaber[5.258] and Jensen and coworkers[5.138] used a Rayleigh–Bénard experiment in mercury subjected to a magnetic field to study quantitative relations in the quasiperiodic route to chaos with fixed winding number. By sweeping the frequency and the amplitude of the magnetic field they mapped Arnold tongues. They were able to observe locked states with denominators larger than 200 and thus the scaling properties of the routes with a fixed winding number could be tested. They found, in agreement with the theoretical prediction, that the fractal dimension of the critical line $D = 0.86 \pm 3 \%$ and the scaling number $\delta = = 2.8 \pm 10 \%$ for the golden mean winding number. Stavans[5.257] and Glazier and Libchaber[5.80] reviewed theoretical and experimental work on quasiperiodicity.

Fauve, Laroche and Perrin[5.64] used a rotating layer of mercury heated from below, where both the Rayleigh number and the rotation rate were varied, to demonstrate that the observed regimes could be analysed in the framework of the codimension two bifurcation theory. Sullivan and Ahlers[5.260] observed Hopf bifurcation to convection near the codimension-two point in a $^3\text{He}-^4\text{He}$ mixture. Walden *et al.*[5.287] studied a Rayleigh–Bénard convection in alcohol–water mixtures where the diffusion of components opposed convection via the Soret effect and observed both periodic and chaotic travelling waves. Winters, Plesser and Cliffe[5.292] studied by bifurcation techniques the onset of convection in a finite container due to surface tension and buoyancy. Ahlers, Hohenberg and Lücke[5.5] studied a Rayleigh–Bénard convection with an external modulation of the imposed heat current and found that the observations could be semiquantitatively described by a Lorenz type model. Later Meyer and coworkers[5.189] presented data showing that temporal modulation allows control of the pattern of convection near onset.

A review of the results of studies of the transition to turbulence in the Rayleigh–Bénard convection in liquid helium was presented by Behringer[5.27].

Hence an entire plethora of various routes to turbulence has been observed in the Rayleigh–Bénard problem under defined values of parameters and initial and boundary conditions. Yahata[5.294] summarized the routes observed in low aspect ratio systems (the arrow denotes an increase in the Rayleigh number):

periodic (f_1) → period-doubling cascade $(f_1/2, f_1/4, ...,)$ → periodic (f_1) → quasiperiodic (f_1, f_2) → phase-locked state (rational rotation number f_1/f_2) → period-doubling cascade $(f_2/2, f_2/4, ...,)$ → chaos,

periodic (f_1) → quasiperiodic (f_1, f_2) → phase-locked state → destruction of the two-torus → chaos,

periodic (f_1) → quasiperiodic (f_1, f_2) → quasiperiodic (f_1, f_2, f_3) → chaos,

periodic (f_1) → chaos with intermittency,

periodic (f_1) → period-doubling bifurcations with intermittency,

periodic (f_1) → quasiperiodic (f_1, f_2) → mode softening $(f_2 → 0)$ → chaos.

Yahata then compared the above scenarios with the numerical solutions of a model system. Basic equations of motion for a finite low aspect system were transformed by Galerkin approximation into a set of 48 ODEs which was then solved numerically. The computational results were consistent with the above summarized experimental observations.

Sreenivasan and Ramshankar[5.256] examined connections between intermittency routes to chaos and intermittent transition to turbulence in open flows (pipe flows, boundary layer flows). They have found some quantitative connections but imperfect agreement in most situations.

The regime changing in convection models was also studied by Shtern[5.253] and the effect of symmetries on the pattern selection by Golubitsky, Swift and Knobloch[5.87].

5.8.3 Other fluid–dynamic systems

Transition to chaos has also been studied in other fluid – dynamic systems and we shall now briefly discuss several interesting examples. Gorman, Widmann and Robbins[5.88] studied three chaotic flow regimes in a convection loop (a loop of fluid heated with a constant flux on the bottom half and cooled at a constant temperature on the top half) and showed that it could be qualitatively described by the Lorenz model.

Ciliberto and Gollub[5.50] showed that in the Faraday experiment (a cylindrical fluid layer in a container that is subjected to a small vertical oscillation, first studied by Faraday in 1831) the interaction of two modes with different symmetries gave rise to slow oscillations and chaos. Keolian *et al.*[5.143] demonstrated existence of subharmonic sequences and discussed their departure from the period-doubling in the Faraday experiment. Ezerskii, Korotin and Rabinovich[5.62] observed in the same system transition to turbulence via the chaotic modulation of two-dimensional structures on the surface of the liquid.

Bonetti *et al.*[5.37] studied the dynamic behaviour of macroscopic structures observed in a periodically excited air jet. They showed that the weakly turbulent state was characterized by a strange attractor with the correlation dimension computed according to the Grassberger–Procaccia algorithm equal to 2.6, whereas the motion of the fully developed turbulent flow was not restricted to a low dimensional attractor. Prasard, Meneveau and Sreenivasan[5.219] presented measurements confirming multifractal nature of the dissipation field of passive concentration and temperature fluctuations in turbulent jets. Lauterborn and Cramer[5.161] discussed a subharmonic route to chaos in an acoustical experiment. Kozlov and coworkers[5.154] studied spatial development of turbulence in a boundary layer on a plate in a wind tunnel, defined correlation dimension of the flow and observed its increase downstream. Perminov and coworkers[5.211] studied numerically difference approximation of the Navier-Stokes equations with boundary conditions corresponding to the flow in flat channel. They found exponential spatial mixing of periodic eddies and formation of a strange attractor when small vibrations of the channel walls were considered.

We have demonstrated through the choice of over three hundred references the ubiquitous character of deterministic chaos, which nevertheless possess a number of common features and indicates further development of the methodology of its studies. Future observations will doubtlessly include an even broader range of nonlinear phenomena in nature. Ito *et al.*[5.132] studied the role of deterministic chaos in great earthquakes and Brand, Kai and Wakabayashi[5.38] studied the effects of noise level on the onset of spatial turbulence in the electrohydrodynamic instability in nematic liquid crystals.

REFERENCES

5.1   Abraham N. B. A new focus on laser instabilities and chaos. *Laser Focus* **19** (1983) 73.

5.2   Abraham N. B., Chyba T., Coleman M., Gioggia R. S., Halas N. J., Hoffer L. M., Lin S. N., Maeda M. and Wesson J. C. Experimental evidence for self—pulsing and chaos in CW—excited lasers. Proc. 3rd New Zealand Symp. in *Laser Physics*, ed. D. Walls and J. Harvey, Berlin, Springer Series in Physics, Springer, 1983.

5.3   Ahlers G. and Behringer R. Evolution of turbulence from the Rayleigh—Bénard instability. *Phys. Rev. Lett.* **40** (1978) 7112.

5.4   Ahlers G. and Walden R. V. Turbulence near onset of convection. *Phys. Rev. Lett.* **44** (1980) 445.

5.5   Ahlers G., Hohenberg P. C. and Lucke M. Externally modulated Rayleigh—Bénard convection: experiment and theory. *Phys. Rev. Lett.* **53** (1984) 48.

5.6   Andereck C. D., Liu S. S. and Swinney H. L. Flow regimes in a circular Couette system with independently rotating cylinders. Preprint, Austin, Dept. of Physics, University of Texas at Austin, 1985.

5.7   Anishchenko V. S. *Stochastic Oscillations in Radiophysical Systems.* Saratov, Publ. House of the Saratov State University, 1985, 1986 (in Russian).

5.8   Anishchenko V. S. Destruction of quasiperiodic oscillations and chaos in dissipative systems. *JTP* **56** (1986) 225 (in Russian).

5.9   Anishchenko V. S. and Astakhov V. V. Experimental investigation of the mechanism of formation and the structure of a strange attractor in a generator with inertial nonlinearity. *Radiophysics and Electronics* **5** (1983) 1109 (in Russian).

5.10 Anishchenko V. S. and Safonova M. A. Bifurcation of attractors in the presence of fluctuations. *JTF* **58** (1988) 641 (in Russian).

5.11 Anishchenko V. S., Astakhov V. V. and Letchford T. E. Multiple—frequency and stochastic oscillations in a generator with inertial nonlinearity. *Radiophysics and Electronics* **10** (1982) 1972 (in Russian).

5.12 Anishchenko V. S., Astakhov V. V., Letchford T. E. and Safonova M. A. On bifurcations in a three—dimensional two—parameter autonomous oscillatory system with a strange attractor. *Rep. of Inst. of Higher Education, sect. Radiophysics* **26** (1983) 169 (in Russian).

5.13 Anishchenko V. S., Letchford T. E. and Safonova M. A. Phase locking effects and bifurcations of phase locking and quasiperiodic oscillations in a nonautonomous generator. *Rep. of Inst. of Higher Education, sect. Radiophysics* **28** (1985) 1112 (in Russian).

5.14 Aoki K. and Yamamoto K. Firing wave instability of the current filaments in a semiconductor. An analogy with neurodynamics. *Phys. Lett.* **98A** (1983) 72: see also ibid **98A** (1983) 217.

5.15 Aoki K., Ikezawa O. and Yamamoto K. Chaotic behaviors of the current filaments in a model of firing wave instability. *J. Phys. Soc. Japan* **53** (1984) 5.

5.16   Aoki K., Miyamae K., Kobayashi T. and Yamamoto K. Phase transition and chaos in electrical avalanche breakdown caused by weak photoexcitation at 4.2 K in n−GaAS. *Physica* **117/118B** (1983) 570.

5.17   Arecchi F. T., Mencli R., Puccioni G. and Tredici J. Experimental evidence of subharmonic bifurcations, multistability, and turbulence in a Q−switched gas laser. *Phys. Rev. Lett.* **49** (1982) 1217.

5.18   Aris R. *The Mathematical Theory of Diffusion and Reaction in Permeable Catalysts.* Vols. I. and II. Oxford, Clarendon Press, 1975.

5.19   Arneodo A., Coullet P., Tresser C., Libchaber A., Maurer J. and d'Humieres D. On the observation of an uncompleted cascade in a Rayleigh−Bénard experiment. *Physica* **6D** (1983) 385.

5.20   Babloyantz A. and Destexhe A. Low dimensional chaos in an instance of epilepsy. *Proc. Natl. Acad. Sci. USA* **83** (1986) 3513.

5.21   Babloyantz A. and Destexhe A. The Creutzfeld−Jacob disease in hierarchy of chaotic attractors. In *From Chemical to Biological Organization*, ed. by M. Markus, S. Muller, G. Nicolis, Berlin, Springer (1988).

5.22   Babloyantz A., Salazar J. M. and Nicolis C. Evidence of chaotic dynamics of brain activity during the sleep cycle. *Phys. Lett.* **111A** (1985) 152.

5.23   Bagley R. J., Mayer−Kress G. and Farmer D. J. Mode locking, the Belousov−Zhabotinski reaction and one−dimensional mappings. Preprint, Los Alamos, NM, Center for Nonlinear Studies, Los Alamos Nat. Lab., 1985.

5.24   Barone A. and Paterna G. *Physics and Applications of the Josephson Effect.* New York, Wiley, 1982.

5.25   Beasley M. R. and Huberman B. A. Chaos in Josephson junctions. *Comments on Solid State Phys.* **10** (1982) 155.

5.26   Beckert S., Schock U., Schulz C. D., Weidlich T. and Kaiser F. Experiments on the bifurcation behaviour of a forced nonlinear pendulum. *Phys. Lett.* **107A** (1985) 347.

5.27   Behringer R. P. Rayleigh−Bénard convection and turbulence in liquid helium. *Rev. Mod. Phys.* **57** (1985) 657.

5.28   Belmonte A. L., Vinson M. J., Glazier J. A., Gunaratne G. H. and Kenny B. G. Trajectory scaling functions at the onset of chaos: Experimental results. *Phys. Rev. Lett.* **61** (1988) 539.

5.29   Belyaev Yu. N., Monakhov A. A., Shcherbakov S. A. and Yavorskaya I. M. Onset of turbulence in rotating fluids. *Letters to Soviet Physics JETP* **29** (1979) 329.

5.30   Belyaev Yu. N., Monakhov A. A., Shcherbakov S. A. and Yavorskaya I. M. Some Routes to Turbulence in Spherical Couette Flow. In IUTAM Symp. *Laminar − Turbulent Transition*, Berlin, Springer, 1985, p. 669.

5.31   Ben−Jacob E., Goldhirsch I., Imry Y. and Fishman S. Intermittent chaos in Josephson junctions. *Phys. Rev. Lett.* **49** (1982) 1599.

5.32   Bergé P., Dubois M., Manneville P. and Pomeau Y. Intermittency in Rayleigh−Bénard convection. *J. Phys. Lett.* **41** (1980) L341.

5.33   Bezaeva L. G. Kapcov L. N. and Landa P. S. Investigations of chaotic modulation of oscillations in the generator with inertial nonlinearity at parametric external action. *Radiophysics and Electronics* **32** (1987) 647 (in Russian).

5.34   Blombergen N. *Nonlinear Optics.* New York, W. A. Benjamin, 1965.

5.35   Blow K. J. and Doran N. J. Global and local chaos in the pumped nonlinear Schrödinger equation. *Phys. Rev. Lett.* **52** (1984) 526.

5.36   Bohr T., Bak P. and Jensen M. H. Transition to chaos by interaction of resonances in dissipative systems. II. Josephson junctions, charge−density waves, and standard maps. *Phys. Rev.* **A30** (1984) 1970.

5.37   Bonetti M., Meynart R., Boon J. P. and Olivari D. Chaotic dynamics in a periodically excited air jet. *Phys. Rev. Lett.* **55** (1985) 492.

5.38   Brand H. R., Kai S. and Wakabayashi S. External noise can suppress the onset of spatial turbulence. *Phys. Rev. Lett.* **54** (1985) 555.

5.39   Brandstätter A., Swift J., Swinney H. L. and Wolf A. A strange attractor in a Couette – Taylor experiment. In *Turbulence and Chaotic Phenomena in Fluids*, ed. T. Tatsumi, Amsterdam, North – Holland, 1983.

5.40   Brandstätter A., Swift J., Swinney H. L., Wolf A., Farmer J. D., Jen E. and Crutchfield J. P. Low – dimensional chas in a hydrodynamic system. *Phys. Rev. Lett.* **51** (1983) 1442.

5.41   Brorson S. D., Dewey D. and Linsay P. S. Self – replicating attractor of a driven semiconductor oscillator. *Phys. Rev.* **A28** (1983) 1201.

5.42   Brun E., Derighetti B., Meier D., Holzner R. and Ravani M. Observation of order and chaos in a nuclear spin – flip laser. *J. Opt. Soc. Am.* **B2** (1985) 156.

5.43   Bryant P. and Jeffries C. The dynamics of phase locking and points of resonance in a forced magnetic oscillator. *Physica* **25D** (1987) 196.

5.44   Carpenter G. A. Bursting phenomena in excitable membranes. *SIAM J. Appl. Maths.* **36** (1979) 334.

5.45   Cascois J., Diloro R. and Costa A. N. Chaos and reverse bifurcation in a RCL circuit. *Phys. Lett.* **93A** (1983) 213.

5.46   Casperson L. W. Spontaneous pulsations in lasers. In *3rd New Zealand Symposium on Laser Physics*, ed. D. Walls and J. Harvey, Lecture Notes in Physics, Berlin, Springer, 1983, pp 107 – 131.

5.47   Chance B., Pye E. K., Ghosh A. K. and Hess B. (Eds.) *Biological and Biochemical Oscillators.* New York, Academic Press, 1973.

5.48   Chandrasekhar S. *Hydrodynamic and Hydromagnetic Stability.* Oxford, Clarendon Press, 1961.

5.49   Chay T. R. Abnormal discharges and chaos in a neuronal model system. *Biol. Cybern.* **50** (1984) 301.

5.50   Ciliberto S. and Gollub J. P. Pattern competition leads to chaos. *Phys. Rev. Lett.* **52** (1984) 922.

5.51   Croquette V. and Poitou C. Cascade of period doubling bifurcations and large stochasticity in the motions of a compass. *J. Physique* **42** (1981) 537.

5.52   D'Humieres D., Beasley M. R., Huberman B. A. and Libchaber A. Chaotic states and routes to chaos in the forced pendulum. *Phys. Rev.* **A26** (1982) 3483.

5.53   Derighetti B., Ravani M., Stoop R., Meier P. F., Brun E. and Badii R. Period – doubling lasers as small – signal detectors. *Phys. Rev. Lett.* **55** (1985) 1746.

5.54   Di Prima R. C. and Swinney H. L. Instabilities and transition in flow between concentric rotating cylinders. In *Hydrodynamic Instabilities and the Transition to Turbulence*, ed. H. L. Swinney and J. P. Gollub, Berlin, Springer, 1981, p. 139.

5.55   Dolník M., Schreiber I. and Marek M. Periodic and chaotic regimes in forced chemical oscillator. *Phys. Lett.* **100A** (1984) 316.

5.56   Donnelly R. J., Park K., Shaw S. and Walden R. W. Early nonperiodic transitions in Couette flow. *Phys. Rev. Lett.* **44** (1980) 987.

5.57   Dubois M., Rubio M. A. and Bergé P. Experimental evidence of intermittencies associated with subharmonic bifurcation. *Phys. Rev. Lett.* **51** (1983) 1446.

5.58   Durrani T. S. and Greated C. A. *Laser Systems in Flow Measurement.* New York, Plenum Press, 1977.

5.59   Eiswirth M., Krischer K. and Ertl G. Transition to chaos in an oscillating surface reaction. *Surface Science* **202** (1988) 565.

5.60   Epstein I. R. Oscillations and chaos in chemical systems. *Physica* **7D** (1983) 47.

5.61   Epstein I. R. Complex dynamical behviour in 'simple' chemical systems. *J. Phys. Chem.* **88** (1984) 187.

5.62   Ezerskii A. B., Korotin P. I. and Rabinovich M. J. Chaotic automodulation of two — dimensional structures on the liquid surface at parametric excitation. *Letters to Soviet Physics JETP* **41** (1984) 129.

5.63   Ezerskii A. B., Rabinovich M. I., Stepanyants Yu. A. and Shapiro M. F. Stochastic oscillations of a parametrically excited nonlinear chain. *Sov. Phys. JETP* **49** (1979) 500.

5.64   Fauve S., Laroche C. and Perrin B. Competing instabilities in a rotating layer of mercury heated from below. *Phys. Rev. Lett.* **55** (1985) 208.

5.65   Fauve S., Laroche C., Libchaber A. and Perrin B. Chaotic phases and magnetic order in a convective fluid. *Phys. Rev. Lett.* **52** (1984) 1774.

5.66   Fenstermacher P. R., Swinney H. L. and Gollub J. P. Dynamical instabilities and the transition to chaotic Taylor vortex flow. *J. Fluid. Mech.* **94** (1979) 103.

5.67   Field R. J. and Burger M. (Eds.) *Oscillations and Travelling Waves in Chemical Systems.* New York, Wiley, 1985.

5.68   Ganapathisubramanian N. and Noyes R. M. A discrepancy between experimental and computational evidence for chemical chaos. *J. Chem. Phys.* **76** (1982) 1770.

5.69   Gaponov — Grekhov A. V. and Rabinovich M. I. Nonstationary structures — chaos and order. In *Synergetics of Brain*, ed. H. Haken, Berlin, Springer, 1983.

5.70   Gaponov — Grekhov A. V., Rabinovich M. I. and Starobinets I. M. Formation of multidimensional chaos in active lattices. Preprint, Gorky, Inst. of Applied Physics, USSR Acad. Sci., 1984.

5.71   Gaponov — Grekhov A. V., Rabinovich M. I. and Starobinets I. M. The onset and development of chaotic structures in dissipative media. In *Autowaves in Biology, Chemistry and Physics*, ed. V. Krinsky, Berlin, Springer, 1984.

5.72   Gibbs H. M., Hopf F. A., Kaplan D. L. and Shoemaker R. L. Observation of chaos in optical bistability. *Phys. Rev. Lett.* **46** (1981) 474.

5.73   Gibson G. and Jeffries C. Observation of period doubling and chaos in spin — wave instabilities in yttrium iron garnet. *Phys. Rev.* **A29** (1984) 811.

5.74   Giglio M., Musazzi S. and Perini V. Transition to chaotic behaviour via a reproducible sequence of period — doubling bifurcation. *Phys. Rev. Lett.* **47** (1981) 243.

5.75   Giglio M., Musazzi S. and Perini V. Low — dimensionality turbulent convection. *Phys. Rev. Lett.* **53** (1984) 2402.

5.76   Gioggia R. S. and Abraham N. B. Routes to chaotic output from a single — mode, dc — excited laser. *Phys. Rev. Lett.* **51** (1983) 650.

5.77   Glass L. and Mackey M. C. Pathological conditions resulting from instabilities in physiological control systems. *Ann. N.Y. Acad. Sci.* **316** (1979) 214.

5.78   Glass L., Goldberger A. L., Courtemanche M. and Shrier A. Nonlinear dynamics, chaos and complex cardiac arrhythmias. *Proc. R. Soc. London* **A413** (1987) 9.

5.79   Glass L., Guevara M., Belair J. and Shrier A. Global bifurcation of a periodically forced biological oscillator. *Phys. Rev.* **A29** (1984) 1348.

5.80   Glazier J. A. and Libchaber A. Quasi — periodicity and dynamical systems: An experimentalist's view. *IEEE Transactions on Circuits and Systems* **35** (1988) 790.

5.81   Glazier J. A., Gunaratne G. and Libchaber A. f(α) curves: Experimental results. *Phys. Rev.* **A37** (1988) 523.

5.82   Goldberger A. L., Bhargava V. and West B. J. Nonlinear dynamics of the heartbeat II. Subharmonic bifurcations of the cardiac interbeat interval in sinus node disease. *Physica* **17D** (1985) 207.

5.83   Gollub J. P. and Benson S. V. . Many routes to turbulent convection. *J. Fluid. Mech.* **100** (1980) 449.

5.84   Gollub J. P. and Steinman J. F. External noise and the onset of turbulent convection. *Phys. Rev. Lett.* **45** (1980) 551.

5.85   Gollub J. P. and Swinney H. L. Onset of turbulence in a rotating fluid. *Phys. Rev. Lett.* **35** (1975) 927.

5.86   Gollub J. P., Romer E. J. and Socolar J. E. Trajectory divergence for coupled relaxation oscillators: measurements and models. *J. Stat. Phys.* **23** (1980) 321.

5.87   Golubitsky M., Swift J. W. and Knobloch E. Symmetries and pattern selection in Rayleigh – Bénard convection. *Physica* **10D** (1984) 249.

5.88   Gorman M., Widmann P. J. and Robbins K. A. Chaotic flow regimes in a convection loop. *Phys. Rev. Lett.* **52** (1984) 2241.

5.89   Grabec I. Chaos generated by the cutting process. *Phys. Lett.* **117A** (1986) 184.

5.90   Grabec I. Chaos generated by the plastic flow in the cutting process. Preprint, Stuttgart, GAMM Meeting, 1987.

5.91   Grüner G. Nonlinear transport due to driven collective modes. *Physica* **23D** (1986) 145.

5.92   Gubankov V. N., Konstantinyan K. I., Kosheletz V. P. and Ovsyannikov G. A. Chaos in Josephson tunnel junctions. *IEEE Trans. Mag.* **19** (1983) 637.

5.93   Gubankov V. N., Zhiglin S. L., Konstantinyan K. J., Koshelec V. P. and Ovsyannikov G. A. Stochastic oscillations in tunnel Josephson junctions. *JETP* **86** (1984) 343 (in Russian).

5.94   Guevara M. R. and Glass L. Phase locking, period doubling bifurcations and chaos in a mathematical model of a periodically driven oscillator: A theory for the entrainment of biological oscillators and the generation of cardiac disrhythmias. *J. Math. Biol.* **14** (1982) 1.

5.95   Guevara M. R., Glass L. and Shrier A. Phase locking, period – doubling bifurcations, and irregular dynamics in periodically stimulated cardiac cells. *Science* **214** (1981) 1350.

5.96   Gwinn E. G. and Westervelt R. M. Scaling structure of attractors at the transition from quasiperiodicity to chaos in electronic transport in Ge. *Phys. Rev. Lett.* **59** (1987) 157.

5.97   Haken H. Laser theory. In *Encyclopedia of Physics* **XXV/2c**, New York, Springer, 1970.

5.98   Haken H. Analogy between higher instabilities in fluids and lasers. *Phys. Lett.* **53A** (1975) 77.

5.99   Harrison R. G., Firth W. J. and Al – Saidi I. A. Observation of bifurcation to chaos in an all – optical Fabry – Perot resonator. *Phys. Rev. Lett.* **53** (1984) 258.

5.100   Harrison R. G., Firth W. J., Emshary C. A. and Al – Saidi I. A. Observation of period-doubling in an all – optical resonator containing NH_3 gas. *Phys. Rev. Lett.* **51** (1983) 562.

5.101   Hayashi H., Ishizuka S. and Hirakawa K. Transition to chaos via intermittency in the Onchidium pacemaker neuron. *Phys. Lett.* **98A** (1983) 474.

5.102   Hayashi H., Ishizuka S. and Hirakawa K. Chaotic response of the pacemaker neuron. *J. Phys. Soc. Japan* **57** (1985) 2337.

5.103   Hayashi H., Ishizuka S., Ohta M. and Hirakawa K. Chaotic behaviour in the Onchidium giant neuron under sinusoidal stimulation. *Phys. Lett.* **88A** (1982) 435.

5.104   Hayashi H., Nakao M. and Hirakawa K. Chaos in the self – sustained oscillations of an excitable biological membrane under sinusoidal stimulation. *Phys. Lett.* **88A** (1982) 265.

5.105   Hayashi H., Nakao M. and Hirakawa K. Entrained, harmonic, quasiperiodic and chaotic responses of the self – sustained oscillation of Nitella to sinusoidal stimulation. *J. Phys. Soc. Japan* **52** (1983) 344.

5.106   He D., Yeh W. J. and Kao Y. H. Transition from quasiperiodicity to chaos in a Josephson – junction analog. *Phys. Rev.* **B30** (1984) 172.

5.107   Held G. A. and Jeffries C. Spatial and temporal structures of chaotic instabilities in an electron – hole plasma in Ge. *Phys. Rev. Lett.* **55** (1985) 887.

5.108   Held G. A., Jeffries C. and Haller E. E. Turbulence in electron-hole plasma. Proc. of 17th Internat. Conf. on the *Physics of Semiconductors*, eds. Chadi J. D. and Harrison W. A., New York, Springer, 1985.

5.109   Held G. A. and Jeffries C. D. Characterization of chaotic instabilities in an electron – hole plasma in germanium. In *Dimensions and Entropies in Chaotic Systems*, ed. Mayer – Kress G., Berlin, Springer, 1986 ,p. 158.

5.110 Held G. A. and Jeffries C. Quasiperiodic transitions to chaos of instabilities in an electron-hole plasma excited by AC perturbations at one and two frequencies. *Phys. Rev. Lett.* **56** (1986) 1183.

5.111 Held G. A., Jeffries C. and Haller E. E. Observation of chaotic behavior in an electron – hole plasma in Ge. *Phys. Rev. Lett.* **52** (1984) 1037.

5.112 Hess B. and Markus M. The diversity of biochemical time patterns. *Ber. Bunsenges. Phys. Chem.* **89** (1985) 642.

5.113 Hiller M. and Schnelle K. – P. Chaotic dynamics in mechanisms. Preprint, Stuttgart, GAMM Meeting, 1987.

5.114 Hoffer L. M., Chyba T. H. and Abraham N. B. Spontaneous pulsing, period doubling, and quasiperiodicity in an unidirectional, single – mode, inhomogeneously broadened ring laser. *J. Opt. Soc. Am.* **B2** (1985) 102.

5.115 Hoffnagle J., DeVoe R. G., Reyna L. and Brewer R. G. Order – chaos transition of two trapped ions. *Phys. Rev. Lett.* **61** (1988) 255.

5.116 Holden A. V. and Ramadan S. M. The response of a molluscan neuron to a cyclic input: entrainment and phase locking. *Biol. Cybern.* **43** (1981) 157.

5.117 Holden A. V., Winlow W. and Haydon P. G. The induction of periodic and chaotic activity in a molluscan neuron. *Biol. Cybern.* **43** (1982) 169.

5.118 Holmes P. J. A nonlinear oscillator with a strange attractor. *Phil. Trans. Roy. Soc.* **A292** (1979) 419.

5.119 Holmes P. J. and Moon F. C. Strange attractors and chaos in nonlinear mechanics. *J. of Appl. Mech.* **50** (1983) 1021.

5.120 Hong – jun Zhang, Jian – hua Dai, Peng – ye Wang, Fu – lai Zhang, Guang Xu and Shi-ping Yang Chaos in liquid crystal opical bistability. In *Directions in Chaos*, ed. Hao Bai – lin, Singapore, World Scientific, 1988, p. 46.

5.121 Hongler M. – O. and Streit L. On the origin of chaos in gearbox models. *Physica* **29D** (1988) 402.

5.122 Hopf E. A., Kaplan D. L., Gibbs H. M. and Shoemaker R. L. Bifurcations to chaos in optical bistability. *Phys. Rev.* **A25** (1982) 2172.

5.123 Huberman B. A., Crutchfield J. P. and Packard N. H. Noise phenomena in Josephson – junctions. *Appl. Phys. Lett.* **37** (1980) 756.

5.124 Hudson J. L. and Mankin J. C. Chaos in the Belousov – Zhabotinsky reaction. *J. Chem. Phys.* **74** (1981) 6171.

5.125 Hudson J. L., Hart M. and Marinko D. An experimental study of multiple peak periodic and non – periodic oscillations in the Belousov – Zhabotinsky reaction. *J. Chem. Phys.* **71** (1979) 1601.

5.126 Iansiti M., Hu Q., Westervelt R. M. and Tinkham M. Noise and chaos in a fractal basin boundary regime of a Josephson – junction. *Phys. Rev. Lett.* **55** (1985) 746.

5.127 Ikeda K. Multiple – valued stationary state and its instability of the transmitted light by a ring cavity system. *Opt. Commun.* **30** (1979) 257.

5.128 Ikeda K. and Kondo K. Successive higher – harmonic bifurcations in systems with delayed feedback. *Phys. Rev. Lett.* **49** (1982) 1467.

5.129 Ikeda K., Daido H. and Akimoto D. Optical turbulence chaotic behaviour of transmitted light from a ring cavity. *Phys. Rev. Lett.* **48** (1980) 617.

5.130 Ilinze J. O. *Turbulence.* New York, McGraw – Hill, 1975 (2nd edn.).

5.131 Isomäki H. M., Von Boehm J. and Räty R. Devil's attractors and chaos of a driven impact oscillator. *Phys. Lett.* **107A** (1985) 343.

5.132 Ito K., Oono Y., Yamazaki H. and Hirakawa K. Chaotic behaviour in great earthquakes – coupled relaxation oscillator model, billiard model and electronic circuit model. *J. Phys. Soc. Japan* **49** (1980) 43.

5.133 Jefferies D. J., and Usher A. Frequency division using diodes in resonant systems. *Phys. Lett.* **99A** (1982) 356.

5.134 Jeffries C. and Perez J. Observation of a Pomeau–Manneville intermittent route to chaos in a nonlinear oscillator. *Phys. Rev.* **A26** (1982) 2117.

5.135 Jeffries C. and Perez J. Direct observation of crises of the chaotic attractor in a nonlinear oscillator. *Phys. Rev.* **A27** (1983) 601.

5.136 Jeffries C. and Wiesenfeld K. Observations of noisy precursors of dynamical instabilities. *Phys. Rev.* **31A** (1985) 1077.

5.137 Jensen K. S., Mosekilde E. and Holstein-Rathlou N.-H. Self–sustained oscillations and chaotic behaviour in kidney pressure regulation. Preprint, Univ. of Copenhagen, 1987.

5.138 Jensen M. H., Kadanoff L. P., Libchaber A. J., Procaccia I. and Stavans J. Global universality at the onset of chaos: Results of a forced Rayleigh–Bénard experiment. *Phys. Rev. Lett.* **55** (1985) 2798.

5.139 Joseph D. D. *Stability of Fluid Motions.* Vols. I and II. Berlin, Springer, 1976.

5.140 Katz V. A. Formation and evolution of chaos in distributed generator with a time delay. *Rep. of Inst. of Higher Education, sect. Radiophysics* **28** (1985) 161 (in Russian).

5.141 Katz V. A. and Trubeckov D. I. Formation of chaos and the destruction of quasiperiodic regimes and transition via intermittency in the distributed generator with a delay. *Lett. to JETP* **39** (1984) 116 (in Russian).

5.142 Keener J. P. Chaotic cardiac dynamics. In *Mathematical Aspects of Physiology.* (AMS Lectures in applied mathematics **19**), ed. Hoppenstadt F. C., Providence, RI, AMS, 1981, p. 299.

5.143 Keolian R., Turkevich L. A., Putterman S. J., Rudnick I. and Rudnick J. A. Subharmonic sequences in the Faraday experiment: departures from period–doubling. *Phys. Rev. Lett.* **47** (1981) 1133.

5.144 Kerr W. C., Williams M. B., Bishop A. R., Fesser K., Lomdahl P. S. and Trullinger S. E. Symmetry and chaos in the motion of the damped driven pendulum. *Z. Phys. B: Condensed Matter.* **B59** (1985) 103.

5.145 Khandokhin P. A. and Khanin Ya. I. Autostochastic regime of generation of solid state ring laser with low–frequency periodic modulation of losses. *Quantum Electronics* **11** (1984) 1483 (in Russian).

5.146 Khandokhin P. A. and Khanin Ya. I. Instabilities in a solid–state ring laser. *J. Opt. Soc. Amer.* **B2** (1985) 226.

5.147 Kiashko S. V., Pikovskii A. S., Rabinovich M. I. Autogenerator of radiorange with stochastic behaviour. *Radiotechnics and Electronics* **25** (1980) 336 (in Russian).

5.148 Kim S. H. and Ostlund S. Simultaneous rational approximations in the study of dynamical systems. *Phys. Rev.* **A34** (1986) 3426.

5.149 Klische W. and Weiss C. O. Instabilities and routes to chaos in a homogeneously broadened one– and two–mode ring laser. *Phys. Rev.* **A31** (1985) 4049.

5.150 Koch B. P. and Leven R. W. Subharmonic and homoclinic bifurcations in a parametrically forced pendulum. *Physica* **16D** (1985) 1.

5.151 Kornev V. K. and Semenov V. K. Chaotic and stochastic phenomena in superconducting quantum interferometers. *IEEE Trans. Mag.* **19** (1983) 633.

5.152 Körös R. M., Bischoff K. B. and Keane T. R. *Chemical Reaction Engineering.* Oxford, Pergamon Press, 1986.

5.153 Koumar J. and Marek M. On the mathematical model of the Zhabotinskii reaction. *Kinetics and Catalysis* **15** (1974) 1333 (in Russian).

5.154 Kozlov V. V., Rabinovich M. I., Ramazanov M. P., Reiman A. M. and Sushchik M. M. Correlation dimension of the flow and spatial development of dynamical chaos in a boundary layer. *Phys. Lett.* **128A** (1988) 479.

5.155 Kücükay F. and Pfeiffer F. Über Rasselschwingungen in Kfz — Schaltgetrieben. *Ingenieur — Archiv* **56** (1986) 25.

5.156 Kurz T. and Lauterborn W. Bifurcation structure of the Toda oscillator. *Phys. Rev.* **A37** (1988) 1029.

5.157 L'vov V. S. and Predtechenskii H. A. On Landau and stochastic attractor pictures in the problem of transition to turbulence. *Physica* **2D** (1981) 38.

5.158 L'vov V. S., Predtechenskii H. A. and Chernykh A. I. Bifurcation and chaos in a system of Taylor vortices: a natural and numerical experiment. *Soviet Physics JETP* **53** (1981) 562.

5.159 Landa P. S., Perminov S. M., Shatalova G. G. and Damgov V. N. Stochastic oscillations in the generator with additional delay feedback. *Radiotechnics and Electronics* **31** (1986) 730 (in Russian).

5.160 Landsberg P. T., Schöll E. and Shukla P. A simple model for the origin of chaos in semiconductors. *Physica* **30D** (1988) 235.

5.161 Lauterborn W. and Cramer E. Subharmonic route to chaos observed in acoustics. *Phys. Rev. Lett.* **47** (1981) 1445.

5.162 Lauterborn W. and Steinhoff R. Bifurcation structure of a laser with pump modulation. *J. Opt. Soc. of America* **B5** (1988) 1097.

5.163 Leonardo da Vinci. *Work in Natural Philosophy.* Moscow, Nauka, 1955 (in Russian).

5.164 Leven R. W., Pompe B., Wilke C. and Koch B. P. Experiments on periodic and chaotic motions of a parametrically forced pendulum. *Physica* **16D** (1985) 371.

5.165 Libchaber A. Convection and turbulence in liquid helium I. *Physica* **B & C 109 − 110** (1982) 1583.

5.166 Libchaber A. and Maurer J. A Rayleigh − Bénard experiment: Helium in a small box. In *Nonlinear Phenomena at Phase Transitions and Instabilities,* ed. T. Riste, New York, Plenum Press 1982, see also *J. Phys. (Paris) Colloq.* **41** (1982) C3 − 51.

5.167 Libchaber A., Laroche C. and Fauve S. Period − doubling cascade in mercury: a quantitative measurement. *J. Phys. Lett.* **43** (1982) L211.

5.168 Liepmann H. W. The rise and fall of ideas in turbulence. *American Scientist* **67** (1979) 221.

5.169 Linsay P. S. Period doubling and chaotic behaviour in a driven anharmonic oscillator. *Phys. Rev. Lett.* **47** (1981) 1349.

5.170 Linsay P. S. Approximate scaling of period doubling windows. *Phys. Lett.* **108A** (1985) 431.

5.171 Ludford G. S. S. (Ed.) *Reacting Flows: Combustion and Chemical Reactors.* AMS Lectures in Applied Mathematics **24**, Providence, RI, AMS, 1986.

5.172 Mackey M. C. and Glass L. Oscillations and chaos in physiological control systems. *Science* **197** (1977) 287.

5.173 Marek M. Dissipative structures in chemical systems. In *Far from Equilibrium Instabilities and Structures,* ed. A. Pacault, C. Vidal, Berlin, Springer, 1979.

5.174 Marek M. Turing structures, periodic and chaotic regimes in coupled cells. In *Modelling of Patterns in Space and Time.* Lecture Notes in Biomathematics, Berlin, Springer, 1984, p. 214.

5.175 Marek M. Periodic and aperiodic regimes in forced chemical oscillations. In *Temporal Order,* ed. L. Rensing and N. I. Jaeger, Springer Series on Synergetics **29**, Berlin, Springer, 1984.

5.176 Marek M. and Stuchl I. Synchronization in two interacting oscillatory systems. *Biophys. Chem.* **3** (1975) 241.

5.177 Marek M. and Svobodová E. Nonlinear phenomena in oscillatory systems of homogeneous reactions — experimental observations. *Biophys. Chem.* **3** (1975) 263.

5.178 Markus M. and Hess B. Transitions between oscillatory modes in a glycolytic model system. *Proc. Natl. Acad. Sci. U.S.A.* **81** (1984) 235.

5.179 Markus M., Kuschmitz D. and Hess B. Chaotic dynamics in yeast glycolysis under periodic substrate input flux. *FEBS Lett.* **172** (1984) 235.

5.180 Martin S. and Martienssen W. Spatio − temporal electrical instabilities in Barium − Sodium − Niobate single crystals. *Physica* **23D** (1986) 195.

5.181 Martin S. and Martienssen W. Circle maps and mode locking in the driven electrical conductivity of barium sodium niobate crystals. *Phys. Rev. Lett.* **56** (1986) 1522.

5.182 Martin S., Leber H. and Martienssen W. Oscillatory and chaotic states of the electrical conduction in barium sodium niobate crystals. *Phys. Rev. Lett.* **53** (1984) 303.

5.183 Maselko J. and Swinney H. A complex transition sequence in the Belousov – Zhabotinskii reaction. *Physica Scripta* **T9** (1985) 35.

5.184 Maselko J. and Swinney H. L. A Farey triangle in the Belousov – Zhabotinskii reaction. *Phys. Lett.* **119A** (1987) 407.

5.185 Matsumoto G., Aihara K., Ichika M. and Tasaki A. Periodic and nonperiodic responses of membrane potential in squid giant axons under sinusoidal current stimulation. *J. Theor. Neurobiol.* **3** (1984) 1.

5.186 Matsumoto T., Chua L. O. and Komuro M. The double scroll. *IEEE Transactions on Circuits and Systems* **32** (1985) 798.

5.187 Maurer J. and Libchaber A. Rayleigh – Bénard experiment in liquid helium; frequency locking and the onset of turbulence. *J. Phys. Lett.* (Paris) **40** (1979) L419.

5.188 Mayer J. H., Risken H. and Vollmer D. Periodic and chaotic breathing of pulses in a ring laser. *Opt. Commun.* **36** (1981) 480.

5.189 Meyer Ch. W., Cannell D. S., Ahlers G., Swift J. B. and Hohenberg P. C. Pattern competition in temporally modulated Rayleigh – Bénard convection. *Phys. Rev. Lett.* **61** (1988) 947.

5.190 Miracky R. F., Clarke J. and Koch R. H. Chaotic noise observed in resistively shunted self – resonant Josephson tunnel junction. *Phys. Rev. Lett.* **50** (1983) 856.

5.191 Mischenko E. F. and Rozov N. *Differential Equations with a Small Parameter and Relaxation Oscillators.* Moscow, Nauka, 1975 (in Russian).

5.192 Moloney J. V. Self – focusing – induced optical turbulence. *Phys. Rev. Lett.* **53** (1984) 556.

5.193 Monin A. S. and Yaglom A. M. *Statistical Fluid Mechanics.* Cambridge MA, MIT Press, 1971, 1975.

5.194 Moon F. C. Fractal boundary for chaos in two – state mechanical oscillator. *Phys. Rev. Lett.* **53** (1984) 962.

5.195 Moon F. C. and Guang – Xuan Li The fractal dimension of the two – well potential strange attractor. *Physica* **17D** (1985) 99.

5.196 Moon F. C. and Holmes P. J. A magnetoelastic strange attractor. *J. Sound. Vib.* **65** (1979) 275; ibid **69** (1979) 339.

5.197 Moon F. C., Cusumano J. and Holmes P. J. Evidence for homoclinic orbits as a precursor to chaos in a magnetic pendulum. *Physica* **24D** (1987) 383.

5.198 Mukai T. and Otsuka K. New route to optical chaos: successive – subharmonic – oscillation cascade in a semiconductor laser coupled to an external cavity. *Phys. Rev. Lett.* **55** (1985) 1711.

5.199 Nagashima H. Experiments on chaotic response of forced Belousov – Zhabotinski reaction. *J. Phys. Soc. Japan* **51** (1982) 21.

5.200 Nicolis J. S. The role of chaos in reliable information processing. *J. Frankl. Inst.* **317** (1984) 289.

5.201 Nöldeke Ch. Chaotic Josephson scenario in an $Sn/Sn_xO_y/Sn$ tunnel junction at 10 GHz. *Phys. Lett.* **112A** (1985) 178.

5.202 Nöldeke Ch. and Seifert H. Different types of intermittent chaos in Josephson junctions. Manifestations in the I – V characteristics. *Phys. Lett.* **109A** (1985) 401.

5.203 Olsen L. F. An enzyme reaction with a strange attractor. *Phys. Lett.* **94A** (1983) 454.

5.204 Olsen L. F. and Degn H. Chaos in an enzyme reaction. *Nature* **267** (1977) 177.

5.205 Onken H. U. and Wicke E. Statistical fluctuations of temperature and conversion at the catalytic CO oxidation in an adiabatic packed bed reactor. *Ber. Bunsenges. Phys. Chem.* **90** (1986) 976.

5.206 Orbán M. and Epstein I. R. Complex periodic and aperiodic oscillations in the chlorite – thiosulfate reaction. *J. Phys. Chem.* **86** (1983) 3907.

5.207 Otnes R. K. and Enochson L. *Digital Time Series Analysis.* New York, Wiley, 1972.

5.208 Peinke J., Mühlbach A., Huebener R. P. and Parisi J. Spontaneous oscillations and chaos in p – germanium. *Phys. Lett.* **108A** (1985) 407.

5.209 Peinke J., Mühlbach A., Röhricht B., Wassely B., Mannhart J., Parisi J. and Huebener R. P. Chaos and hyperchaos in the electric avalanche breakdown of p – germanium at 4.2K. *Physica* **23D** (1986) 176.

5.210 Perez J. and Jeffries C. Effects of additive noise on a nonlinear oscillator exhibiting period doubling and chaotic behaviour. *Phys. Rev.* **B26** (1982) 3460.

5.211 Perminov S. M., Dobrovoskii V. A., Anikeev G. I. and Genkin M. D. Process of mixing of liquid in a flat channel. *JTP* **57** (1987) 171 (in Russian).

5.212 Pfeiffer F. Chaos in Getriebe. Preprint, Stuttgart, GAMM Meeting, 1987.

5.213 Pfeiffer F. Seltsame Attraktoren in Zahnradgetrieben. *Ingenieur* – *Archiv* **58** (1988).

5.214 Pikovsky A. S. A dynamical model for periodic and chaotic oscillations in the Belousov – Zhabotinsky reaction. *Phys. Lett.* **85A** (1981) 13.

5.215 Pikovsky A. S. and Rabinovich M. I. Stochastic oscillations in dissipative systems. *Physica* **2D** (1981) 8.

5.216 Pikovsky A. S. and Rabinovich M. I. A simple self – sustained oscillator with stochastic behaviour. Preprint, Gorky, Inst. of Applied Physics, USSR Acad. Sci.,1984.

5.217 Pocheau A., Croquette V. and Le Gal P. Turbulence in a cylindrical container of Argon near threshold of convection. *Phys. Rev. Lett.* **55** (1985) 657.

5.218 Pomeau Y., Roux J. C., McCornick W. D. and Swinney H. L. Intermittent behaviour in the Belousov – Zhabotinsky reaction. *J. Phys. Lett.* **42** (1981) 271.

5.219 Prasard R. R., Meneveau C. and Sreenivasan K. R. Multifractal nature of the dissipation field of passive scalars in fully turbulent flows. *Phys. Rev. Lett.* **61** (1988) 74.

5.220 Puccioni G. P., Poggi A., Gadoruski W., Tredicci J. R. and Arecchi F. T. Measurement of the formation and evolution of a strange attractor in a laser. *Phys. Rev. Lett.* **55** (1985) 339.

5.221 Rabinovich A. and Thierberger R. Time series analysis of chaotic signals. *Physica* **28D** (1987) 409.

5.222 Rapp P. E. Oscillations and chaos in cellular metabolism. In *Chaos, An Introduction*, ed. A. V. Holden, Manchester, Manchester Univ. Press, 1987.

5.223 Raschman P., Kubíček M. and Marek M. Waves in distributed chemical systems: Experiments and computations. In *New Approaches to Nonlinear Problems in Dynamics*, ed. P. J. Holmes, Philadelphia, SIAM Publ., 1980, p. 271.

5.224 Razón L. F., Chang S. M. and Schmitz R. A. Chaos during the oxidation of carbon monoxide on platinum – experiments and analysis. *Catal. Rev. Sci. Eng.* **28** (1986) 89.

5.225 Reiner G., Belic' M. R. and Meystre P. Optical turbulence in phase – conjugate resonators. *J. Opt. Soc. Am.* **B5** (1988) 1193.

5.226 Richetti P. and Arneodo A. The periodic – chaotic sequences in chemical reactions: a scenario close to homoclinic conditions. *Phys. Lett.* **109A** (1985) 359.

5.227 Richetti P., Roux J. C., Argoul F. and Arneodo A. From quasiperiodicity to chaos in the Belousov – Zhabotinskii reaction; I. Experiment, II. Modelling and Theory. *J. Chem. Phys.* **86** (1987), 3325, 3339.

5.228 Ringland J. and Turner J. S. One – dimensional behaviour in a model of the Belousov – Zhabotinskii reaction. *Phys. Lett.* **105A** (1984) 93.

5.229 Robbins K. A. A new approach to subcritical instability and turbulent transitions in a simple dynamic. *Math. Proc. Camb. Phil. Soc.* **82** (1977) 309.

5.230 Rössler O. and Wegmann K. Chaos in the Zhabotinsky reaction. *Nature* **271** (1978) 89.

5.231 Roux J. C. Experimental studies of bifurcations leading to chaos in the Belousov – Zhabotinsky reaction. *Physica* **7D** (1983) 57.

5.232   Roux J. C. and Swinney H. L. Topology of chaos in a chemical reaction. In *Nonlinear Phenomena in Chemical Dynamics*, ed. C. Vidal and A. Pacault, Berlin, Springer, 1981, p. 33.

5.233   Roux J. C., Rossi A., Bachelart S. and Vidal C. Representation of a strange attractor from an experimental study of chemical turbulence. *Phys. Lett.* **77A** (1980) 391.

5.234   Roux J. C., Rossi A., Bachelart S. and Vidal C. Experimental observation of complex dynamical behaviour during a chemical reaction. *Physica* **2D** (1981) 395.

5.235   Roux J. C., Simoyi R. H. and Swinney H. L. Observation of a strange attractor. *Physica* **8D** (1983) 257.

5.236   Roux J. C., Turner J. S., McCormick W. D. and Swinney H. L. Experimental observation of complex dynamics in a chemical reaction. In *Nonlinear Problems, Present and Future*, ed. A. R. Bishop, D. K. Campbell and B. Nicolaenko, Amsterdam, North Holland, 1982, p. 409.

5.237   Ruelle D. Some comments on chemical oscillations. *Trans. N.Y. Acad. Sci.* **35** (1973) 66.

5.238   Sakai K. and Yamaguchi Y. Nonlinear dynamics of a Josephson oscillator with a cos term driven by dc− and ac− current sources. *Phys. Rev.* **B30** (1984) 1219.

5.239   Sargent M., Scully M. O. and Lamb W. E. *Laser Physics*. London, Addison−Wesley, 1974.

5.240   Sbitnev V. I. The origin and evolution of a strange attractor in the system of coupled oscillators. *Biofizika* **29** (1984) 113, (in Russian).

5.241   Schmidt G. Universality of dissipative systems. In *Directions in Chaos*, ed. Hao Bai−lin, World Scientific, Singapore, 1988, p. 11.

5.242   Schmitz R. A. Multiplicity, stability and sensitivity of state in chemically reacting systems. *Adv. Chem. Ser.* **148** (1975) 154.

5.243   Schmitz R. A., Graziani K. R. and Hudson J. L. Experimental evidence of chaotic states in the Belousov−Zhabotinsky reaction. *J. Chem. Phys.* **67** (1977) 3040.

5.244   Schmitz R. A., Renola G. T. and Ziondas A. P. Strange oscillations in chemical reactions. In *Dynamics and Modelling of Reactive Systems*, ed. P. Rabinowitz, New York, Academic Press, 1980.

5.245   Scholz M. J., Yamada T., Brand H. and Graham R. Intermittency and chaos in a laser system with modulation inversion. *Phys. Lett.* **82A** (1981) 321.

5.246   Seeger K. *Semiconductor Physics*. New York, Springer, 1982 (2nd ed.).

5.247   Seifert H. Intermittent chaos in Josephson junctions represented by stroboscopic maps. *Phys. Lett.* **98A** (1983) 213.

5.248   Seifert H. Structures on Josephson current steps, bifurcation and chaos. *Phys. Lett.* **101A** (1984) 230.

5.249   Seiler D. G., Littler C. L., Justice R. J. and Milonni P. W. Nonlinear oscillations and chaos in n−InSb. *Phys. Lett.* **108A** (1985) 462.

5.250   Ševčíková H. and Marek M. Chemical waves in electric field − modelling. *Physica* **21D** (1986) 61.

5.251   Shaw S. W. and Wiggins S. Chaotic dynamics of a whirling pendulum. *Physica* **31D** (1988) 190.

5.252   Sherwin M., Hall R. and Zettl A. Chaotic ac conductivity in the charge−density−wave state of $(TaSe_4)_2$ I. *Phys. Rev. Lett.* **53** (1984) 1387.

5.253   Shtern V. N. Onset and properties of chaos in simple models of a heat convection. Preprint, Novosibirsk, Institute of Thermal Physics, 1983.

5.254   Simoyi R. H., Wolf A. and Swinney H. L. One−dimensional dynamics in a multicomponent chemical reaction. *Phys. Rev. Lett.* **49** (1982) 245.

5.255   Smith J. M. and Cohen R. J. Simple finite−element model accounts for wide range of cardiac disrhythmias. *Proc. Natl. Acad. Sci. U.S.A.* **81** (1984) 233.

5.256   Sreenivasan K. R. and Ramshankar R. Transition intermittency in open flows, and intermittency routes to chaos. *Physica* **23D** (1986) 246.

5.257   Stavans J. Experimental study of quasiperiodicity in a hydrodynamical system. *Phys. Rev.* **A35** (1987) 4314.

5.258 Stavans J., Heslot F. and Libchaber A. Fixed winding number and the quasiperiodic route to chaos in a convective fluid. *Phys. Rev. Lett.* **55** (1985) 596.

5.259 Stelter P. and Popp K. Chaotic motion of a driven self−sustained oscillator with dry friction. Preprint, GAMM Meeting, Stuttgart, 1987.

5.260 Sullivan T. S. and Ahlers G. Hopf bifurcation to convection near the codimension−two point in a 3He−4He mixture. *Phys. Rev. Lett.* **61** (1988) 78.

5.261 Swinney H. L. Observations of order and chaos in nonlinear systems. *Physica* **7D** (1983) 3.

5.262 Swinney H. L. Observations of complex dynamics and chaos. In *Fundamental Problems in Statistical Mechanics* **VI**, ed. E. G. D. Cohen, Amsterdam, North−Holland, 1984.

5.263 Swinney H. L. and Gollub J. P. (Eds.) *Hydrodynamic Instabilities and the Transition to Turbulence.* Berlin, Springer, 1981.

5.264 Takens F. Detecting strange attractors in turbulence. In *Lecture Notes in Mathematics* **98**, ed. D. A. Rand and L. S. Young, Berlin, Springer, 1981, p. 366.

5.265 Tanaka S., Matsumoto T. and Chua L. D. Bifurcation scenario in a driven R−L−diode circuit. *Physica* **28D** (1987) 317.

5.266 Teitsworth S. W. and Westervelt R. M. Subharmonic and chaotic response of periodically driven extrinsic Ge photoconductors. *Phys. Rev. Lett.* **56** (1986) 516.

5.267 Teitsworth S. W. and Westervelt R. M. Nonlinear current−voltage characteristics and spontaneous current oscillations in p−Ge. *Physica* **23D** (1986) 181.

5.268 Teitsworth S. W., Westervelt R. M. and Haller E. E. Nonlinear oscillations and chaos in electrical breakdown in Ge. *Phys. Rev. Lett.* **51** (1983) 825.

5.269 Testa J. S., Perez J. and Jeffries C. Evidence for universal chaotic behviour of a driven nonlinear oscillator. *Phys. Rev. Lett.* **48** (1982) 714 ; see also Hunt E. R., Comment on a driven nonlinear oscillator. *Phys. Rev. Lett.* **49** (1982) 1054.

5.270 Thompson J. M. T. and Virgin L. N. Spatial chaos and localization phenomena in nonlinear elasticity. *Phys. Lett.* **A126** (1988) 491.

5.271 Thompson J. M. and Ghaffari R. Chaotic dynamics of an impact oscillator. *Phys. Rev.* **A27** (1983) 1741.

5.272 Tomita K. Structure and macroscopic chaos in biology. *J. Theor. Biol.* **99** (1982) 111.

5.273 Tomita K. The significance of the concept chaos. *Prog. Theor. Phys.* **79** (1984) 1.

5.274 Tomita K. and Daido H. Possibility of chaotic behaviour and multi−basins in forced glycolytic oscillations. *Phys. Lett.* **79A** (1980) 133.

5.275 Tomita K. and Tsuda I. Chaos in the Belousov−Zhabotinsky reaction in a flow system. *Phys. Lett.* **71A** (1979) 489.

5.276 Tomita K. and Tsuda I. Towards the interpretation of Hudson's experiments on the Belousov−Zhabotinsky reaction. *Prog. Theor. Phys.* **64** (1980) 1138.

5.277 Tsuda I. A hermeneutic process of the brain. *Prog. Theor. Phys.* **79** (1984) 241.

5.278 Turner J. S., Roux J. C., McCormick W. D. and Swinney H. L. Alternating periodic and chaotic regimes in a chemical reaction: experiment and theory. *Phys. Lett.* **85A** (1981) 9.

5.279 Tyson J. J. On the appearance of chaos in a model of the Belousov reaction. *J. Math. Biol.* **5** (1978) 351.

5.280 Ueda Y. and Akamatsu N. Chaotically transitional phenomena in the forced negative−resistance oscillator. *IEEE Trans. on Circuits and Systems* **28** (1981) 217.

5.281 Van−Buskirk R. and Jeffries C. Observation of chaotic dynamics of coupled nonlinear oscillations. *Phys. Rev.* **A31** (1985) 3332.

5.282 Velarde M. G. and Antoranz J. C. Strange attractor (optical turbulence) in a model problem for the laser with saturable absorber and the two−component Benard convection. *Prog. Theor. Phys.* **66** (1981) 717.

5.283 Vidal C., Roux J. C., Bachelart S. and Rossi A. Experimental study of the transition to turbulence in Belousov−Zhabotinsky reaction. *Ann. N.Y. Acad. Sci.* **357** (1980) 377.

5.284 Walden R. W. Some new routes to chaos in Rayleigh – Bénard convection. *Phys. Rev.* **A27** (1983) 1255.

5.285 Walden R. W. and Donnely R. J. Reemergent order of chaotic circular Couette flow. *Phys. Rev. Lett.* **42** (1979) 301.

5.286 Walden R. W., Kolodner P., Passner A. and Surko C. M. Nonchaotic Rayleigh – Bénard convection with four and five incommensurate frequencies. *Phys. Rev. Lett.* **53** (1984) 242.

5.287 Walden R. W., Kolodner P., Passner A. and Surko M. Traveling waves and chaos in convection in binary fluid mixtures. *Phys. Rev. Lett.* **55** (1985) 496.

5.288 Wegman K. and Rössler O. Different kinds of chaotic oscillations in the Belousov – Zhabotinsky reaction. *Z. Naturwiss.* **33A** (1979) 1179.

5.289 West B. J., Goldberger A. L., Rovner G. and Bhargava V. Nonlinear dynamics of the heart beat I. The AV junction passive conduit or active oscillator? *Physica* **17D** (1985) 198.

5.290 Westervelt R. W. and Teitsworth S. W. Nonlinear dynamics and chaos in extrinsic photoconductors. *Physica* **23D** (1986) 187.

5.291 Wiesenfeld K. and McNamara B. Period – doubling systems as small – signal amplifiers. *Phys. Rev. Lett.* **55** (1985) 13.

5.292 Winters K. H., Plesser Th. and Cliffe K. A. The onset of convection in a finite container due to surface tension and buoyancy. *Physica* **29D** (1988) 387.

5.293 Wolf A., Swift J. B., Swinney H. L. and Vastano J. A. Determining Lyapunov exponents from a time series. *Physica* **16D** (1985) 285.

5.294 Yahata H. Onset of chaos in the Rayleigh – Bénard convection. *Progr. Theor. Phys. Suppl.* **79** (1984) 26.

5.295 Yamada T. and Graham R. Chaos in a laser under modulated external field. *Phys. Rev. Lett.* **45** (1980) 1322.

5.296 Yamazaki H., Oono Y. and Hirakawa K. Experimental study on chemical turbulence I and II. *J. Phys. Soc. Japan* **44** (1978) 335; ibid **46** (1979) 721.

5.297 Yeh W. J. and Kao Y. H. Universal scaling and chaotic behviour of a Josephson – junction analog. *Phys. Rev. Lett.* **49** (1982) 1888.

5.298 Yeh W. J. and Kao Y. H. Intermittency in Josephson junctions. *Appl. Phys. Lett.* **42** (1983) 299.

5.299 Yorke J. A., Grebogi C., Ott E. and Tedeschini – Lalli L. Scaling behaviour of windows. *Phys. Rev. Lett.* **54** (1985) 1095.

5.300 Zeldovich Ya. B., Barenblat G. I., Librovich V. B. and Machviladze G. M. *Mathematical Theory of Combustion and Explosion.* Moscow, Nauka, 1974 (in Russian).

5.301 Zettl A. Electronic and elastic mode locking in charge density wave conductors. *Physica* **23D** (1986) 155.

5.302 Zhabotinsky A. M. *Concentration Oscillations.* Moscow, Nauka, 1974 (in Russian).

5.303 Special issue on instabilities in active optical media. *J. Opt. Soc. Am.* **B2** (1985).

6

Forced and coupled chemical oscillators – a case study of chaos

In this chapter we discuss in detail two examples of dynamical systems. The first example shows the use of one-dimensional maps for the interpretation of chaotic and periodic regimes observed in experiments with a periodically forced chemical oscillator. The second example illustrates the use of numerical techniques for the investigation of chaotic phenomena in a model of coupled reaction diffusion cells.

6.1 Periodically forced continuous stirred tank reactor

There are several reasons for the interest in periodically forced chemical reactors. First, advances in electronics and computer technology make the controlled operation of chemical systems under periodically varying conditions feasible. Periodically operated processes might possess advantages over steady state processes as discussed by Bailey[6.5]. Second, forced chemical systems serve as simple experimental analogues of various dynamical systems arising at different levels of organization of living systems, for example synthesis of proteins, cell cycles, tissue growth, excitatory tissues, physiological control systems, etc.[6.4, 6.24, 6.27, 6.28, 6.31, 6.56]. Third, chemical systems, particularly oscillatory reactions, serve as prototype experimental systems for studies of transition to chaos in dissipative systems[6.34, 6.47].

A particularly interesting and important case arises when undamped oscillations of the limit cycle type occur in the flow-through reaction cell.

This type of dynamical systems is called an *oscillator*. Another important case of chemical dynamical systems is an *excitable system* or *excitator*[6.13]. Excitators are characterized by a non oscillatory behaviour in the absence of external forcing, but they can be excited and become oscillatory under superthreshold periodic perturbations. Forcing by a pulse-like or continuous periodic variation of inlet concentrations of various reaction components not only gives detailed information about dynamical behaviour of the studied system and thus helps to differentiate between alternative reaction mechanisms (or mathematical models) but it also models the actual processes occurring on various levels of living systems. It is still very difficult to study the behaviour of various neurotransmitters *in vitro*, for example, choline, acetylcholine and acetylcholinesterase system. Before the enzymatic parts of such systems become available for experimental studies from efficient production processes based on genetically manipulated cells or tissue cultures, a methodical base for classification of the behaviour of such systems can be developed with various organic or inorganic analogues available.

The system studied in the greatest detail is the *Belousov-Zhabotinski* (BZ) *reaction* (oxidation of malonic acid by bromate in the presence of Ce^{3+}/Ce^{4+} catalyst). It is possible to observe undamped limit cycle oscillations of reaction components and of the redox potential with the oscillation period constant within 1.5 %, if the reaction is studied in a flow-through well-stirred reaction cell[6.16, 6.17, 6.53]. As has already been discussed in Chapter 5, proposed detailed reaction mechanisms can include more than 20 reaction components, hence the dimension of the corresponding state space is quite high. However, even these rather complicated dynamical models do not describe the results of the concentration forcing experiments. On the other hand, we shall now illustrate in detail how both periodic and chaotic dynamics of the continuous stirred reaction cell with a periodic pulse concentration forcing can be understood using iterated one-dimensional phase transition curves (i.e. one-dimensional mappings) determined in single pulse experiments[6.16].

Similar experimental results were obtained on biological material by Glass *et al.*[6.22, 6.23]. Effects of sinusoidal forcing on glycolytic oscillations were reported by Markus and Hess[5.178]. Buchholtz, Freund and Schneider[6.12] studied effects of sinusoidal concentration forcing of the BZ system. Ruoff [6.46] discussed single pulse perturbations of the BZ reaction in a closed system (batch reactor) with pulses of Br^- and Ag^+ ions and by HOBr.

The results of the single pulse experiments are usually expressed in the form of *phase transition curves* (PTC) defined in the following way, see Ref. 6.58. It is assumed that the unperturbed oscillator is described by the flow $g^t(\mathbf{x})$, $\mathbf{x} \in R^n$, *which generates an attracting periodic orbit* γ *with a period T*. To each point $\mathbf{x} \in \gamma$ we can uniquely assign a phase φ defined on a circle S^1 of length 1 in such a way that

$$\frac{d\varphi}{dt} = \frac{1}{T}, \qquad (6.1)$$

i.e. one circulation of the phase point $\mathbf{x}$ around γ implies one circulation of the phase φ around S^1. Suppose that the periodically oscillating system is perturbed by a pulse perturbation applied at a particular phase φ into the point $\mathbf{x}_a \notin \gamma$ and then returns, that is $g^t(\mathbf{x}_a) \to \gamma$. The asymptotic oscillations are shifted by

$\bar{\varphi} - \varphi$ where the new phase $\bar{\varphi}$ is by definition equal to the phase $\varphi(\mathbf{x})$ of a point $\mathbf{x} \in \gamma$ such that

$$\lim_{t \to \infty} d[\mathbf{g}^t(\mathbf{x}_a), \, \mathbf{g}^t(\mathbf{x})] = 0 , \qquad (6.2)$$

where $d[\cdot, \cdot]$ denotes the Euclidean distance in R^n. To each phase φ at the instant of the pulse we can assign an asymptotic phase $\bar{\varphi}$. The function $\bar{\varphi} = \vartheta(\varphi)$ is called the PTC. The difference $\vartheta(\varphi) - \varphi$ is called the *phase resetting curve* (PRC). To measure the phase it is necessary to choose a point $\mathbf{x}_0 \in \gamma$ corresponding to a *reference event* and set there $\varphi = 0$ (mod 1). First we shall present examples of experimental results on responses to single pulse perturbations in the form of PTCs and then examples of dynamic regimes observed in the periodically forced system. The use of PTCs for the explanation of periodic and chaotic dynamics will be then discussed.

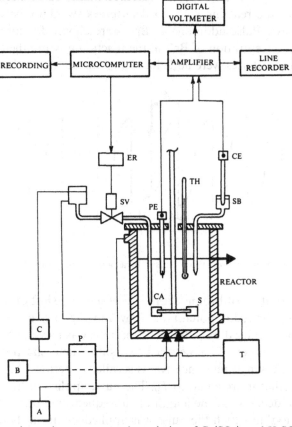

Fig. 6.1. Experimental arrangement: A – solution of $Ce(SO_4)_2$ and H_2SO_4; B – malonic acid and $KBrO_3$; C – KBr; P – pump; T – thermostat; SV – solenoid valve; ER – electromagnetic relay; CA – capillary; PE – platinum electrode; CE – calomel electrode; SB – salt bridge; TH – thermometer; S – stirrer.

6.1.1 Single pulse experiments – phase transitions curves

The flow-through stirred tank reactor used in the experiments is shown in Fig. 6.1. The glass reaction cell contains a platinum electrode and a capillary for introduction of the pulse perturbation. All experiments were conducted at the constant temperature 35.0 ± 0.1 °C. The redox potential corresponding to the change of $\ln (Ce^{4+}/Ce^{3+})$ was measured by platinum and calomel electrodes. The solution of 0.001 M $Ce(SO_4)_2$, 1.5 M H_2SO_4, 0.05 M $KBrO_3$ and 0.05 M malonic acid was fed into the reactor. The residence time was 18 min. A microcomputer was used for the control of reaction conditions, pulse addition and data acquisition.

The BZ reaction exhibits, under the above conditions (temperature, concentrations, residence time), strongly relaxational oscillations of basic period $T_B \approx 50$ s. Bromide ions (Br^-) form an important reaction intermediate and the level of Br^- concentration in the reacting solution determines switching between important reaction pathways. Pulse additions of Br^- were chosen for forcing; immediate increase of the concentration of Br^- in the reaction mixture then determines the pulse amplitude A. Pulses were added and well mixed in less than 1 s.

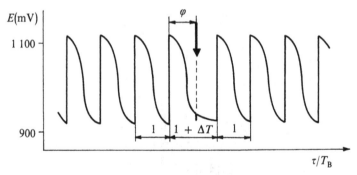

Fig. 6.2. Phase shift of oscillations caused by a pulse of Br^- ions, $E \sim \ln Ce^{4+}/Ce^{3+}$.

Single pulse addition of bromide ions causes a phase shift of oscillations. The time variation of the redox potential is shown in Fig. 6.2. The measured phase shift of oscillations caused by pulse perturbations was used for the construction of PTCs. The time interval δT between two subsequent sharp increases of the redox potential chosen as reference events is called a 'period'. All 'periods' of the unperturbed oscillator are equal to T_B. The pulse applied at the phase φ causes an increase or a decrease in the length of subsequent 'periods'. Only the length of the first 'period' in which the pulse was applied significantly differs from T_B under the experimental conditions used[6.16]. This rapid return to periodic oscillations suggests that the system is strongly dissipative. However, under different experimental conditions the pulse affects several 'periods'[6.53].

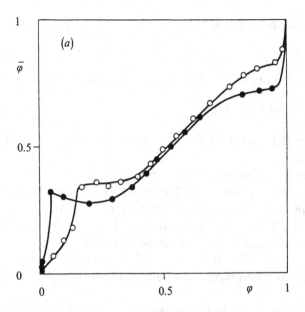

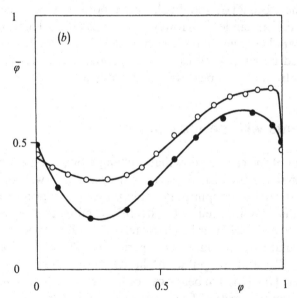

Fig. 6.3. Experimentally measured phase transition curves (PTC); reaction conditions –
see text: (*a*) PTC of type '1', concentration of Br$^-$ ions $A = 1.43 \times 10^{-5}$ M – O, $A =$
5×10^{-5} M – ●; (*b*) PTC of type '0', concentration of Br$^-$ ions $A =$
$= 1.4 \times 10^{-4}$ M – O, $A = 5 \times 10^{-4}$ M – ●.

Defining the relative change of the first 'period',

$$\Delta T = (\delta T - T_B)/T_B ,\qquad (6.3)$$

the PTC can be written as (note that $-\Delta T$ is equivalent to the PRC)

$$\bar{\varphi} = \varphi - \Delta T(\varphi) \;(\text{mod } 1) = \vartheta(\varphi). \qquad (6.4)$$

The shape of the PTC is dependent on the amount of Br^- added, i.e. on the amplitude A of the pulse.

Very small phase shifts were observed for low amplitudes $(A \lesssim 1 \times 10^{-6}\,\text{M})$, hence $\bar{\varphi} \approx \varphi$ in this case. Pulse additions with amplitudes $1 \times 10^{-6}\,\text{M} \lesssim A \lesssim 1 \times 10^{-5}\,\text{M}$ caused phase shifts with nondecreasing PTCs of type $1^{6.58}$, (i.e. of topological degree 1, see Section 3.3.6). When $A \gtrsim 1 \times 10^{-5}\,\text{M}$, the PTCs are nonmonotonic. A transition of type 1 PTC to the type 0 PTC occurs between the pulse amplitudes $5 \times 10^{-5}\,\text{M}$ and $1 \times 10^{-4}\,\text{M}$. Several examples are shown in Fig. 6.3, while more complete experimental results are given in Refs 6.16, 6.17, 6.37.

Regression polynomials were used to fit the data for numerical simulations. It was considered that the BZ reaction system can be described by a smooth dynamical system and therefore PTCs should also be smooth provided they are defined at every point of the circle$^{6.20}$. However, the relaxational nature of oscillations can lead to rapid changes of $d\vartheta/d\varphi$, see Fig. 6.3. Thus for simplicity, the PTCs were composed of several regression polynomials such that the regression curves were continuous and piecewise differentiable on the circle.

6.1.2 Experiments with periodic pulses

The natural rhythm of the chemical oscillator changes by periodic forcing with Br^-. The forced system can also be seen as a system of two coupled oscillators with a one-directional coupling. Periodic pulses correspond to the driving oscillator and the reaction cell with the BZ reaction to the driven oscillator. Let the unperturbed oscillator be characterized by the period T_B and the sequence of the external periodic pulses by a period T_F. The dimensionless forcing period $T = T_F/T_B$ and the pulse amplitude A are parameters that specify the overall system. The dynamic behaviour of the periodically perturbed system was analysed by power spectra of the sequence $\{\delta T_k\}$ of 'periods'. The sequence was considered periodic with period N and *phase-synchronized* with the external forcing in the ratio $N : M$ if

$$M \cdot T_F = \sum_{k=1}^{N} \delta T_k \text{ and } \delta T_k \approx \delta T_{k+N} . \qquad (6.5)$$

It follows from the single pulse experiments that the system relaxes after a pulse perturbation in a time less than T_B. That is if we choose $T \gtrsim 1$ then two sequences of 'periods' observed at two values of T differing by an integer will be identical when the 'periods' with no external pulse are deleted. Hence it is sufficient to vary T within a unit interval.

The experiments are inevitably subjected to a certain level of experimental noise, reflecting nonisotropicity of the hydrodynamic flow (imperfect mixing) and fluctuations of temperature and flow-rates. The noise-induced fluctuations of 'periods' in the absence of an external forcing were never higher than 1.5 %.

Table 6.1 *Comparison of experimental results with simulation for* $A = 5 \times 10^{-5}$ M

T	Experiment		Eq. (6.7)		Eq. (6.12)	
	ϱ_{exp}	regime	ϱ	regime	ϱ	regime
2.10	2.00	2 : 1	2.00	2 : 1	2.00	2 : 1
2.17	2.06	aper.	2.00	2 : 1	2.02	aper.
2.27	2.23	aper.	2.25	9 : 4	2.24	aper.
2.31	2.33	7 : 3	2.33	7 : 3	2.33	7 : 3
2.34	2.33	7 : 3	2.33	7 : 3	2.33	7 : 3
2.38	2.44	aper.	2.40	aper.	2.40	aper.
2.45	2.50	5 : 2	2.50	5 : 2	2.50	5 : 2
2.51	2.50	5 : 2	2.50	5 : 2	2.50	5 : 2
2.61	2.50	5 : 2	2.50	5 : 2	2.50	5 : 2
2.64	2.60	aper.	2.60	13 : 5	2.60	aper.
2.67	2.62	aper.	2.66	8 : 3	2.65	aper.
2.74	3.00	3 : 1	2.75	11 : 4	2.75	aper.
2.77	3.00	3 : 1	3.00	3 : 1	3.00	3 : 1
2.98	3.00	3 : 1	3.00	3 : 1	3.00	3 : 1

Table 6.2. *Comparison of experimental results with simulation for* $A = 1.43 \times 10^{-3}$ M

T	Experiment		Eq. (6.7)		Eq. (6.12)	
	ϱ_{exp}	regime	ϱ	regime	ϱ	regime
2.07	2.00	2 : 1	2.00	2 : 1	2.00	2 : 1
2.14	2.00	2 : 1	2.00	2 : 1	2.00	2 : 1
2.23	2.00	2 : 1	2.00	2 : 1	2.00	2 : 1
2.30	2.00	2 : 1	2.00	2 : 1	2.00	2 : 1
2.38	2.00	2 : 1	2.00	2 : 1	2.00	2 : 1
2.65	2.40	aper.	2.00	4 : 2	2.42	aper.
2.85	2.50	5 : 2	2.50	5 : 2	2.50	5 : 2
3.00	3.00	3 : 1	3.00	3 : 1	3.00	3 : 1

Examples of observed dynamic regimes at several values of T, classified as $N : M$ phase-synchronized or aperiodic are given for the forcing amplitude $A = = 5 \times 10^{-5}$ M (corresponding to the PTC of type 1) in Table 6.1 and for the amplitude $A = 1.43 \times 10^{-3}$ M (corresponding to the PTC of type 0) in Table 6.2 (for definition of ϱ and $\varrho_{\exp}$ cf. Eqs (6.9) and (6.10), respectively). We can infer from the tables that the aperiodic regimes are more frequent at the lower amplitude while the periodic regimes are common for both levels of amplitude.

Periodic regimes with values of $M = 1, 2$ and 3 were observed for the lower forcing amplitude. The fact that periodic regimes with higher values of M were not observed follows from both the effects of noise and the very narrow regions of values of T where such regimes exist. When the higher forcing amplitude was

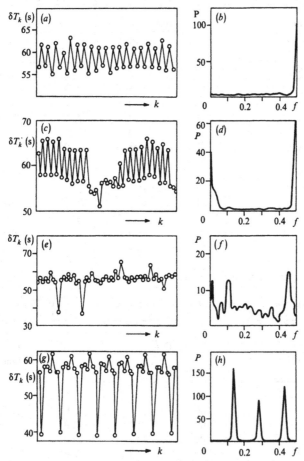

Fig. 6.4. Periodic perturbations, sequence of periods δT_k with corresponding power spectra, $A = 5 \times 10^{-5}$ M: (a), (b) $T = 2.10$, periodic regime 2 : 1; (c), (d) $T = 2.17$, aperiodic regime – intermittency; (e), (f) $T = 2.27$, aperiodic regime; (g), (h) $T = 2.31$, periodic regime 7 : 3.

used, periodic regimes with the values of $M = 1$ and 2 were observed. The results of numerical simulations confirm that regimes with higher values of M do not exist for this forcing amplitude, see Section 6.1.5.

Examples of the sequences $\{\delta T_k\}$ and of the power spectra corresponding to $A = 5 \times 10^{-5}$ M are given in Figs. 6.4(a)–(h). At $T = 2.1$ (Fig. 6.4(a), (b)) we observe a regime phase-synchronized at $N : M = 2 : 1$. The effects of experimental noise may be seen in both the sequence $\{\delta T_k\}$ and the power spectra. When $T = 2.17$ (Fig. 6.4(c), (d)) the behaviour is typically intermittent. The laminar phase corresponds to the phase synchronization at $N : M = 2 : 1$. The alternation of the laminar and the turbulent phase causes the appearance of a marked low-frenquency component in the power spectrum. Chaotic oscillations with broad-band noise in the power spectrum are observed at $T = 2.27$ (Fig. 6.4(e), (f)). Finally the phase-synchronized regime $(N : M = 7 : 3)$ is observed at $T = 2.31$ (Fig. 6.4(g), (h)).

The situation is simpler for $A = 1.43 \times 10^{-3}$ M. An aperiodic regime was observed at the transition from the 2 : 1 regime to the 5 :2 regime, see Table 6.2, while all other regimes are periodic.

6.1.3 Geometrical models for PTCs

The trajectories constructed from experimentally observed aperiodic oscillations of the unperturbed BZ system can be located in three-dimensional space without mutual intersections[5.235]. This fact indicates the presence of low-dimensional deterministic chaos and points to the existence of three most significant modes in the dynamics of the reaction system studied. For the sake of simplicity let us start with the assumption that the unperturbed experimental system is described by the flow $g^t(\mathbf{x})$, $\mathbf{x} \in \mathbf{R}^3$. Let this system have a stable periodic orbit γ of period T_B with basin of attraction W. The perturbations will be modelled as pulses that will shift $\mathbf{x}$ jumpwise off γ.

If one of coordinate axes in $\mathbf{R}^3$ corresponds to the concentration of bromide ions, then each pulse of amplitude A can be visualized as a shift of size A parallel to this axis. When the state space point is not removed from W due to the pulse, then almost all trajectories tend back to γ and the system returns to the periodic regime. Only a set of zero volume will not be attracted to γ. Let us take the simplest example of such a set, a saddle stationary point $\mathbf{S}$ and its stable manifold W_S^s of dimension one (i.e. its codimension is 2), see Fig. 6.5(a), (b). Winfree[6.58] calls this set a *'phaseless set'*, because a point located in this set never returns to γ (it approaches $\mathbf{S}$) and thus the asymptotic phase is not defined. The above assumptions are satisfied, for example, by the model *Oregonator*, one of the simplest and until now most often used models of the BZ reaction[6.19]. A generalization to more-dimensional models is straightforward under the assumption that the codimension of W_S^s is equal to two.

The perturbations shift every point on γ in the direction parallel to the Br$^-$ axis by the amplitude A and thus give rise to a shifted cycle $\bar{\gamma}$. Provided all points on the shifted cycle $\bar{\gamma}$ are attracted to γ, i.e. when $g^t(\bar{\gamma}) \rightarrow \gamma$, two qualitatively

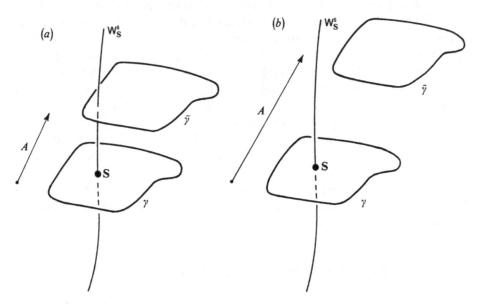

Fig. 6.5. Pulse perturbation in the phase space R^3, schematically: (a) weak perturbation, stable manifold W$_S^s$ passes 'inside' the shifted periodic orbit $\bar{\gamma}$ – PTC of the type '1'; (b) strong perturbation – W$_S^s$ is located 'outside' $\bar{\gamma}$ – PTC of the type '0'.

different situations can occur. When the pulse amplitude A is low, then the phaseless set passes through the 'inside' of the shifted cycle $\bar{\gamma}$ see Fig. 6.5(a), when A is high, it passes 'outside', see Fig. 6.5 (b).

The asymptotic phase $\bar{\varphi}$ of the points on $\bar{\gamma}$ is given according to Eq. (6.2). The PTC, $\bar{\varphi} = \vartheta(\varphi)$, maps the circle to itself. The *topological degree* of ϑ describes how many times $\bar{\varphi}$ winds the circle in the course of one revolution of φ. Because $g^t(\bar{\gamma})$ is at a given time a closed curve, which cannot pass through the phaseless set, the topological degree of ϑ for lower values of amplitudes will be equal to one (PTC of type 1) and for high amplitudes will be equal to zero (PTC of type 0); this is in agreement with the experimental observations.

The transition from the PTC of type 1 to the PTC of type 0 when the amplitude A increases is indicated by a singular state, where $\bar{\gamma} \cap (S \cup W_S^s)$ has a singular point at which the asymptotic phase is not defined. The existence of the singularity is reflected in the screw-like character of the graph of $\bar{\varphi} = \vartheta(\varphi, A)$ in $(\bar{\varphi}, \varphi, A)$ space[6.58]. A simple graphical representation of this surface based on equiscalar curves of the function $\bar{\varphi} = \vartheta(\varphi, A)$ determined from experiments is given in Fig. 6.6. The curves $\bar{\varphi} = $ const wind on the singular point and a screw-like surface is thus formed. A critical value of the forcing amplitude at which the topological character of the PTC changes can be estimated as $A^* \approx \approx 7.0 \times \times 10^{-5}$ M. The corresponding value of the critical phase is $\varphi^* \approx 0$ (mod. 1).

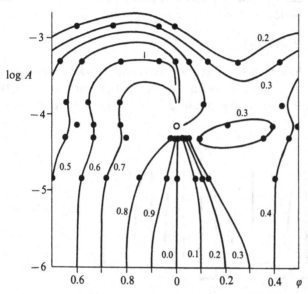

Fig. 6.6. Equiscalar curves of the function $\bar{\varphi} = \vartheta(\varphi, A)$, experimental results for the pulse perturbations of the BZ reaction.

The oscillations could, theoretically, be stopped by applying a pulse at the critical amplitude and phase. However, this was not observed in the experiments. A natural explanation is that the phaseless set has zero volume in the state space and thus the probability of hitting it is also zero. The stopping of oscillations could be realized, for example, in cases where the limit cycle oscillations would coexist with a stable steady state[6.14, 6.15, 6.38] (which is not so in this case).

The system, perturbed at values of amplitude and phase close to critical, approaches the stationary point **S** along its stable manifold but then is repelled away from **S**. The approach of the state point to γ may last a long time; it would then cause large time shifts between the oscillations before and after the perturbation. This phenomenon was not observed in the experiments. We can explain it in two ways: either the interval of amplitudes where this effect is observable is very narrow and was missed in the experiments, or an accelerated repulsion from the stationary point is caused by random noise inherent in the experimental system.

6.1.4 Iterated PTCs – model for periodic forcing

The standard method of description of the dynamics of an oscillating system periodically forced by short pulses makes use of PTCs. The construction of a simple model in the form

$$\phi_{k+1} = T + \phi_k + \Delta T(\phi_k), \qquad (6.6)$$

where $T = T_F/T_B$ is dimensionless forcing period and $\Delta T(\phi_k)$ is dimensionless change of the 'period' affected by the pulse, see Eq. (6.3), is shown schematically in Fig. 6.7. The integer part of the phase ϕ in Eq. (6.6) counts the number

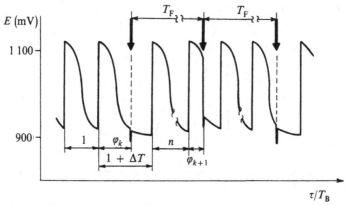

Fig. 6.7. Construction of the 1D map.

of oscillations of the driven oscillator starting from a chosen reference event. The difference equation (6.6) has the real axis as its state space. If we set $\varphi = \phi \pmod 1$, $\vartheta = \phi + \Delta T(\phi) \pmod 1$, we can write down the corresponding equation on a circle,

$$\varphi_{k+1} = T + \vartheta_A(\varphi_k) \pmod 1 = f(\varphi_k). \qquad (6.7)$$

The experimentally determined function ϑ_A is dependent on the amplitude A but not on the period T. The dynamics of one-dimensional maps is described in Section 3.3. In particular, an orbit $\gamma \equiv \{f^k(\varphi_0)\}$ of Eq. (6.7) can be characterized by its Lyapunov exponent (LE) λ,

$$\lambda = \lim_{k \to \infty} \frac{1}{k} \log \left| \frac{\mathrm{d}f^k(\varphi_0)}{\mathrm{d}\varphi} \right|. \qquad (6.8)$$

The attractors of one-dimensional maps are in general chaotic ($\lambda > 0$), periodic ($\lambda < 0$) or neutral ($\lambda = 0$). For circle maps a special case of the neutral attractor, called a quasiperiodic attractor, coincides with the circle.

If the orbit γ is periodic of period q, then its LE is

$$\lambda = \frac{1}{q} \log |\sigma| \, ,$$

where $\sigma = [\mathrm{d} f^q(\varphi_0)]/\mathrm{d}\varphi$ is the multiplier of the orbit. If $|\sigma| < 1$ then γ is locally stable, $\sigma = 1$ and $\sigma = -1$ are bifurcation values. The bifurcation structure of degree one circle maps was described in Sections 3.3.6 and 3.3.8.

Let d be the topological degree of f, see Eq. (3.29). A convenient characteristic of an orbit γ of f with $d = 1$ is the *rotation number*

$$\varrho(\varphi, f) = \lim_{k \to \infty} \frac{F^k(\phi) - \phi}{k} \, , \tag{6.9}$$

where F is the lifting of f, see Eq. (3.30).

This limit, if exists, depends on φ and f. If f is continuous with a continuous inverse (a homeomorphism) then ϱ does not depend on φ. Different orbits of a nonmonotonic map f may have different ϱs which are generally contained in a closed interval called the *rotation interval* [6.6, 3.61]. There are orbits for which the limit set of the sequence $(F^k(\phi) - \phi)/k$ is not a unique point but rather a closed subinterval of the rotation interval [6.6]. Thus the dynamics of f with $d = 1$ may be very complicated and, in particular, multiple attractors can be expected.

The rotation number can be used for a characterization of the phase-synchronized periodic orbits. If a q-periodic orbit of Eq. (6.7) exists then

$$F^q(\phi) = \phi + p$$

for all points on the orbit and p is an integer. The corresponding rotation number is $\varrho = p/q$. The experimental interpretation is that during q pulses the system oscillates p-times; see Section 6.1.2 for a comparison with the forced **BZ** oscillator where $N \equiv p$ and $M \equiv q$. If ϱ is irrational and f is a homeomorphism then the dynamics is quasiperiodic [6.29].

Unfortunately, if f is of degree zero which corresponds to the PTC of type 0 then Eq. (6.9) always gives $\varrho = 0$. In addition, if $d > 1$ or $d < 0$, ϱ tends to infinity. Thus, ϱ is useful only for maps of degree one. However, when describing the experimental regimes we can define ϱ_{exp} analogously to ϱ, independent of the type of the PTC,

$$\varrho_{exp} = \frac{N_e}{N_F} \tag{6.10}$$

where N_e is the number of reference events, and N_F is the number of forcing pulses within an experimental run. To be able to compare ϱ_{exp} with an appropriate theoretical counterpart we need a definition of the rotation number for maps of degree different from one.

The idea is to change the lifting function F so that the limit in (6.9) will become finite and nonzero regardless of d, see Refs 6.8, 6.17. Let a critical phase $\varphi_c = \phi_c \pmod 1$ be given; ϕ_c is unique up to an integer which can be arbitrarily chosen, e.g. so that $0 \leqq \phi_c < 1$. Restricting F to the interval $[\phi_c, \phi_c + 1)$ we can define a *modified lifting* $\bar{F}$ as follows:

$$\bar{F}(\phi) = F(\bar{\phi}) + n, \qquad \phi = \bar{\phi} + n, \qquad \bar{\phi} \in [\phi_c, \phi_c + 1),$$

$$n = 0, \pm 1, \pm 2, \ldots . \qquad (6.11)$$

By definition, $\bar{F}(\phi + 1) = \bar{F}(\phi) + 1$, and the rotation number according to (6.9) with F replaced by $\bar{F}$ will become finite and generally nonzero for any topological degree of f. As before, ϱ depends on φ, f and in addition on φ_c. A convenient property of $\bar{F}$ is that the rotation number of a q-periodic orbit of f is rational for any degree, $\varrho = p/q$.

We note that $\bar{F}$ is equal to F for $d = 1$ but if $d \neq 1$, $\bar{F}$ has discontinuities of width $d - 1$ at integer multiples of ϕ_c, i.e.

$$\bar{F}(n\phi_c +) = \bar{F}(n\phi_c -) + d - 1, \qquad n = 0, \pm 1, \pm 2, \ldots .$$

The difference between F and $\bar{F}$ for $d = 0$ is shown in Fig. 6.8. The discontinuity of $\bar{F}$ for $d \neq 1$ implies that given a one-parameter family of periodic orbits, the rotation number as a function of the parameter is not a constant but rather a piecewise constant function with discontinuous jumps, see Guevara and Glass[6.26], Keener and Glass[6.33] and Belair and Glass[6.8].

We consider now the case of $d = 0$ and $d = 1$ which is relevant to our experimental observations. The map f depends on two parameters, the period T

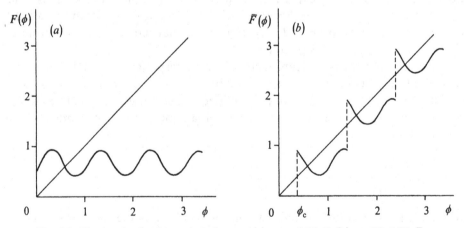

Fig. 6.8. The graph of a degree 0 circle map: (a) normal lift F; (b) modified lift $\bar{F}$

and the amplitude A of the pulses. It is well known that for maps of degree one the set of points in the parameter plane corresponding to a periodic point of f with a given rotation number form a cusp-shaped region called the Arnold tongue[2,4]. Arnold tongues may overlap if f is noninvertible, giving rise to rotation intervals. For A fixed and f an invertible map the dependence of ϱ on T is a stepwise nondecreasing continuous function called the *devil's staircase*[6,29]. When f is noninvertible then the end points of the rotation interval depend on T in a similar way[3,61]. These functions are constant if ϱ is rational and strictly increasing if ϱ is irrational.

For maps of degree zero the dependence of ϱ (or of the end points of the rotation set) on T is discontinuous. Moreover, the rotation number is dependent on the discontinuity point φ_c which in turn is given by the original continuous time system. Experiments suggest that the choice $\varphi_c = 0 \pmod 1$ is appropriate to put correspondence between ϱ_{exp} obtained from (6.10) and its theoretical counterpart ϱ according to (6.9) and (6.11) for PTCs of both types.

6.1.5 Numerical computations and comparison with experiments

Every experimental run was characterized by the sequence of 'periods' $\{\delta T_k\}$ and its power spectrum, by the sequence of phases $\{\varphi_k\}$ and by the value of the rotation number ϱ_{exp}. The experimental results were compared with the mathematical model. To distinguish between periodic and aperiodic regimes in the model we also computed LE. The effect of noise on the dynamics of the forced system was modelled with a simple Langevin type equation

$$\varphi_{k+1} = f(\varphi_k) + \varepsilon X_k,\tag{6.12}$$

where X is a random number with a uniform distribution on the interval $[-1, 1]$ and ε is the noise amplitude. A 'periodic orbit' of (6.12) with period q will be formed by q separate intervals; the phase variable will cycle regularly among these intervals and the rotation number will be the same as in the deterministic limit.

Comparison of the rotation numbers determined from the experiments with those from the deterministic and stochastic models is made in Table 6.1 for $A = 5 \times 10^{-5}$ M and in Table 6.2 for $A = 1.43 \times 10^{-3}$ M. It is evident from the tables that in several cases where the experimental regime was aperiodic, the deterministic model predicted a periodic regime and only the introduction of noise brought agreement with the experimental data. This is an example of noise-induced aperiodicity. Thus the prediction power of the deterministic model was improved by the introduction of noise. The dependence of rotation numbers and LEs on T for both deterministic and stochastic models are given in Fig. 6.9(a) to (d) for $A = 5 \times 10^{-5}$ M and in Fig. 10(a) to (d) for $A = 1.43 \times \times 10^{-3}$ M.

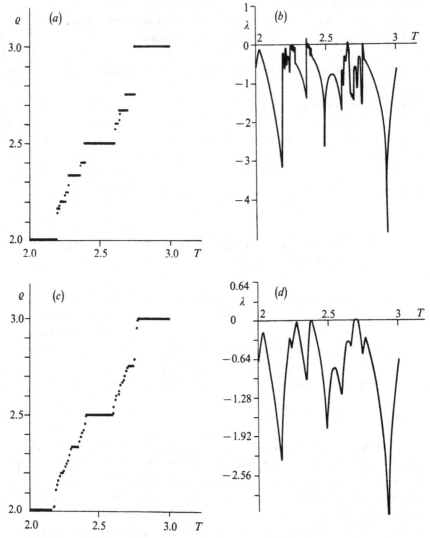

Fig. 6.9. Rotation number ϱ and Lyapunov exponent λ in dependence on T, $A = 5 \times 10^{-5}$ M: (a), (b) Eq. (6.7); (c), (d) Eq. (6.12), $\varepsilon = 0.02$.

A multiplicity was observed when $A = 5 \times 10^{-5}$ M. However, the rotation numbers as well as LEs computed for different orbits reveal observable differences only on a much finer scale than that used in Fig. 6.9. There was no multiple attractor behaviour observed when $A = 1.43 \times 10^{-3}$ M, see Fig. 6.10.

In the first case (the corresponding PTC is of type 1) there exist many different phase-synchronized regimes seen as intervals with constant value of ϱ in Fig. 6.9(a). Chaotic regimes indicated by a positive value of λ (see

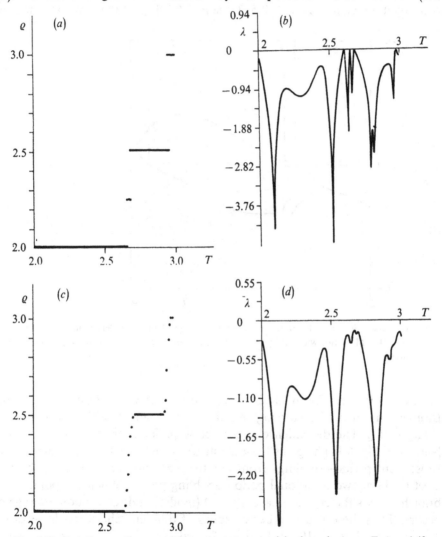

Fig. 6.10. Rotation number ϱ and Lyapunov exponent λ in dependence on T, $A = 1.43 \times$ $\times\ 10^{-3}$ M: (a), (b) Eq. (6.7); (c), (d) Eq. (6.12), $\varepsilon = 0.02$.

Fig. 6.9(b)) are scattered among resonances. The random noise causes narrowing of the resonance windows (see Fig. 6.9(c)); at the same time the values of λ are shifted to negative ones (see Fig. 6.9(d)), particularly in cases where λ is positive in the deterministic limit. This interesting behaviour may be due to the presence of nearby resonances as noted by Kaneko[6.30].

In the second case (corresponding PTCs are of type zero) the rotation number can vary discontinously along the branch of periodic solutions; this follows from the discontinuity of the map $\bar{F}$. There are only four possible values of the rotation number, $\varrho = 2, 2.25, 2.5, 3$, see Fig. 6.10($a$). At the same time, four zero

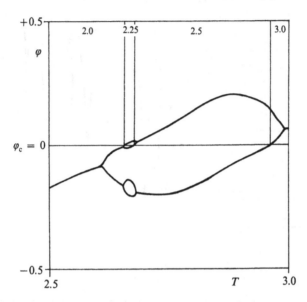

Fig. 6.11. Solution diagram generated by Eq. (6.7) in dependence on T, $A = 1.43 \times$ $\times 10^{-3}$ M; zones with different rotation numbers are marked in the upper part of the figure.

values of LEs, see Fig. 6.10(b), indicate points at which bifurcations appear. Bifurcation values of the forcing period T do not coincide with points of discrete changes of ϱ. The dependence of attracting periodic orbits of Eq. (6.7) on T (shown in Fig. 6.11) helps us to understand the actual situation. There are two subsequent period-doublings from left to right and the bifurcating branches meet again at two additional period-doubling points. When the point on any branch crosses the critical phase $\varphi_c = 0 \pmod 1$ a discontinuous change of ϱ occurs. Thus there is a number of zones of different phase synchronizations, 2: 1, 4 :2, 8 : 4, 9 : 4, 10 : 4, 5 : 2, 6 : 2 and 3 : 1.

The addition of random noise decreases the width of synchronization zones and also decreases the values of LEs as in the previous case. The lack of phase synchronization is observed close to bifurcation points as expected. The rotation number $\varrho = 2.25$ which corresponds to the synchronization 9 : 4 disappears due to the effect of noise (see Figs. 6.10 (a), (c)). These observations are essentially consistent with experimental results, which give regimes with period one and two and an aperiodic regime in between, see Table 6.2.

The description of both chemical and biological systems forced periodically by short pulses via Eq. (6.7) may be inadequate for two reasons. First, a slow relaxation to the limit cycle after a perturbation causes the next pulse to be realized at the moment when the oscillator is still off the limit cycle. Thus the necessary assumption of the validity of a one-dimensional model is a sufficiently fast relaxation back to the limit cycle. The accuracy of the one-dimensional approximation increases with the length of the forcing period. When the forcing period is decreased below a certain critical value the accuracy of the model sharply decreases and additional modes have to be included. The stronger the dissipation, the lower the critical value of the forcing period. The return to the original limit cycle is usually very fast in the case of relaxation oscillations but a slower return may occur as well[6.53].

Second, the presence of random noise in principle requires a stochastic description. Thus the deterministic model (6.7) does not reflect dynamics of a 'noisy' experimental system. In this case a simple stochastic model (6.12) can be used provided the stochastic fluctuations are not very large.

6.2 Coupled reaction–diffusion cells

Reaction cells with mutual mass exchange are standard model systems for the study of transformation and transport processes in living cells, tissues, networks of neurons, various compartmental representations of physiological systems, ecological models, and of all forms of chemical, biochemical and biological reactors[6.3, 6.5, 6.9, 6.10, 6.24, 6.40, 6.42, 6.44, 6.45, 6.50, 6.52, 6.54, 6.56, 6.57]. A system of well-stirred flow-through reaction cells with mutual mass exchange through common walls have also been used for experimental studies of various nonhomogeneous steady states and time-dependent regimes – dissipative structures. Coexisting nonhomogeneous steady states in two cells[5.176, 6.14], combinations of nonhomogeneous steady states (Turing structures) in a system of up to seven cells[5.174] as well as periodic and aperiodic time-dependent regimes in two coupled cells have been reported[6.48, 6.49, 6.51].

In this section we discuss in detail results of systematic studies of inter-relations of steady states, periodic regimes, quasiperiodic regimes and various forms of chaotic behaviour in a simple model of two coupled reaction-diffusion cells with the well-known Brusselator kinetic scheme. Behaviour of large linear and cyclic arrays of coupled cells are then discussed briefly. Here we show how the standard numerical techniques based on the continuation methods discussed in Chapter 4 can be used in the analysis of dynamical systems of two coupled reaction–diffusion cells.

6.2.1 Model of two coupled reaction–diffusion cells

The model of two well mixed reaction cells with linear diffusion coupling and the *Brusselator reaction kinetic scheme* is used as a standard model system in studies of dissipative structures in nonlinear chemical systems[6.36, 6.42] and similarly as the *Lorenz model* serves for the studies of chaotic behaviour in simple models of turbulence[1.44].

In the reaction scheme the initial components A and B are transformed into products D and E via the reaction intermediates X and Y

$$A \xrightarrow{k_1} X,$$

$$B + X \xrightarrow{k_2} Y + D,$$

$$2X + Y \xrightarrow{k_3} 3X,$$

$$X \xrightarrow{k_4} E.$$

When the concentrations A, B are considered constant (the components are supplied from the environment) and the concentrations of all components are made dimensionless so that they include the rate constants, then the model (see Fig. 6.12) can be written in the form

$$\frac{dX}{dt} = v(X),\tag{6.13a}$$

$$X = \begin{bmatrix} x_1 \\ y_1 \\ x_2 \\ y_2 \end{bmatrix}, \quad v(X) = \begin{bmatrix} A - (B+1)\,x_1 + x_1^2 y_1 + D_1(x_2 - x_1) \\ Bx_1 - x_1^2 y_1 \hspace{2.2cm} + D_2(y_2 - y_1) \\ A - (B+1)\,x_2 + x_2^2 y_2 + D_1(x_1 - x_2) \\ Bx_2 - x_2^2 y_2 \hspace{2.2cm} + D_2(y_1 - y_2) \end{bmatrix}.$$

$$\tag{6.13b}$$

Here A, B are constant parameters, x_i, $y_i (i = 1, 2)$ are dimensionless concentrations of the reaction intermediates X and Y in the first and second cells. The parameters D_1 and D_2 define the intensity of mass exchange between the cells. In the numerical computations we shall set $A = 2$, $D_1/D_2 = q = 0.1$ and study solutions of the system (6.13) in dependence on two parameters $B > 0$, $D_1 \geq$ ≥ 0. In the following, the system (6.13) with $D_1 = D_2 = 0$ and $x_1 = x_2$, $y_1 = y_2$ will be called a *decoupled system*.

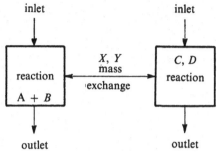

Fig. 6.12. Two reaction-diffusion cells with mutual mass exchange.

The choice of two identical cells (with respect to values of the parameters in the kinetic model) is reflected in the inherent symmetry of the vector field v:

$$v(S(\mathbf{X})) = S(v(\mathbf{X})), \qquad S^2 = \text{id}, \tag{6.14}$$

where S can be represented by a permutation matrix

$$S = \begin{bmatrix} 0 & 0 & 1 & 0 \\ 0 & 0 & 0 & 1 \\ 1 & 0 & 0 & 0 \\ 0 & 1 & 0 & 0 \end{bmatrix}, \tag{6.15}$$

exchanging the first component of the vector $\mathbf{X}$ with the third one and the second component with the fourth one. This corresponds to the exchange of cells. Hence, the orbits of (6.13) are mutually symmetric with respect to the symmetry plane Δ in the phase space $\mathbb{R}^4$, given by

$$\Delta = \{\mathbf{X} \in \mathbb{R}^4; \ x_1 = x_2, y_1 = y_2\}.$$

The orbits located in Δ will be called *homogeneous*, the others *nonhomogeneous*.
The following qualitatively different types of orbits exist:
(a) two distinct nonhomogeneous *asymmetric orbits* Γ and $\bar{\Gamma}$ such that $\bar{\Gamma} =$ $= S(\Gamma)$ and $\Gamma = S(\bar{\Gamma})$;
(b) a single orbit Γ invariant with respect to S, where two possibilities arise:
  (i) a nonhomogeneous Δ-*symmetric orbit*, i. e. $\Gamma = S(\Gamma)$, $\Gamma \not\subset \Delta$ (orbits of this type cannot be steady state solutions)
  (ii) a homogeneous orbit, i.e. $\Gamma = S(\Gamma)$, $\Gamma \subset \Delta$ (orbits of this type are identical with those of the decoupled system (6.13).

6.2.2 Stationary solutions

Stationary solutions satisfy

$$v(\mathbf{X}) = 0 . \tag{6.16}$$

We have

$$x_1 + x_2 = 2A , \tag{6.17a}$$

and for

$$u = x_1 - x_2$$

we obtain either

$$u = 0 , \tag{6.17b}$$

or u satisfies a biquadratic equation

$$q\omega u^4 + 4(B - 2A^2 q\omega + 2D_1\omega) u^2 + 16(A^4 q\omega - A^2 B + 2A^2 D_1\omega) = 0 , \tag{6.17c}$$

where $\omega = (D_1 + 0.5)/D_1$. Then

$$y_1 = \frac{2A(2B + q\omega u^2)}{4A^2 + u^2} - \frac{q\omega u}{2} , \tag{6.17d}$$

$$y_2 = \frac{2A(2B + q\omega u^2)}{4A^2 + u^2} + \frac{q\omega u}{2} . \tag{6.17e}$$

It follows from Eqs. (6.17b) and (6.17c) that either one, three or five steady state solutions exist. The homogeneous solution S_{H}: $x_1 = x_2 = A$; $y_1 = y_2 = = B/A$ lies in $\varDelta$ and exists for all values of parameters. The nonhomogeneous solutions can be found in asymmetric pairs S_{N}^1, $\bar{S}_{\mathrm{N}}^1$ and S_{N}^2, $\bar{S}_{\mathrm{N}}^2$, see Table 6.3. All four homogeneous solutions exist for parameter values satisfying

$$\omega(4A\sqrt{D_1 q} - 2D_1) < B < \omega(A^2 q + 2D_1), \qquad 4D_1 < A^2 q . \tag{6.18a}$$

Two nonhomogeneous solutions exist for

$$B > \omega(A^2 q + 2D_1) . \tag{6.18}$$

Table 6.3 *Types of stationary and periodic solutions*

Symbol	Type of solution	Homogeneity	Symmetry	Solution symmetric to the original one	Mutual phase relations in two cells
S_H	stationary	homogeneous	$S_H \subset \varDelta$	$S_H = S(S_H)$	–
S_N	stationary	nonhomogeneous	asymmetric	$\bar{S}_N = S(S_N)$	–
P_H	periodic	homogeneous	$P_H \subset \varDelta$	$P_H = S(P_H)$	synchronized
P_{NA}	periodic	nonhomogeneous	$\varDelta$-symmetric	$P_{NA} = S(P_{NA})$	anti-phase
P_{NI}	periodic	nonhomogeneous	asymmetric	$\bar{P}_{NI} = S(P_{NI})$	in-phase
P_{NO}	periodic	nonhomogeneous	asymmetric	$\bar{P}_{NO} = S(P_{NO})$	out-of-phase

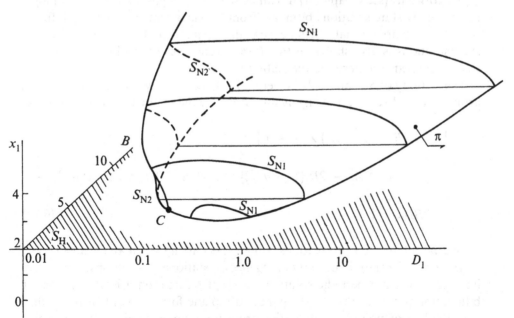

Fig. 6.13. Steady state solutions in (x_1, B, D_1) space; (nonhomogeneous solutions lie on the surface π which is symmetric with respect to the plane of homogeneous solutions $x_1 = A = 2$ (only the upper part of the solution surface is shown here). The solutions S_N^2 and $\bar{S}_N^2$ arise at the cusp point C.

The dependence of these nonhomogeneous solutions on B and D_1 is shown in Fig. 6.13. The solutions are characterized by means of the coordinate x_1; all points on the surface π in the space (x_1, B, D_1) represent the nonhomogeneous steady state solutions. The symmetry of solutions with respect to Δ corresponds

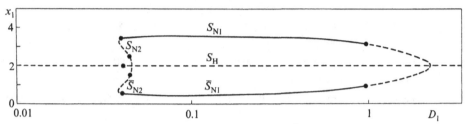

Fig. 6.14. Solution diagram of steady state solutions (dependence of x_1 on D_1), $B = 5.9$: full line – stable solution; dashed line–unstable solution; periodic solutions branch off at the Hopf bifurcation points denoted by ●, see Fig. 6.17.

to the symmetry of π with respect to the plane $x_1 = A$. A cross-section of the surface π with the plane $B = 5.9$ is shown in Fig. 6.14. This graph is called a *solution diagram*, which can also be generated by means of standard continuation algorithms (see Chapter 4). It can be seen from Fig. 6.14, that nonhomogeneous steady state solutions bifurcate from S_H via symmetry breaking bifurcations, i.e. in pairs mutually symmetric with respect to Δ. The stability of steady state solutions is also shown in Fig. 6.14 (eigenvalues of the linearized system were evaluated to determine the stability).

The stability of S_H can easily be determined analytically. On linearizing (6.13) around S_H and solving the eigenvalue problem, the characteristic equation is

$$\left[\lambda^2 - (B - 1 - A^2)\,\lambda + A^2\right]\left[\lambda^2 - \right.$$

$$- (B - 1 - A^2 - 2D_1(1 + q^{-1}))\,\lambda + (1 + 2D_1)(A^2 + 2D_1 q^{-1}) -$$

$$\left. - 2BD_1 q^{-1}\right] = 0 . \tag{6.19}$$

Of special interest are bifurcations from S_H, occurring when an eigenvalue λ is either zero (formation of nonhomogeneous stationary solutions) or purely imaginary (when a periodic solution via Hopf bifurcation arises). The loci of bifurcation points in the (B, D_1) parametric plane form curves which together with the loci of points of bifurcation from nonhomogeneous stationary solutions are shown in the *bifurcation diagram* in Fig. 6.15. The number and stability of stationary solutions in various regions of the parametric plane (B, D_1) can be learned from the diagram.

The condition for the symmetry breaking bifurcation from S_H is obtained from (6.19) when one eigenvalue λ is set to zero, i.e.

$$B = \omega(A^2 q + 2D_1), \qquad (6.20a)$$

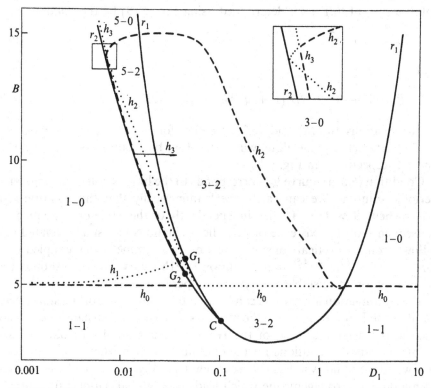

Fig. 6.15. Bifurcation diagram of steady state solutions in the parametric plane (B, D_1). The plane is divided by the bifurcation curves into regions denoted by $m - n$, where m is the total number of solutions and n is the number of stable solutions: full lines – limit points (curve r_2) or symmetry breaking bifurcations (curve r_1); dashed lines – Hopf bifurcations from stable steady state solutions; dotted lines – Hopf bifurcations from unstable steady state solutions. Curves of Hopf bifurcations as well as degenerate bifurcation points G_1, G_2 are described in the text.

see curve r_1 in Fig. 6.15. The condition for the coalescence of S_N^1 with S_N^2 (and $\bar{S}_N^1$ with $\bar{S}_N^2$) at limit points can be obtained from Eq. (6.17c) by putting the discriminant equal to zero, i.e.

$$B = \omega(4A\sqrt{D_1 q} - 2D_1), \qquad (6.20b)$$

see curve r_2 in Fig. 6.15.

Now we examine the case of a purely imaginary eigenvalue λ (Hopf bifurcation). The loci of all Hopf bifurcation points form smooth curves located on the surface of stationary solutions in (X, B, D_1) space. Projections of these curves into the plane (B, D_1) are shown in Fig. 6.15.

The Hopf bifurcation points for the family of homogeneous steady state solutions S_H of (6.13) are determined using Eq. (6.19) by the relations

$$B = 1 + A^2 \tag{6.21a}$$

and

$$B = 1 + A^2 + 2D_1(1 + q^{-1}), \qquad B > (2D_1 + qA^2). \tag{6.21b}$$

The relations (6.21a) and (6.21b) define for fixed A and q curves in the (B, D_1) parametric plane along which the Hopf bifurcations occur, see curves h_0 and h_1, respectively, in Fig. 6.15.

Condition (6.21a) (curve h_0) corresponds to the bifurcation of a homogeneous periodic solution. We can easily check numerically that this solution always exists when $B > 1 + A^2$ (i.e. independently of the value of D_1) which is the consequence of the existence of periodic oscillations in a single isolated cell. It follows from theoretical considerations on two symmetrically coupled oscillators[6.32, 6.35, 6.55, 4.13, 4.40, 4.58] that condition (6.21b) (curve h_1) corresponds to the bifurcation of an unstable Δ − symmetric periodic solution.

The continuation algorithms can be used to find periodic orbits starting at the point of the Hopf bifurcation; the methods of asymptotic expansions[3.40] may be used in the neighbourhood of the Hopf bifurcation point to obtain starting points on periodic solutions for the continuation algorithm.

If we admit equality in (6.21b) (see point G_1 in Fig. 6.15) then S_H is degenerate with a double zero eigenvalue which leads to a 'global' Hopf bifurcation[6.18].

Hopf bifurcation curves also exist on the families of nonhomogeneous solutions S_N. The curves h_2 and h_3 in Fig. 6.15 are composed of two parts, corresponding to a Hopf bifurcation either from a stable or an unstable solution. Both parts of the curve h_2 start from points of a degenerate bifurcation with one zero and two purely imaginary eigenvalues. The unstable part goes through the S_N^2 branch, then reaches the S_N^1 branch via the limit point and meets the stable part at the point with two pairs of purely imaginary eigenvalues. This point is the point of intersection of h_2 and h_3 and divides h_3 into stable and unstable parts as well. The curve h_3 extends on the S_N^1 branch very close to the curve r_2 of limit points and reaches this curve at the point G_2 with a double zero eigenvalue. Both h_2 and h_3 correspond to the bifurcation of asymmetric periodic solutions. In addition, there exists a mirror image to each point on h_2 and h_3 with respect to Δ for a given B, D_1. Although the projection in Fig. 6.15 cannot distinguish between a solution and its mirror image, they can be differentiated with the help of Fig. 6.13.

It follows from stability analysis that stable nonhomogeneous stationary solutions may exist for a range of parameter values where the homogeneous solution S_H is not stable and homogeneous oscillations may be expected. Similar behaviour was observed in experiments with the Belousov–Zhabotinski reaction in two coupled cells[6.14].

6.2.3 Classification of periodic solutions and their bifurcations

According to the classification of orbits given above, periodic solutions can be either nonhomogeneous (divided into asymmetric and Δ-symmetric solutions) or homogeneous. Note that the Δ-symmetric periodic oscillations with period T imply the following phase relations,

$$x_1\left(t + \frac{T}{2}\right) = x_2(t),$$

$$y_1\left(t + \frac{T}{2}\right) = y_2(t), \tag{6.22}$$

i.e. both cells oscillate in opposite phases and hence the Δ–symmetric solution may also be called an 'anti-phase' solution.

It is, moreover, useful to differentiate between the two types of asymmetric solutions according to the character of the mass exchange between the cells. The flux of the component X (or Y) can be either unidirectional or it can alternate, i.e. sgn $(x_1 - x_2)$ (or sgn $(y_1 - y_2)$) is either constant or it alternates.

Based on the symmetry and character of the mass flux between cells the following classification of periodic solutions (P) can be made (see Table 6.3):

(a) Homogeneous solution (P_H) with the orbit located in Δ (zero mass flux between the cells).

(b) Nonhomogeneous solutions (P_N):

    (i) Δ-symmetric *or anti-phase solution* (P_{NA}); the corresponding closed orbit is self-symmetric with respect to Δ and thus (6.22) holds; here sgn$(x_1 - x_2) = 1$ in one half of the period and sgn $(x_1 - x_2) = -1$ in the other one, hence the flux alternates.

    (ii) *in-phase asymmetric* solutions $(P_{NI}$ if $x_1 > x_2$ and $\bar{P}_{NI}$ if $x_1 < x_2)$; two closed orbits P_{NI} and $\bar{P}_{NI}$ are mutually symmetric with respect to Δ; the flux between cells is unidirectional.

    (iii) *out of-phase asymmetric* solutions $(P_{NO}$ if $x_1 > x_2$ in the larger part of the period and $\bar{P}_{NO}$ otherwise); in contrast to the in-phase solutions the flux between the cells alternates.

The periodic solutions (both stable and unstable) can be found numerically, using an algorithm based on the transformation of the system (6.13) into a boundary value problem with mixed boundary conditions. This algorithm combined with an algorithm for the continuation of solutions in dependence on a parameter (see Section 4.1) was used for the computation of a one-parameter family of periodic solutions.

The stability of the computed periodic solutions is determined on the basis of multipliers σ computed along the branch of periodic solutions; the computation is easily realized in combination with the above-mentioned continuation algorithm. One multiplier is always equal to 1, because the system is autonomous, see Section 2.3.

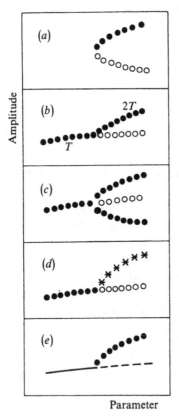

Fig. 6.16. Local bifurcations involving periodic solutions: (a) type $(+1)$ – limit point; (b) type (-1) – period doubling; (c) type (SB) – symmetry breaking; (d) type (T) – torus bifurcation; (e) type (H) – Hopf bifurcation.
Stable (unstable) steady state solutions are denoted by full (dashed) lines, stable (unstable) periodic solutions by full (empty) circles and tori by asterisks. Only bifurcations of stable solutions are shown, but bifurcations of unstable solutions from both stable and unstable branches may occur as well.

The stability may change at the bifurcation points, where one of the multipliers computed along the branch of solutions crosses the unit circle (see Section 3.2 for a more detailed description). We observed the following types of local bifurcations (see Fig. 6.16):

(a) type $(+1)$: limit point on the dependence of periodic solutions on a parameter. The number of solutions changes by two when a parameter is varied. Either a stable branch joins an unstable one at this point (Fig. 6.16 (a)) or both branches are unstable;

(b) type (-1): period doubling bifurcation point. A branch of solutions with double period branches off the original branch of solutions (Fig. 6.16 (b));

(c) type (SB): symmetry breaking bifurcation point. A pair of solutions mutually symmetric with respect to $\varDelta$ bifurcates from a $\varDelta$-symmetric or a homogeneous periodic solution. This bifurcation can generally arise only in systems with inherent symmetry (Fig. 6.16 (c));

(d) type (T): bifurcation into an invariant torus (see Fig. 6.16 (d)); (Bifurcations at points of strong resonance were also observed);

(e) type (H): Hopf bifurcation, i.e. a bifurcation of the branch of periodic solutions from the branch of steady state solutions (Fig. 6.16 (e)).

The above types were found numerically in the system of two coupled '*Bruselator*' cells . However, system (6.13) may also possess global bifurcations, see Section 6.2.5.

One-parameter systems of periodic orbits form 'tubes' in the product space $R^4 \times R$ consisting of pairs $(\mathbf{X}, \alpha)$; the parameter α can be either D_1 or B. If we assign to each periodic orbit a suitably chosen norm (e.g. a diameter) then the 'tubes' are represented by smooth curves in a solution diagram. Let us define a *branch of periodic orbits* as a maximal piece of a 'tube' which can be parametrized by the parameter α so that the norm of corresponding periodic orbits is a smooth single valued function of α.

Thus the bifurcation points of type $(+1)$ or (-1) separate two different branches, the bifurcation point of type (SB) separates three different branches. The *family of periodic orbits* is the union of all branches that are separated by a limit point. The family may have two ends at points of local bifurcations except the limit point or at a point of a global bifurcation where the period and/or amplitude tend to infinity. The family can also be formed by a closed cycle of branches, the first and the last branch being joined at the limit point. A single branch may also form a family.

Applying the same definitions to stationary solutions we have five branches which we denote S_H, S_N^1, S_N^2, $\bar{S}_N^1$, $\bar{S}_N^2$ and three families S_H, $S_N^1 \cup S_N^2$, $\bar{S}_N^1 \cup \bar{S}_N^2$ (see Fig. 6.14). This definition of the family can be extended to a higher-dimensional parameter space and will be used also in the case of a simultaneous dependence of periodic solutions on B and D_1.

6.2.4 Dependence of periodic solutions on D_1

Let us take $A = 2$, $q = 0.1$, $B = 5.9$ and follow the dependence of periodic solutions on D_1. Under these conditions the corresponding single cell system has an unstable stationary solution (corresponding to S_H) and a stable periodic solution (corresponding to P_H).

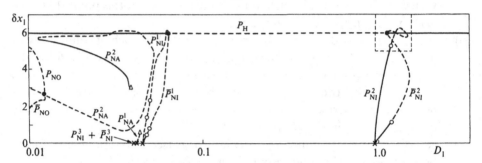

Fig. 6.17. Dependence of the amplitude δx_1 of several periodic solutions on D_1, $B =$ $= 5.9$: full lines – stable solutions; dashed lines – unstable solutions; × – Hopf bifurcations; ● – symmetry breaking bifurcations; ○ – period-doubling bifurcations; △ – heteroclinic loops; no bifurcations occur at the other intersections of the lines. The complicated structure of solutions in the rectangle is shown in Fig. 6.19.

From Fig. 6.14 we can infer that one point of the Hopf bifurcation from S_H and three symmetric pairs of the Hopf bifurcation points from S_N exist. The branches of periodic solutions are depicted in dependence on D_1 in the solution diagram in Fig. 6.17. The amplitude δx_1 of the concentration x_1 is taken as the norm (T denotes period),

$$\delta x_1 = \max_{t \in [0,\,T]} x_1(t) - \min_{t \in [0,\,T]} x_1(t). \tag{6.23}$$

The following solutions arise at the points of Hopf bifurcation:

(a) $D_1 \approx 0.040\ 9$: unstable periodic solution P_{NA}^1 branches off the solution S_H and continues in the direction of increasing D_1.

(b) $D_1 \approx 0.038\ 2$: from the solution S_N^1 $(\bar{S}_N^1)$ an unstable periodic solution P_{NI}^3 $(\bar{P}_{NI}^3)$ bifurcates to the right (see Fig. 6.21). These solutions are of the in-phase type.

(c) $D_1 \approx 0.044\ 6$: an unstable solution P_{NI}^1 $(\bar{P}_{NI}^1)$ branches off the solution S_N^2 $(\bar{S}_N^2)$ to the right.

(d) $D_1 \approx 0.954\ 3$: a stable solution P_{NI}^2 $(\bar{P}_{NI}^2)$ branches off the solution S_N^1 $(\bar{S}_N^1)$ to the right.

The homogeneous periodic solution P_H has arisen by a Hopf bifurcation at $B = 5$, see Eq. (6.21b), and its amplitude is independent of D_1, see Fig. 6.17.

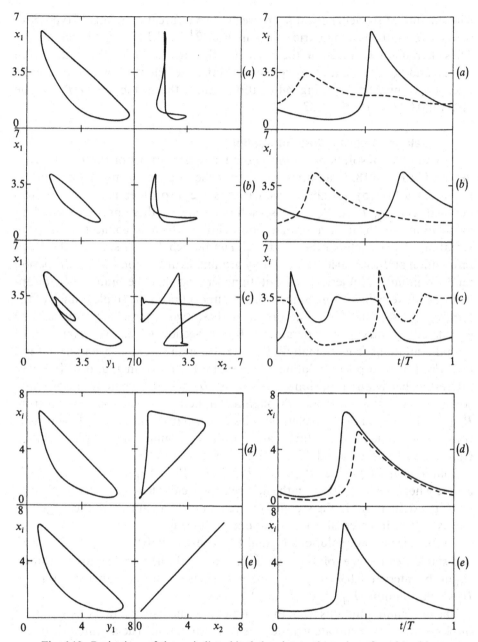

Fig. 6.18. Projections of the periodic orbits belonging to the region of weak and intermediate interaction into the planes (x_1, y_1), (x_1, x_2) and the time dependence of x_1 (full line) and x_2 (dashed line);
$B = 5.9$: (a) P_{NO}, $D_1 = 0.01$, $T = 5.331\ 02$; (b) P_{NA}^1, $D_1 = 0.01$, $T = 4.450\ 35$; (c) P_{NA}^2, $D_1 = 0.03$, $T = 12.499\ 91$; (d) P_{NI}^1, $D_1 = 0.062\ 5$, $T = 4.933\ 83$; (e) P_H, D_1 arbitrary, $T = 4.986\ 52$.

The stability of P_H depends on D_1; there are two bifurcation points of type (SB), where two mutually symmetric solutions P_{NI}^1, $\bar{P}_{NI}^1$ and P_{NI}^2, $\bar{P}_{NI}^2$ branch off P_H. These families have arisen on the other side through the Hopf bifurcation from S_N^2, $\bar{S}_N^2$ and S_N^1, $\bar{S}_N^1$. Hence they connect steady state behaviour with the periodic one. The symmetry-breaking bifurcations cause the instability of P_H in the interval $0.063 \lessgtr D_1 \lessgtr 1.122$.

Weak and intermediate interaction

Examples of solutions for low and intermediate values of $D_1(D_1 \lessgtr 0.1)$ are depicted in Fig. 6.18. In addition to the branches arising through Hopf bifurcations (cases (a), (b), (c) above) we can observe also other periodic solutions denoted as P_{NO}, $\bar{P}_{NO}$, P_{NA}^2. These solutions arise either through secondary bifurcations or through a primary bifurcation which is connected with the variation of parameters other than D_1. The branch P_{NA}^1 arising via the Hopf bifurcation at $D_1 \approx 0.040\,9$ apparently terminates at the point where the period tends to infinity. Numerical computations suggest that the branch P_{NA}^1 disappears at a heteroclinic loop. Similar behaviour occurs on the stable branch of the family P_{NA}^2, see Fig. 6.17. The origin of these two heteroclinic loops can be elucidated by varying the parameter B, see Section 6.2.5. Two examples of periodic solutions from the family P_{NA}^2 are shown in several projections in Figs. 6.18 (*b*), (*c*). Anti-phase solutions in this family, shown in Fig. 6.18 (*c*), have interesting behaviour at the limit $D_1 \to 0$. At $D_1 \approx 0.012$ a pair of out-of-phase solutions P_{NO}, $\bar{P}_{NO}$ bifurcates through a symmetry breaking bifurcation from P_{NA}^2. With decreasing D_1, the amplitudes of oscillations in both cells approach those in decoupled cells on this branch of the P_{NA}^2 family but the phase shift is equal to half of the period. The entire branch is unstable.

Solutions in the families P_{NO}, $\bar{P}_{NO}$, (see Fig. 6.18 (*a*)) approach, for $D_1 \to 0$, a state where the oscillations in the first cell come close to single cell oscillations, while the oscillations in the second cell are damped and approach the single cell steady state. However, all these solutions are unstable.

Two stable periodic solutions P_H and P_{NA}^2 coexist for $0.011\,2 \lessgtr D_1 \lessgtr 0.038\,3$ and stable steady state solutions S_N^1, $\bar{S}_N^1$ coexist with the stable periodic solution P_H in the interval $0.038\,16 \lessgtr D_1 \lessgtr 0.063$. Hence in a small range of values of D_1 stable solutions P_H, P_{NA}^2, S_N^1 and $\bar{S}_N^1$ coexist.

The remaining solutions from the range of intermediate interactions are of the in-phase type and bifurcate via the type (H) bifurcation. The asymmetric orbits P_{NI}^1, $\bar{P}_{NI}^1$, see Fig. 6.18 (*d*), bifurcate to the right and finally annihilate at the point of the type (SB) bifurcation. Both families are unstable and each contains two points of the type (-1) bifurcation (newly bifurcating branches were not followed). The unstable families P_{NI}^3, $\bar{P}_{NI}^3$ branch off to the right and almost immediately terminate at two mutually symmetric homoclinic orbits, see later.

Strong interaction

A very complicated system of in-phase nonhomogeneous periodic solutions exists in the interval $0.954\,2 \lesssim D_1 \lesssim 1.472\,4$ (the interaction is an order of magnitude stronger now). The corresponding solution diagram (a part of the solution diagram from Fig. 6.17) is shown in Fig. 6.19. Let us denote each family

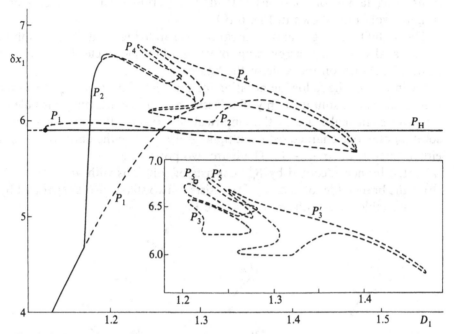

Fig. 6.19. Dependence of the amplitude δx_1 of periodic solutions $P_1, \ldots, P_5$ and P_H in the region of strong interaction: full lines – stable solutions; dashed lines – unstable solutions; families of solutions are connected via period doubling bifurcations with the exception of type (SB) (denoted by ●) bifurcation connecting P_1 and P_H. Isolated families are shown separately.

in Fig. 6.19 by a symbol P_m, where m gives a number of local maxima on any coordinate of $x(t)$ within one period. In this notation the basic family P_{NI}^2 arising via the type (H) bifurcation at $D_1 \approx 954\,2$ is equivalent to P_1.

At the point $D_1 \approx 1.172\,0$ the family P_2 with a double period bifurcates from the basic family P_1; from the family P_2 bifurcates the family P_4 with a four-fold period (related to P_1) and so on. The intervals between the bifurcation values of D_1 for the subsequently bifurcating periodic solutions decrease geometrically with the universal quotient δ. The sequence of the double period bifurcation points is oriented in the direction of increasing D_1.

A similar sequence of period-doubling bifurcations begins at the point $D_1 \approx$ $\approx 1.470\,2$, close to the limit point on P_1; the subsequent bifurcation points are oriented in the direction of decreasing D_1. The family P_2 bifurcating at $D_1 \approx$

$\approx 1.470\,2$ finally joins the family P_2 bifurcating at $D_1 \approx 1.72\,0$. Two families P_4 bifurcate on this common P_2 family; both P_4 families return to the P_2 family. Altogether four families of P_8 solutions bifurcate from the two P_4 families, etc.

Several solution families (e.g. P_3 and P_5 in Fig. 6.19) form closed loops-*isolas*; cascades of period doubling bifurcations may again start from the isolas. Any bifurcating family of solutions with double period terminates again on the original isola (not shown in Fig. 6.19).

The stability changes and bifurcations on individual families of solutions occur usually in a very narrow range of values of the parameter D_1 and thus they cannot be shown on the scale of the figure.

We introduce the following notation for branches. Let $B_m^s \subset P_m$ be a branch containing stable solutions only. To classify branches containing unstable solutions we shall make use of the empirical observation that for each unstable solution only one characteristic multiplier σ lies outside the unit circle and thus either $\sigma > 1$ or $\sigma < -1$. There are two possibilities:

(a) the branch (denoted by B_m^+) consists of solutions with $\sigma > 1$,

(b) the branch (denoted by B_m^{s-s}) consists of two stable pieces separated by an unstable piece with $\sigma < -1$, see Fig. 6.20.

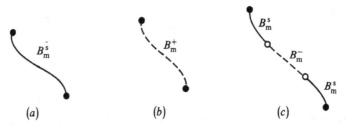

Fig. 6.20. Types of branches forming the families P_m: (a) stable branch B_m^s; (b) unstable branch B_m^+; (c) combined branch B_m^{s-s} with two period doubling bifurcation points; O period-doubling bifurcation point; ● points where the branches terminate (i.e., Hopf bifurcation points, limit points, symmetry breaking bifurcation points, period doubling bifurcation points).

Each family P_m can be constructed combining B_m^s, B_m^+ and B_m^{s-s}. For example, the families P_2 and P_4 in Fig. 6.19 (i.e. the solutions which do not form isolas) arise via the combination $B_m^{s-s} \cup B_m^+ \cup B_m^{s-s}$, the isolas P_3, P_3', and P_5' can be formed via the combination $B_m^{s-s} \cup B_m^+ \cup B_m^{s-s} \cup B_m^+$ and P_5 via the combination $B_m^+ \cup B_m^s$.

We may expect that the behaviour of families P_m for higher values of m will be similar.

Periodic solutions for $D_1 = 1.26$ from the branches B_1^{s-s}, B_2^{s-s}, B_3^{s-s}, B_4^{s-s}, B_5^{s-s} and B_6^{s-s} are depicted in Figs 6.21 (a)–(f).

The structure of periodic solutions in the region of the strong interaction is very complex (it appears that an infinite number of solutions exist here). On the

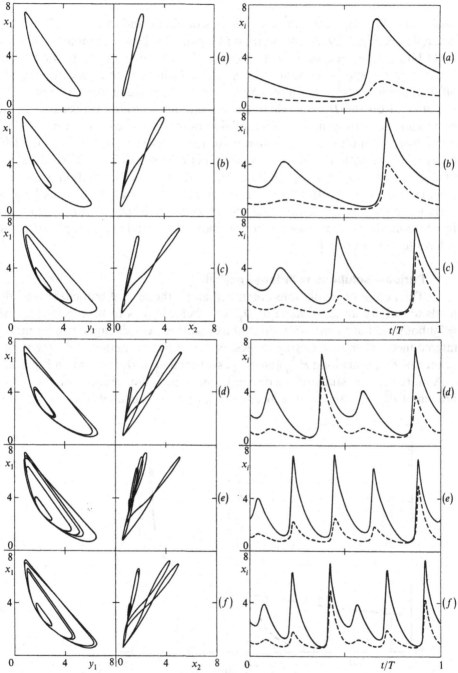

Fig. 6.21. Projections of the periodic orbits belonging to the region of strong interaction into the planes (x_1, y_1), (x_1, x_2) and the time dependence of x_1 (full line) and x_2 (dashed line); $B = 5.9$, $D_1 = 1.26$: (a) P_1, $T = 4.707\,83$; (b) P_2, $T = 8.403\,20$; (c) P_3, $T = 12.319\,18$; (d) P_4, $T = 16.610\,83$; (e) P_5, $T = 21.828\,97$; (f) P_6, $T = 24.938\,97$.

other hand, the way in which new solutions originate and change their stability is simple, see Fig. 6.19. At the limit point a pair of solutions bifurcates through $(+1)$ bifurcation; one solution is stable and the second one is unstable. If a bifurcation of type (-1) occurs on the stable branch, then a cascade of (-1) bifurcations is expected to follow and terminate at the accumulation point[3.23, 6.2]. An interval on the D_1-axis where stable periodic solutions between the limit point and the corresponding accumulation point exist is called a window. The numerical computations suggest that in the range $1.193 \lesssim D_1 \lesssim 1.470$ infinitely many windows with infinitely many stable periodic solutions exist. At the same time it appears that at many values of D_1 no stable periodic solutions exist and complicated chaotic attractors are observed[6.48, 6.51]. Chaos generated by a period-doubling sequence repeatedly occurs between the neighbouring windows. In addition, the windows sometimes overlap, which leads to multiple attractor behaviour, see Section 6.2.7.

Periodic solutions with large period

The period of oscillations computed along the several branches evidently tends to infinity at some value of D_1. This behaviour is to be expected in the neighbourhood of a *homoclinic* orbit or a closed loop consisting of two (or more) interconnected *heteroclinic* trajectories. For example, the period along the stable branch of P_{NA}^2 and along P_{NI}^3 and $\bar{P}_{NA}^3$ as a function of D_1 is shown in Fig. 6.22.

A more detailed study of the development of periodic solutions on the stable branch of P_{NA}^2 reveals that the increase in the period is caused by two gradually

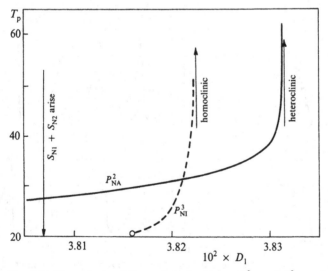

Fig. 6.22. Dependence of the period of solutions P_{NA}^2 and P_{NI}^3 on D_1, $B = 5.9$; stable branch P_{NA}^2 approaches the heteroclinic loop, unstable branch P_{NI}^3 approaches the homoclinic orbit.

lengthening phases, containing phase points which are close to the stationary states S_N^2 and $\bar{S}_N^2$, see Fig. 6.23 (a). At first the phase point on the corresponding orbit stays close to $\bar{S}_N^2$ for almost half the period. Then it rapidly jumps close to S_N^2, and behaves in the same way due to the symmetry given by (6.14).

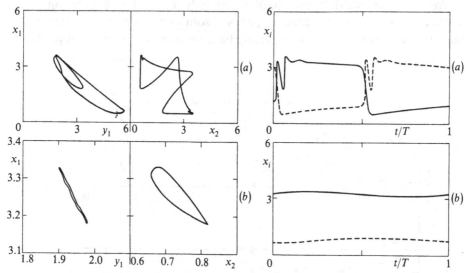

Fig. 6.23. Projections of the periodic orbits P_{NA}^2, P_{NI}^3, close to the heteroclinic loop and the homoclinic orbit, respectively, into the planes (x_1, y_1), (x_1, x_2) and the time dependences of x_1 (full line) and x_2 (dashed line), $B = 5.9$: (a) P_{NA}^2, $D_1 = 0.038\ 316\ 51$, $T = 62.087\ 70$; (b) P_{NI}^3, $D_1 = 0.038\ 220\ 41$, $T = 42.085\ 32$.

Although it is not possible to continue P_{NA}^2 further for numerical reasons, we may assume that the solution P_{NA}^2 approaches a heteroclinic loop between unstable steady states S_N^2 and $\bar{S}_N^2$. The same behaviour occurs on the unstable branch P_{NA}^1. A further description involves two parameter families of solutions, see Section 6.2.5.

Numerical computations suggest that the families P_{NI}^3 and $\bar{P}_{NI}^3$ terminate at homoclinic orbits connected with steady states S_N^2 and $\bar{S}_N^2$. These families arise through the Hopf bifurcation at $D_1 = 0.038\ 162$, see Fig. 6.22; at $D_1 = 0.038\ 204$ the period of the solution is already very high and the solution cannot be numerically continued. The P_{NI}^3 and $\bar{P}_{NI}^3$ orbits have a very small amplitude and they are nearly planar, see Fig. 6.23(b).

6.2.5 Behaviour of periodic solutions in the parameter plane (B, D_1)

A numerical computation of two parameter families of periodic solutions was performed sequentially by choosing several fixed values of the first parameter and continuing the second parameter families. Thus the global picture is obviously incomplete and therefore it is presented only in a qualitative way, see Fig. 6.24 (*a*) to (*d*). Nevertheless, with the exception of the possibly complicated behaviour arising near several codimension two bifurcation points, we have here a rather clear picture of the behaviour of solutions.

We note that two different families may only intersect at bifurcation points, see Section 6.2.3. Hence families of periodic solutions can be divided into groups such that the families within one group have no common bifurcation point with

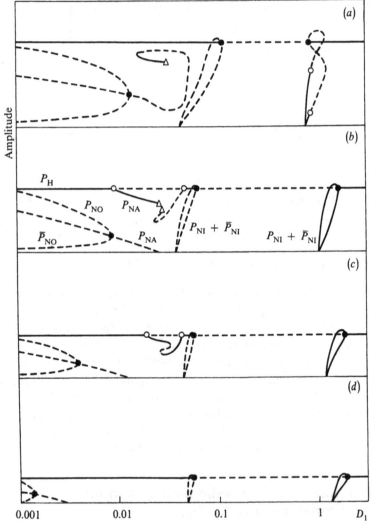

Fig. 6.24. Qualitative behaviour of periodic solutions in the parameter plane (B, D_1) depicted as a sequence of the dependences of an amplitude on D_1 for different values of B; $\bigcirc$ – period-doubling bifurcations; $\bullet$ – symmetry breaking bifurcations; $\triangle$ –heteroclinic loops: (*a*) $B \approx 6.3$; (*b*) $B \approx 5.5$; (*c*) $B \approx 5.3$; (*d*) $B \approx 5.1$.

the families of another group. For example, one-parameter families in Fig. 6.17 can be divided into five groups (1) P^2_{NA}, P_{NO}, $\bar{P}_{NO}$, (2) P^3_{NI}, (3) $\bar{P}^3_{NI}$, (4) P^1_{NA}, (5) P_H, P^1_{NI}, $\bar{P}^1_{NI}$, P^2_{NI}, $\bar{P}^2_{NI}$. However, when the solutions are seen as dependent on B and D_1 simultaneously, then the first, fourth and fifth groups merge together as will be explained below.

All families within one group are mutually connected through bifurcations. Thus starting from a chosen solution for given B and D_1 we can reach any solution in the same group by choosing a continuous path through the group. Let a *primary family* be that one that starts from a steady state solution via the Hopf bifurcation; as a *secondary family* we shall denote the family that bifurcates from a primary family, etc.

Classification of families of solutions

The behaviour of two-parameter families P^3_{NI} and $\bar{P}^3_{NI}$ is relatively simple. For given B and D_1 the solutions from P^3_{NI} and $\bar{P}^3_{NI}$ are mutual images under Δ (see Eq. (6.14)) and though both families form two distinct groups, they can be treated simultaneously. They originate at the curve of the Hopf bifurcation points (see curve h_3 in Fig. 6.15) from the nonhomogeneous solutions S^1_N, $\bar{S}^1_N$, respectively, (two curves of the Hopf bifurcation points in the space $R^4 \times R^2$ merge together in the projection into the (B, D_1) plane). The curves terminate at the critical point G_2 ($B \approx 5.382\,76$; $D_1 \approx 0.045\,43$) where two purely imaginary eigenvalues vanish. The results of Bogdanov[3.8] are directly applicable and we may conclude that a one-parameter family of homoclinic orbits (and its mirror image) arises at G_2. Numerical computations show that this curve extends very near to the Hopf bifurcation curve h_3 forming a very narrow cusp-shaped region of the existence of P^3_{NI} and $\bar{P}^3_{NI}$. The width of this region at $B = = 5.9$ can be seen in Fig. 6.22.

All remaining families of solutions are interconnected via loci of bifurcations and form the third group. This group includes a very complicated structure of solution families originating from primary ones which themselves originate at three curves of the type (H) bifurcation points (see curves h_0, h_1 and h_2 in Fig. 6.15).

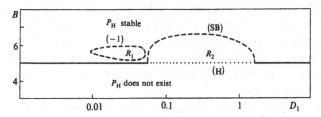

Fig. 6.25. Bifurcation diagram of the homogeneous periodic solution P_H in the (B, D_1) plane.

The first Hopf bifurcation curve is given by Eq. (6.21a) and is associated with the appearance of the homogeneous solution P_H (see curve h_0 in Fig. 6.15). The bifurcation diagram for P_H in Fig. 6.25 shows two regions R_1 and R_2 of an instability of P_H bounded by bifurcation curves. The boundary of R_1 is formed by a closed curve of type (-1) bifurcation points. The symmetry of the system implies that an anti-phase solution P_{NA} with a double period will bifurcate along this curve, see Fig. 6.24(*b*) and (*c*).

The one-parameter family originating at a point on the boundary of R_1 is shown in Fig. 6.26 as a family depending on B for $D_1 = 0.02$. Using Fig. 6.17 we can conclude that this family is of the type P_{NA}^2. However, a chosen solution

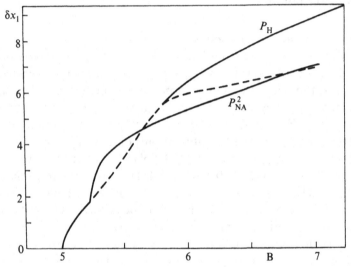

Fig. 6.26. Dependence of the amplitude δx_1 of the solutions P_H and P_{NA}^2 on B, $D_1 = = 0.02$.

from the family P_{NA}^1 (see Fig. 6.17) continued in B for fixed D_1 leads again to P_{NA}^2. It follows that P_{NA}^2 and P_{NA}^2 form a unique family (denoted by P_{NA}) when considered as dependent on B and D_1 simultaneously. This family bifurcates from P_H via a period-doubling bifurcation (see Fig. 6.25) and simultaneously from S_H via the Hopf bifurcation (see curve h_1 in Fig. 6.15). Hence P_{NA} is a primary family. The situation is schematically shown in Fig. 6.24 (*a*)–(*d*) see also Fig. 6.17. Two heteroclinic loops described for $B = 5.9$ come close together if B is decreased and finally merge and disappear. The disappearance of one of both heteroclinic loops with increasing B is associated with the degenerate bifurcation point G_1, see Fig. 6.15.

The boundary of R_2 (see Fig. 6.25) is formed by two parts, the linear part of the type (H) bifurcation points being at the same time part of the line given by Eq. (6.21a), and the curve of the type (SB) bifurcation points. The symmetry of

Eq. (6.13) implies that a pair of the in-phase solution families P_{NI}, $\bar{P}_{NI}$ will bifurcate along the (SB) curve, see Fig. 6.24 (a)–(d).

Numerical computations show that these families themselves bifurcate (or terminate) from two symmetric curves of Hopf bifurcation points on a branch of nonhomogeneous steady states (see curve h_2 in Fig. 6.15). All solutions arising via Hopf bifurcation from a nonhomogeneous steady state must necessarily be of the in-phase type and this is in agreement with the branching of the in-phase solution from the (SB) curve. Thus P_{NI} and $\bar{P}_{NI}$ are primary families and they meet each other and at the same time intersect P_H at the (SB) curve of the boundary of R_2. The intersection is limited to $5 \lesssim B \lesssim 6.7$, see Fig. 6.25. Comparing these results with the solutions studied earlier for $B = 5.9$, we are led to the conclusion that P_{NI}^1 $(\bar{P}_{NI}^1)$ from the region of intermediate interactions and P_{NI}^2 $(\bar{P}_{NI}^2)$ from the region of strong interactions meet for $B \gtrsim 6.7$ and form a unique family (denoted by P_{NI} $(\bar{P}_{NI})$) when considered as two-parametric systems.

Period doubling and tori bifurcations

A continuation algorithm based on a *simple shooting* procedure was used in cases where the modulus of the largest multiplier was not too high. For very unstable orbits (modulus approximately higher than 10^8) this method is not efficient and a *multiple shooting* method must be used, see Section 4.1.

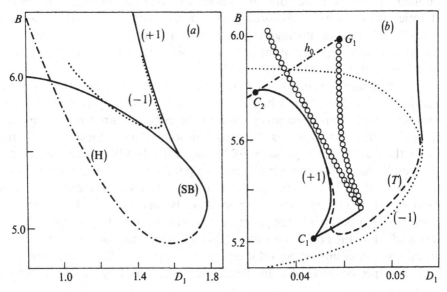

Fig. 6.27. Bifurcation diagram in the plane (B, D_1) for the solutions P_{NI} (Fig. 6.27 (a)) and P_{NA} (Fig. 6.27 (b)): .... – period doubling bifurcation curves; – – – – – torus bifurcation curve; –·–·–·– –Hopf bifurcation curves; ——— – limit point or symmetry breaking bifurcation curves; ∘∘∘∘∘∘ – heteroclinic bifurcation curves.

It appears that only small regions of stable nonhomogeneous periodic solutions exist for $B \gtrsim 8$ and thus we shall not follow this region in more detail. Instead, we shall study two regions of the (B, D_1) plane containing two types of chaos[6.48, 6.49].

Fig. 6.27(a) contains a bifurcation diagram of the solution in the (B, D_1) plane (see also Fig. 6.24 (a)–(d), Fig. 6.17 and Fig. 6.19). P_{NI} is bounded by curves of type (SB), (H) and $(+1)$ bifurcation points and a bifurcating family with a double period is bounded by a type (-1) bifurcation curve. Isolated families as well as potentially chaotic behaviour exist inside the region bounded by the curve of accumulation points of the period-doubling sequences close to the type (-1) curve (not shown in the figure). Note that there is a region in the (B, D_1) plane such that stable nonhomogeneous oscillations exist for $B < 1 + A^2$, i.e. the interaction between the cells can lead to an oscillatory state even when the isolated cells possess no periodic solution[6.1].

The second region of interest involves the bifurcation of tori from P_{NA}, see Fig. 6.27 (b). The region of existence of P_{NA} is bounded by a curve of type (-1) bifurcation points. In the cusp region bounded by two curves of type $(+1)$ bifurcations there exist three different solutions (see Fig. 6.24 (c); two of them are stable and one is unstable close to the cusp point C_1. However, there exists another bifurcation changing the stability of P_{NA} – a bifurcation of type (T). The family P_{NA} is unstable inside the region formed by the (T) curve, and a large number of new periodic solutions associated with the resonant tori and with chaotic attractors arise. According to general results[2.4], the global picture of periodic solutions in this region is expected to be very complicated. A detailed numerical study might be of interest as very similar behaviour including symmetries has been observed in experiments[6.11].

To understand the global behaviour of various branches of the P_{NA} family we also included curves along which the heteroclinic loops appear though they are computed with limited accuracy. One of the curves of limit points emanating from C_1 splits into two curves of heteroclinic loops. One of these curves terminates at the point $G_1 \equiv (B \approx 5.979\ 79, D_1 \approx 0.044\ 54)$ where the stationary solution S_H has two zero eigenvalues. This point coincides with the end point of the Hopf bifurcation curve h_0 (see Fig. 6.15) and we may expect behaviour similar to that in the vicinity of the point G_2. However, due to the symmetry, the point G_1 does not fall into the generic cases considered by Bogdanov[3.8]. This case was studied by Fiedler[6.18] who shows that a limiting case of periodic orbits with an infinite amplitude or period exists in the vicinity of G_1. In the case studied here the expected curve of homoclinic orbits is replaced by a curve of heteroclinic loops.

The second curve of limit points starting at C_1 terminates at the point C_2 which again lies on the curve h_0. The periodic solutions appearing along h_0 change the direction of the bifurcation at C_2.

6.2.6 Period-doublings and chaotic attractors

Typical chaotic oscillations generated by Eqs. (6.13) are shown in Fig. 6.28. We shall use *Poincaré mapping* which reduces Eqs. (6.13) into a discrete time system which has three-dimensional state space and we shall follow transitions from the periodic regimes into more complex ones.

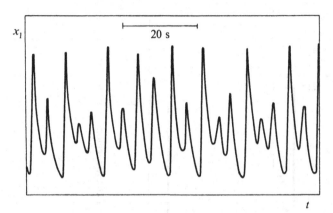

Fig. 6.28. Temporal oscilations of the concentration x_1: $A = 2$; $B = 5.9$; $q = 0.1$; $D_1 = 1.21$

The Poincaré mapping Z will be defined on a hypersurface of codimension one chosen in such a way that the symmetry of the original system is preserved. We choose $\Sigma = \{(x_1, y_1, x_2, y_2) \in R^4; \; x_1 - y_1 + x_2 - y_2 + c = 0\}$. In fact, Σ is equivalent to a three-dimensional Euclidean space. If a point $X_0 \equiv (x_1^0, y_1^0, x_2^0, y_2^0) \in \Sigma$ then also the point $\bar{X}_0 = S(X_0) \equiv (x_2^0, y_2^0, x_1^0, y_1^0) \in \Sigma$. To study the asymptotic behaviour of Eqs. (6.13) it is useful to set $c = 2(B/A - A)$. This choice implies that the homogeneous stationary point S_H (i.e. the point with coordinates $(A, B/A, A, B/A)$) is contained in Σ. Subsequent intersections (provided they exist) of the trajectory starting at $X_0 \in \Sigma$ are given by iterates of Z, i.e. $X_k = Z^k(X_0)$. The projection of this discrete orbit into the plane (x_1, x_2) is sufficient to differentiate between asymmetric and symmetric orbits and will be used for graphical representation of limit sets.

As above we shall set $A = 2$, $q = 0.1$ and investigate two chaotic regions in the (B, D_1) plane. Transitions to chaos will be described, dependent on the coupling strength between the cells, i.e. B will be fixed and D_1 will be changed. There is a period-doubling route to chaos for $B = 5.9$ while a transition to chaos via a torus occurs for $B = 5.5$.

First, we set $B = 5.9$ and return to the one-parameter families of periodic orbits P_n, $n = 1, 2, ...$, which are shown in Fig. 6.19. This figure shows the complicated behaviour of the families; in particular, there are two main period-

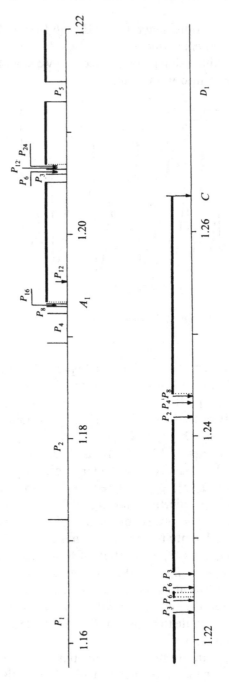

Fig. 6.29. Period-doubling sequences and the window structure in a chaotic region, only large windows are shown.

doubling sequences starting from the P_1 family, one from the left to the right and the other in the opposite direction, which are apparently infinite and accumulate at $D_1 \approx 1.193\,3$ and at $D_1 \approx 1.472\,0$, respectively. Let us denote these points A_1, A_2 respectively. Chaotic behaviour is expected to occur in the interval $A_1 < < D_1 < A_2$. However, a direct numerical integration of Eqs. (6.13) reveals chaotic attractors only in the interval $A_1 < D_1 < C \approx 1.263\,5$. In addition, many periodic windows occur in this region, the largest ones being indicated in Fig. 6.29. Numerical computations suggest that most windows (i.e. intervals of the values of D_1 corresponding to stable branches of periodic orbits connected via successive period-doubling bifurcation points) contain an infinite sequence of period-doubling bifurcations. Two questions arise:

1. What are the asymptotic properties of observed sequences? If we denote a sequence of bifurcation values of D_1 where the branchings $P_{2^i n} \rightarrow P_{2^{i+1} n}$, $i = 0, 1, 2, ...$, occur at $\{D_n^i\}$, and

$$\delta_i = \frac{D_n^{i+1} - D_n^i}{D_n^{i+2} - D_n^{i+1}},$$

then does δ_i converge to Feigenbaum's δ?

2. Do the windows fill out the entire interval $A_1 < D_1 < A_2$ between the accumulation points of the two main period-doubling sequences or are there gaps, i.e. intervals (or at least points) on the D_1 axis, yielding chaotic attractors?

As the bifurcations leading to chaos in this region of the parameter values closely resemble those in one-dimensional maps, we first recall the results for the one-dimensional quadratic mappings of the interval with a single extreme, discussed in Chapter 3: (1) $\lim_{k \to \infty} \delta_k = \delta \approx 4.669$ is a universal constant for all mappings with a quadratic extreme; (2) there is a Cantor set of 'chaotic' parameter values and this set has a positive Lebesgue measure; the chaotic attractor is a finite union of closed intervals, i.e. it does not possess a fractal structure.

In the multidimensional case the answer to the first question was given by Collet, Eckmann and Koch[3.11] who show that Feigenbaum's results on one-dimensional maps can be extended to finite dimensional state spaces and the value of δ remains unchanged. However, rigorous results for the second problem are still lacking. Hence, numerical computations are largely used and we have to accept their limited accuracy. We are thus unable to decide exactly whether the observed chaotic parameter set contains intervals.

The results of the numerical study of (6.13)[6.48] reveal the following information:

1. The first period-doubling sequence from the left (starting at the point $D_1 \approx 1.172\,0$) apparently converges geometrically at a rate given by δ. Several

first terms of the sequence are given in Table 6.4. Presumably, all other infinite period-doubling sequences follow the same behaviour.

Table 6.4 *The rate of convergence of the sequence of period-doubling bifurcations,* $\delta_i = (D_1^{i+1} - D_1^i)/(D_1^{i+2} - D_1^{i+1})$; D_1^i *is the value of* D_1 *at the bifurcation point where branching* $P_{2^i} \rightarrow P_{2^{i+1}}$ *occurs*

i	D_1^i	δ_i
0	1.172 000 0	5.822
1	1.189 399 3	5.822
2	1.192 388 1	4.347
3	1.193 075 6	4.699
4	1.193 221 9	

2. Trajectories for most values of D_1 do not approach any periodic orbit and the resulting asymptotic set has a complex geometric structure typical for chaotic attractors of multidimensional dissipative systems. All larger windows in the interval $A_1 < D_1 < C$ are shown in Fig. 6.29. There is a vast number of other windows but these are very narrow and can be indicated only by very careful computations.

Let us now follow the overall development of attractors starting from the appearance of P_1 (and its mirror image $\bar{P}_1$) at $D_1 \approx 0.954\ 3$ to the disappearance of chaotic attractors at $D_1 \approx 1.263\ 5$. The periodic orbits P_1 and $\bar{P}_1$ are the only two attractors until $D_1 \approx 1.122\ 2$ as the homogeneous periodic orbits P_H as well as all three stationary points are unstable. However, P_H is stable for $D_1 \gtrsim 1.122\ 2$ hence the attractors P_1, $\bar{P}_1$ and P_H coexist there. When $D_1 \approx 1.172\ 0$, P_1 and $\bar{P}_1$ become unstable and P_2 and $\bar{P}_2$ become attractors. The string of bifurcations is repeated until the accumulation point A_1 is reached; symmetry pairs of periodic attractors from the period-doubling sequence are substituted by symmetry pairs of chaotic attractors above A_1. Chaotic attractors are frequently replaced by periodic ones if D_1 is further increased. The coexistence of periodic and chaotic attractor pairs and of two pairs of chaotic attractors were indicated. This sequence of changes is very complicated.

If we follow these changes in the Poincaré section Σ, then the attractor P_n is subsequently represented by one, two, four points and so on; above the accumulation point A_1 an analogous sequence of changes occurs but in the opposite order and instead of the points we now have pieces of chaotic attractors. The pieces are successively joined pairwise and finally at $D_1 \approx 1.20$ a connected attractor arises. This sequence of changes on the chaotic attractor closely resembles the sequence of reverse bifurcations known from one-dimensional systems such as the logistic map (see Chapter 3). The period-doublings may also accumulate with a decreasing parameter unlike those in the logistic map. The reverse bifurcations occur in a certain interval of the parameter D_1 on the chaotic side of each accumulation point; this interval can be called a 'window of reverse

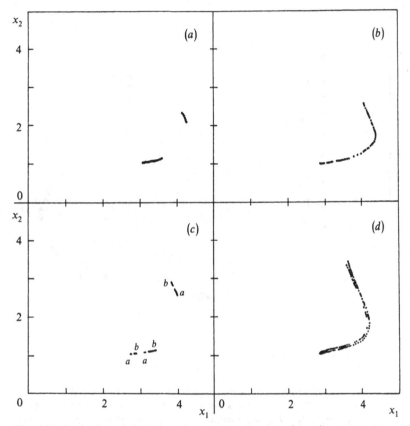

Fig. 6.30. Projection of the Poincaré maps into the plane (x_1, x_2); $A = 2$, $B = 5.9$, $q = 0.1$: (a) $D_1 = 1.194$; (b) $D_1 = 1.21$; (c) $D_1 = 1.224\ 6$, coexistence of two attractors; (d) $D_1 = 1.263$.

bifurcations'. The width of such a window is usually much smaller than the width of the corresponding periodic window. In fact, the reverse bifurcation sequence converges to the accumulation point from the chaotic side and its rate of convergence is given by δ.

Typical Poincaré maps of chaotic attractors are shown in Fig. 6.30 (a)–(d). The Poincaré mapping contracts so much that the attractor appears to be a union of several curves of a finite length in Σ. When we study the attractor carefully, we can observe that it consists of points located along many curves passing close to each other. This complex and fine structure cannot be observed in one-dimensional maps. It can be seen from Fig. 6.30 that the attractor splits into a variable number of pieces with the variation of the parameter. Qualitative differences in the behaviour of 'connected' and 'split' chaotic attractors are best characterized by power spectra, see Section 6.2.8.

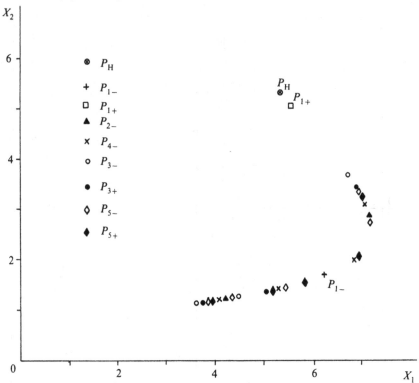

Fig. 6.31. Poincaré map representation of several periodic orbits of Eq. (6.13): $A = 2$; $B = 5.9$; $q = 0.1$; $D_1 = 1.21$. P_H is stable, the other periodic orbits are unstable. P_{1-}, P_{2-}, P_{3-}, P_{3+}, P_{4-}, P_{5-}, P_{5+} are contained in the chaotic attractor (compare Fig. 6.30 (b)) while P_{1+} is located in the boundary of the basin of attraction of the chaotic attractor.

In the following we use the symbols P_{ns}, P_{n-} and P_{n+} for a periodic orbit contained in the family P_n if it is stable, unstable with one multiplier $\sigma < -1$ and unstable with one multiplier $\sigma > 1$, respectively. Let us study in more detail the structure of the attractor in Fig. 6.30(b) at $D_1 = 1.21$. If we choose a point $\mathbf{X} \in \Sigma$ located on the attractor, then subsequent iterates of Z fill out the entire attractor in an apparently random way, but the shape and the orientation of the attractor in the state space can be reconstructed from unstable periodic orbits contained in the attractor. Let us take all Poincaré sections of unstable periodic orbits P_{n-}, P_{n+} which were previously found by a continuation and plot them in the same way as the chaotic attractor. If we compare the result (see Fig. 6.31) with the corresponding chaotic attractor (Fig. 6.30 (b)) we can observe an evident correspondence. In fact, the attractor contains infinitely many unstable periodic orbits which are expected to be dense in it. The construction in Fig. 6.31 can be called a *frame of the attractor*.

In general, the complex motion of the state point within the chaotic attractor is caused by an exponential instability of all orbits in the attractor, i.e. there are expanding and contracting directions in the attractor. Periodic orbits in the attractor are saddles and their stable and unstable manifolds intersect in homoclinic orbits. This arrangement implies that a point which was originally repelled from the periodic orbit, must after a finite time return to its neighbourhood. It can be shown rigorously that if this feedback mechanism exists, then the attractor has to possess a very complex topological structure as well as a sensitive dependence on initial conditions.

6.2.7 Intermittency, multiple attractor behaviour and crises

As in the logistic map, intermittency occurs near limit points which bound all observed periodic windows at least from one side (the other side is bounded either by an accumulation point, if the period-doubling sequence is infinite, or by a limit point). Thus the intermittent transition can occur both for increasing and decreasing parameter values. The intermittent dynamics is analogous to that of the logistic map, see Chapter 3. Later we shall study an interesting intermittent transition in a different parameter region, hence we do not describe it here.

One aspect in which the dynamic behaviour in the parameter interval $A_1 < < D_1 < C$ differs from that of the logistic map is the occurrence of multiple attractors (each of the attractors has its own symmetric image). Two coexisting chaotic attractors are shown in Fig. 6.30c. Their fate if D_1 is changing can be explained in terms of a boundary crisis using the results of the continuation of periodic orbits. Let us follow the behaviour of the system (6.13) in the interval $1.222\,8 \lesssim D_1 \lesssim 1.226\,4$. Previous computations of families of periodic orbits (see Fig. 6.19) reveal that the family P_3 'bends' here twice, hence between the limit points there is a triad of P_3 orbits located close to each other, of which the middle one is the saddle orbit P_{3+}. There are two infinite period-doubling sequences starting near the limit points one increasing and the other decreasing, see Fig. 6.29. Behind the accumulation points exist two reverse windows containing chaotic attractors. These windows overlap and thus the multiplicity of chaotic attractors results. Varying the parameter D_1 we obtain a picture which is schematically given in Fig. 6.32.

Both attractors coexist until one of them collides with the P_{3+} orbit, i.e. a boundary crisis occurs. The topological structure of the attractors varies in the interval of coexistence; the chaotic and the periodic attractors coexist for some values of the parameter (for example at $D_1 = 1.225$). The chaotic attractor ceases to exist at the collision point and the periodic attractor remains (in both cases it is the attractor P_6).

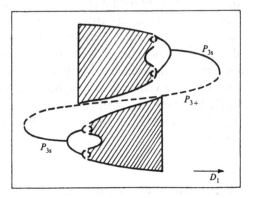

Fig. 6.32. Occurence of multiple attractors, schematically.

The chaotic invariant set which remains after the destruction of the chaotic attractor is not attracting (a chaotic saddle) and long-lived chaotic transients are expected to exist close to this set. Other families of orbits form similar loops such as P_3, hence the multiplicity can be expected for many values of D_1. Seeking the multiplicities is difficult, because the basins of attraction of the coexisting attractors generally intermingle in a very complicated way[6.25].

Though chaotic behaviour should occur for all values of D_1 in the interval between the accumulation points A_1, A_2 of the main period-doubling sequences starting from P_1, chaotic attractors do not occur above the critical point $C \approx \approx 1.263\ 5$ (see Fig. 6.29). The chaotic invariant set still exists in the interval $C < D_1 < A_2$ but is not attracting. However, there are stable periodic orbits different from the homogeneous orbit P_H for certain $D_1 > C$; trajectories approach either such a symmetric orbits or P_H, depending on the initial condition.

At the small right-hand neighbourhood of the critical point C preturbulence (or transient, metastable chaos) can be observed. The trajectories starting in a neighbourhood of the nonattracting chaotic set remain close to it for a long time but finally leave it and approach P_H. The average length of chaotic transients increases when D_1 approaches the critical value C from above in the same way as in the logistic map. To explain this behaviour we have to return to the point $D_1'' \approx 1.122$ where two saddle periodic orbits P_{1+} and $\bar{P}_{1+}$ arise via a symmetry breaking bifurcation. In the following we omit the symmetric images. For $D_1'' < D_1 < C$ at least two attractors always coexist, the homogeneous periodic orbit P_H and an asymmetric attractor which is either periodic or chaotic. The basins of attraction of coexisting attractors have a common boundary. If $D_1'' < D_1 < C$, the orbit P_{1+} lies just in this boundary. In fact, the boundary of both basins of attraction contains the stable manifold of P_{1+} (which is formed by the set of all initial points for which the trajectory of Eqs. (6.13) asymptotically approach P_{1+}). For D_1 not too close to C the asymmetric attractor is well inside its basin

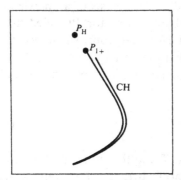

Fig. 6.33. Collision of the chaotic attractor with the unstable periodic orbit P_{1+} at $D_1 = C$, schematically.

of attraction. It develops a fold with increasing D_1 (see Fig. 6.30 (d)) and finally comes into contact with P_{1+} at $D_1 = C$ and a boundary crisis occurs. The situation is shown schematically in Fig. 6.33. This collision results in the loss of attracting character of the chaotic set. For D_1 slightly above C there is a large neighbourhood of the nonattracting chaotic set in which the trajectories are retained close to the chaotic set for a long time until they are ultimately repelled. This neighbourhood shrinks very fast with increasing D_1, hence the chaotic transients are not observable for D_1 not close to C.

6.2.8 Lyapunov exponents, dimensions and power spectra

LE spectra and Lyapunov dimensions computed for several chaotic attractors studied previously are given in Table 6.5. The chaotic attractors have LE spectra of type $(+, 0, -, -)$ and the Lyapunov dimension is slightly larger than 2. The other possible chaotic LE spectrum of type $(+, +, 0, -)$ was not

Table 6.5 *Lyapunov exponents λ_i and Lyapunov dimension D_L for several chaotic attractors associated with the period-doubling route to chaos; $A = 2$, $B = 5.9$, $q = 1$*

D_1	λ_1	λ_2	λ_3	λ_4	D_L
1.10	0.078	0.00	-2.68	-30.87	2.03
1.21	0.107	0.00	-2.69	-30.67	2.04
1.26	0.200	0.00	-2.82	-32.17	2.07

observed in this system. Numerical methods described in Chapter 4 were used in the computations. Two smallest LEs are sufficiently negative to ensure rapid contraction in two directions and this implies the observed almost one-dimensional behaviour of the Poincaré mapping.

The LE spectrum can also be approximated on the basis of a knowledge of only a few periodic orbits contained in the chaotic attractor. Provided that we know several such orbits (e.g. from continuation) we can approximately describe the dynamics on the attractor in such a way that the state space point will switch randomly among the periodic orbits. If we assume a uniform participation of all orbits, then

$$\lambda_i \approx \sum \tilde{\lambda}_i T / \sum T, \qquad i = 1, 2, ..., \tag{6.24}$$

where $\tilde{\lambda}_i$ are LEs of a periodic orbit of period T and the sum is over all known periodic orbits which are contained in the chaotic attractor. For example, using the orbits from Fig. 6.21 we obtain from (6.24) an estimate of the largest LE $\lambda_1 \approx 0.24$ at $D_1 = 1.26$ which is quite comparable with the value from Table 6.5.

Power spectra of chaotic orbits of the system (6.13), computed by the *Fast Fourier Transform method* are shown in Fig. 6.34. The power spectra are generally of two types. The first type contains periodic components (peaks) above a certain level of broad-band noise and the second type is without characteristic periodic components. In practice, the power spectra of chaotic attractors always have at least several orders of magnitudes higher noise levels than those of periodic attractors. The presence of periodic components in a chaotic spectrum expresses a high degree of *phase coherence*; the state space point more or less regularly visits several distinct regions of the attractor. It is clearly visible in Poincaré maps where the attractor consists of several pieces (see Fig. 6.30 (a), (c)) and the state space point visits individual pieces in a fixed order as in one-dimensional maps (see Section 3.3). Thus chaotic attractors from the reverse windows are expected to have power spectra with periodic components. The second type of power spectra does not reveal the superimposed periodic structure. The Poincaré maps of these attractors have a structure without separated pieces, see Fig. 6.30 (b), (d).

We define the mean frequency $\hat{f}$ as the frequency at which the numerically computed power spectrum reaches its global maximum. The time $\hat{T} = 1/\hat{f}$ is a *mean orbital time* of one loop around the attractor. For the periodic attractor P_n with the period T, $\hat{T} = T/n$ and the power spectrum has periodic components at integer multiples of $\hat{f}/n$. The transition $P_n \rightarrow P_{2n}$ is reflected in the spectrum by doubling the number of periodic components. Similarly attractors which form n-pieces in their Poincaré maps have n periodic components at integer multiples of $\hat{f}/n$ and the reverse bifurcation is accompanied by the halving of the number of periodic components in the spectrum. However, the mean

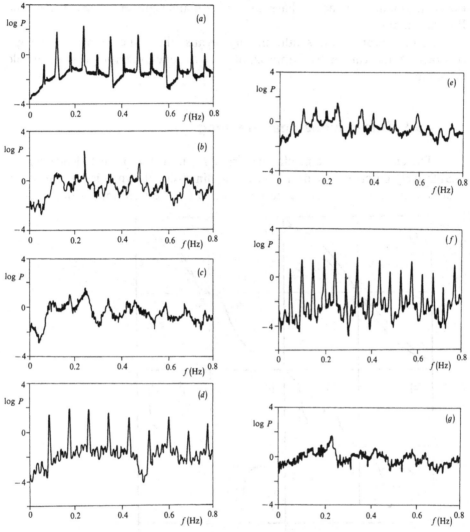

Fig. 6.34. Power spectra of various chaotic attractors. $A = 2$, $B = 5.9$, $q = 0.1$:
(a) $D_1 = 1.194\,0$; (b) $D_1 = 1.2$; (c) $D_1 = 1.21$; (d) $D_1 = 1.225$; (e) $D_1 = 1.235$;
(f) $D_1 = 1.245$; (g) $D_1 = 1.263$.

orbital time $\hat{T}$ does not relate to any specific periodic orbit contained in the chaotic attractor.

Thus the spectra in Fig. 6.34 (a), (b), (d), (f) correspond to phase coherent attractors. The spectrum in Fig. 6.34 (b) is obtained from that in Fig. 6.34 (c) after two reverse bifurcations; the spectra in Fig. 6.34 (d) and 6.34 (f) correspond to attractors with three and five islands in their Poincaré maps, respectively. The remaining spectra in Fig. 6.34 (c), (e) contain only broad-band noise and the corresponding attractors have an approximately connected structure in the Poincaré maps.

Such an interpretation is valid mainly because the Poincaré map is strongly contracting and can be well approximated by a one-dimensional nonivertible mapping.

6.2.9 One-dimensional mapping

The entire sequence of changes from P_{1s} until the preturbulence can be qualitatively described by means of a one-dimensional dynamics. Because the chaotic attractor of the Poincaré mapping Z can be well approximated by a

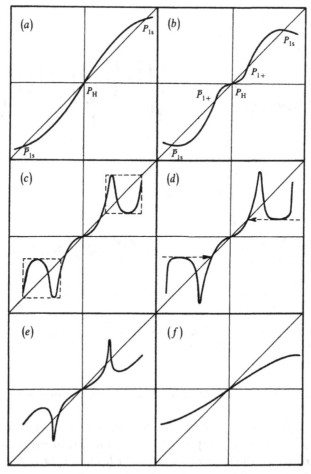

Fig. 6.35. One-dimensional map as a model of the evolution of chaos. See text for the description.

curve in Σ, we can reduce Z only on this curve and after proper transformation of coordinates we get a noninvertible mapping f of the interval to itself. The price for this simplification is the loss of the fine fractal structure which cannot occur in a one-dimensional state space. An advantage of the use of the one-dimensional mapping follows from a simple geometrical interpretation.

If D_1 is changed we obtain a one-parameter family of maps $f_{D_1}: \mathsf{R} \to \mathsf{R}$ which are shown qualitatively in several typical situations in Fig. 6.35. The function f is odd, which follows from the symmetry of two coupled cells. The fixed points of the Poincaré mapping (i.e. the orbits P_H, P_1 and $\bar{P}_1$) correspond to fixed points of f, which are determined by the intersections of the graph of $f(x)$ with the diagonal line. The point P_H is located at the origin, the points P_1 and $\bar{P}_1$ are symmetric with respect to the origin. Fixed points are stable (unstable) if $|df/dx|$ at the given point is less (greater) than 1. If $D_1 < D_1'' \approx 1.122$ (see Fig. 6.35 (a)) then P_H is unstable and forms the boundary of basins (i.e. of intervals) of attraction of stable points P_{1s} and $\bar{P}_{1s}$. P_H is stable for $1.122 \lesssim D_1 \lesssim 1.172$ and points P_{1+} and $\bar{P}_{1+}$ separate the basins of attraction of the points P_H, P_{1s} and $\bar{P}_{1s}$, see Fig. 6.35 (b). For $D_1 > 1.172$ P_{1s} passes into the unstable P_{1-} (similarly for the mirror image orbits) and the sequence of period-doubling bifurcations leads to chaotic region with the window structure. The function f develops two extremes in the chaotic region on both sides of the origin hence a multiple attractor behaviour is possible. A typical graph of the function f in the chaotic region is shown in Fig. 6.35 (c). The situation is similar to that in Fig. 6.35 (b), but in place of P_{1s} and $\bar{P}_{1s}$ we now have chaotic attractors located in the invariant region denoted in the figure. The boundary crisis at $D_1 \approx 1.263\,5$ is shown in Fig. 6.35 (d), chaotic attractors now touch the points P_{1+} and $\bar{P}_{1+}$. Two pairs of fixed points still exist in the interval $1.263 \lesssim D_1 \lesssim 1.472\,4$, but the chaotic attractors cannot occur (see Fig. 6.35 (e)); finally, when $D_1 \gtrsim 1.472\,4$ *then the only fixed point of f* (which is at the same time an attractor) *is the orbit* P_H, see Fig. 6.35 (f).

The discussion based on Fig. 6.35 is only illustrative, not quantitative. It is a question whether a more careful construction of the one-dimensional model can give more detailed information which can be carried over to the original problem.

6.2.10 Transition from torus to chaos

All transitions of periodic oscillations into chaotic ones which have been described can be found in the region of the parameter plane (B, D_1) which is shown in Fig. 6.27 (a). There is another region in the same parameter plane containing a transition from the homogeneous periodic orbit P_H to the Δ-symmetric periodic orbit P_{NA} then to a two-dimensional torus and finally to a chaotic attractor, see Fig. 6.27 (b). Let us describe this transition in more detail.

We again study the evolution of attractors by varying D_1 and keeping B fixed at $B = 5.5$. We are interested in the range $0.052 \lesseqgtr D_1 \lesseqgtr 0.053$, i.e. where the intensity of the interaction between the reaction cells is two orders of magnitude lower than in the previous case.

Table 6.6 *Regions of existence of various types of attractors of* (6.13) *for* $0.05 \lesseqgtr D_1 \lesseqgtr 0.055$; $A = 2$, $B = 5.5$, $q = 0.1$

Range of values of D_1	Type of attractors
$D_1 > 0.052\,95$	P_{H}
$0.052\,95 - 0.052\,46$	quasiperiodic orbit on a torus
$0.052\,46 - 0.052\,44$	periodic orbit on a torus with the rotation number 1/3
$0.052\,44 - 0.052\,39$	quasiperiodic orbit on a torus
$0.052\,38 - 0.052\,06$	chaotic attractor
$D_1 < 0.052\,06$	$S_{N1},\ \bar{S}_{N1}{}^{(+)}$

$^{(+)}$ Stationary points are stable in the entire followed range of D_1 and cause multistability for $D_1 > 0.052\,06$

Typical attractors that emerge as D_1 is decreased are characterized in Table 6.6. A Δ-symmetric solution P_{NA} ('anti-phase' solution) arises via period-doubling bifurcation from the stable periodic solution P_{H}. This orbit becomes unstable when a pair of complex conjugate multipliers crosses the unit circle in the complex plane (a bifurcation of the (T) type) and a stable Δ-symmetric torus is formed. Typical phases of the further evolution of attractors are depicted in Fig. 6.36 by means of the Poincaré mapping. The torus (see Fig. 6.36 (*a*)) becomes resonant (see Fig. 6.36 (*b*)), i.e. a pair of periodic orbits, one stable and the other unstable, arises on the torus via saddle-node bifurcation; both orbits continue to exist in some interval of the parameter and finally coalesce in another saddle-node point. Then folds develop on the attractor (see Fig. 6.36 (*c*)) which indicates that the torus disintegrates and the chaotic attractor is formed. At first the chaotic attractor consists of two disjoint parts (see Fig. 6.36 (*d*)), later both parts approach each other and finally are joined (see Fig. 6.36 (*e*) and (*f*)). The chaotic attractor disappears via a boundary crisis (probably by a collision with the unstable orbit P_{NI}^2) and all trajectories are attracted to one of the stationary points S_{N1} or $\bar{S}_{\mathrm{N1}}$ (as transient chaotic trajectories).

We can summarize further considerations about the structure of this transition to chaos in the following way:

(a)   The torus is represented by two closed curves in the Poincaré map Z which are visited alternately by the evolving trajectory. This topology is implied by the topology of the Δ-symmetric orbit P_{NA}^1 which intersects the hyperplane Σ in a pair of asymmetric points mutually related by the symmetry (6.14).

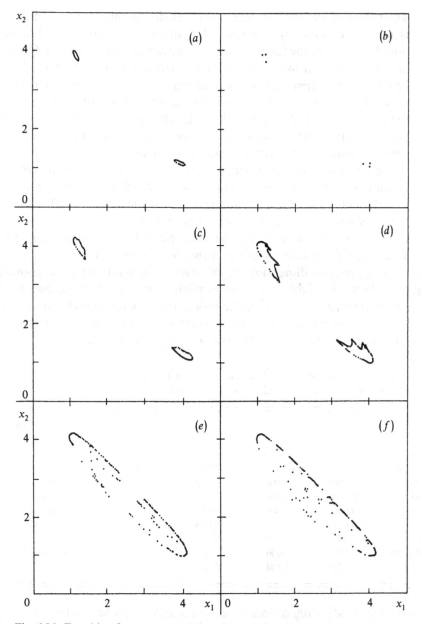

Fig. 6.36. Transition from torus to chaos represented by Poincaré maps. $A = 2$, $B = 5.5$, $q = 0.1$: (a) torus, $D_1 = 0.052\,47$; (b) periodic orbit with the rotation number $\frac{1}{3}$, $D_1 = 0.052\,45$; (c) torus, $D_1 = 0.052\,42$; (d) chaotic attractor, $D_1 = 0.052\,3$; (e) chaotic attractor, $D_1 = 0.052\,2$; (f) chaotic attractor, $D_1 = 0.052\,1$.

(b)   The orbits on the torus can be quasiperiodic, in this case the entire torus is densely covered by a single orbit and it is an attractor. The other possibility is that the torus contains an attracting periodic orbit. Similarly, as in diffeomorphisms of the circle (see Section 3.3), a rotation number can be introduced when we study second iterates of Z restricted to one of the closed curves which represent the torus. The rotation number expresses the average speed of rotation along the closed curve. The torus contains quasiperiodic orbits if the rotation number is irrational or an attracting periodic orbit if the rotation number is rational.

(c)   There are two kinds of periodic orbits on the torus, $\varDelta$-symmetric and asymmetric; the latter ones occur in pairs related by the symmetry (6.14).

(d)   It is difficult to determine numerically the critical value of D_1 where the torus ceases to exist; at this point infinitely many folds are expected to occur. This bifurcation is a global one and is, in general, associated with a lack of differentiability of the torus, see Section 3.4.

LEs and Lyapunov dimension D_L of several attractors near the transition point are shown in Table 6.7. Quasiperiodic torus is characterized by LE spectrum of type $(0, 0, -, -)$ and chaotic attractors have spectra of type $(+, 0, -, -)$. Lyapunov dimension considerably exceeds two indicating that one-dimensional approximation of the Poincaré map is not valid here.

Table 6.7 *Lyapunov exponents λ_i and Lyapunov dimension D_L for several attractors characterizing a transition from torus to chaos;* $A = 2$, $B = 5.5$, $q = 0.1$

D_1	1	2	3	4	D_L	Type of the attractor
0.052 47	0.000	0.000	−0.011	−2.684	2	torus
0.052 45	0.000	−0.010	−0.010	−2.681	1	periodic
0.052 42	0.000	0.000	−0.034	−2.677	2	torus
0.052 39	0.000	0.000	−0.049	−2.673	2	torus
0.052 38	0.009	0.000	−0.063	−2.671	2.14	chaotic
0.052 30	0.043	0.000	−0.172	−2.658	2.25	chaotic
0.052 20	0.068	0.000	−0.42	−2.57	2.16	chaotic
0.052 10	0.075	0.000	−0.58	−2.47	2.13	chaotic

A better understanding of this transition to chaos can be reached by varying two parameters, B and D_1 (see Fig. 6.27 (b)). Numerical computations show that the dynamics is quite similar to that of the simple oscillator periodically perturbed by discrete pulses described in Chapter 2.

Interesting intermittent oscillations without the presence of a chaotic attractor are observed in a different parameter region. As before let $A = 2$, $B = 5.5$ but we now change the parameter q from the value $q = 0.1$ to $q = 2$, i.e. the diffusion of the reaction component X is now slightly preferred. This leads to large changes in the phase portrait. All nonhomogeneous stationary states are unstable and they exist only for much higher values of B than were used for computations. Thus changing the free parameter D_1, we cannot find any stable

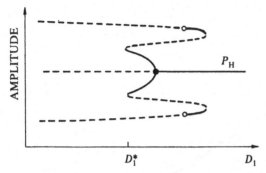

Fig. 6.37. Solution diagram for periodic orbits in the region of the intermittent transition to torus, schematically. Full lines – stable orbits, dashed lines – unstable orbits.

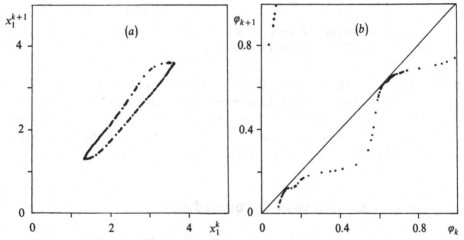

Fig. 6.38. Poincaré maps of the intermittent torus, $A = 2$, $B = 5.5$, $q = 2$, $D_1 = 0.043$: (a) plot of the pairs (x_1^{k+1}, x_1^k); (b) the circle map reconstructed from (a).

stationary state. The only attractor is the homogeneous periodic orbit which exists for D_1 sufficiently large. If we decrease D_1 a symmetry breaking bifurcation is observed (see Fig. 6.37) leading to a pair of stable asymmetric periodic orbits. These orbits lose their stability at a pair of limit points at $D_1 = D_1^*$. More complicated oscillations occur to the left of the limit points.

The usual Poincaré map projected to the (x_1, x_2) plane is not very instructive. So we took the first coordinate x_1 of the points of intersection with Σ and plotted the pairs (x_1^{k+1}, x_1^k) obtained from two successive iterations (Fig. 6.38 (a)). This figure clearly reveals that the attractor is a two-dimensional torus. If D_1 is close to D_1^*, the motion on the torus is intermittent.

The laminar phase is characterized by a local slowing down of rotation along the transversal section of the torus at two points correspondig to the pair of limit

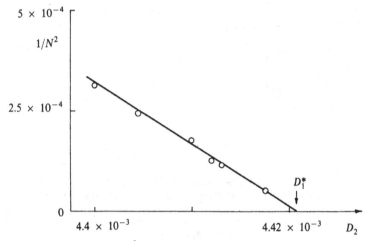

Fig. 6.39. The plot of $1/N^2$ versus D_2; N is the average number of cycles during the laminar phase. The linear dependence confirms the scaling law (3.22).

points. This is clearly visible on the reconstructed circle map in Fig. 6.38 (*b*). There are two laminar channels similar to those described in Section 3.3.4. The 'turbulent' bursts correspond to an accelerated rotation but the overall motion is restricted to the torus and hence it cannot be chaotic. Nevertheless the length of the laminar phase scales according to Eq. (3.22), as shown in Fig. 6.39.

6.2.11 Larger arrays of coupled cells

Complicated stationary and dynamic dissipative structures have been observed in experimental studies in a hexagonal structure of seven mutually coupled reaction cells[5.174]. Studies of analogous model systems using the Brusselator kinetic scheme have also revealed the existence of complex periodic and chaotic regimes[6.5] and we shall now briefly discuss them to illustrate development of these regimes with increasing number of cells.

Let us consider simple unbranched linear or cyclic arrays of cells, where each cell exchanges mass both with the environment and with its neighbours (each cell in a cyclic array has two neighbours; the boundary cells in a linear array have just one neighbour). Let us assume that all cells operate under the same restrictions (equal mass exchange with the environment and the same values of reaction kinetic parameters) and that the coupling between the cells is linear, symmetric, and equal for all coupled cells.

Mass balances for the system of N coupled cells containing n independent reaction components may then be written in the form

$$\frac{dc_i^k}{dt} = f_i(c_1^k, ..., c_n^k) + \sum_{j=1}^{n} d_{ij} \sum_{l=1}^{N} \delta^{lk}(c_j^l - c_j^k),$$

$$i = 1, ..., n, \quad k = 1, ..., N. \quad (6.25)$$

Here c_i^k denotes the concentration of the i-th reaction component in the k-th cell, $f_i: \mathbf{R}^n \to \mathbf{R}$ are functions describing reaction kinetics and mass exchange with the environment. The rate of transport of the i-th component between the cells is proportional to the gradient between the l-th and k-th cell as long as the coupling between these cells exists ($\delta^{lk} = 1$ when the coupling exists, otherwise $\delta^{lk} = 0$). It follows from symmetry that $\delta^{lk} = \delta^{kl}$. When the cross-fluxes between the components are neglected, then the matrix $\mathbf{D}$ of the transport coefficients d_{ij} has all off-diagonal components equal to zero.

If we denote the diagonal elements of $\mathbf{D}$ by D_i, $i = 1, ..., n$, then for one-dimensional arrays with the Brusselator kinetics the mass balances (6.25) can be written as

$$\frac{dx_k}{dt} = A - (B + 1) x_k + x_k^2 y_k + D_1(x_{k-1} + x_{k+1} - 2x_k),$$

$$\frac{dy_k}{dt} = Bx_k - x_k^2 y_k + D_2(y_{k-1} + y_{k+1} - 2y_k), \quad (6.26)$$

$$k = 1, ..., N$$

with zero-flux boundary conditions

$$x_0 = x_1, \quad x_{N+1} = x_N,$$

$$y_0 = y_1, \quad y_{N+1} = y_N,$$

for linear arrays and with periodic boundary conditions

$$x_0 = x_N, \quad x_{N+1} = x_1,$$

$$y_0 = y_N, \quad y_{N+1} = y_1,$$

for cyclic arrays.

Let us again choose $A = 2$, $q = 0.1$ for numerical computations. Various types of attractors exist for this choice, when values of two remaining parameters B and D_1 are chosen properly. The computations show that sustained nonhomogeneous oscillations may exist for a certain values of D_1 if $5 < B \lesssim 7$. When $B \gtrsim 5.8$, a strong hysteresis among the homogeneous and nonhomogeneous regimes exists.

We are interested in the dependence of solutions of (6.26) on two most interesting parameters, D_1 and N, the intensity of interaction (mass exchange) between the cells and the number of cells in an array, respectively. Linear arrays

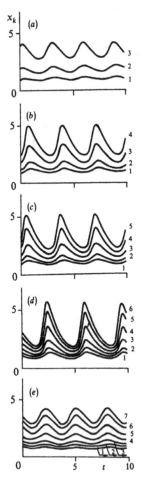

Fig. 6.40. Periodic oscillations of the components x_k, $k = 1, ..., N$. $A = 2$, $B = 5.5$, $q = 0.1$: (a) $N = 3$, $D_1 = 2.6$; (b) $N = 4$, $D_1 = 4.6$; (c) $N = 5$, $D_1 = 7.1$; (d) $N = 6$, $D_1 = 10.6$; (e) $N = 7$, $D_1 = 12.5$.

up to seven cells and cyclic arrays up to six cells were studied. The value of D_1 was varied from zero up to values for which only homogeneous structures are possible due to the high intensity of interaction.

Nonhomogeneous structures

For $B > 1 + A^2$ we always find a homogeneous periodic orbit P_H of the system (6.26). When the value of $B = 5.5$ is fixed, then numerical simulations show that P_H is stable both for sufficiently low and sufficiently high values of D_1. Thus, P_H is stable for $D_1\{(0, D_1') \cup (D_1'', \infty)\}$. Let us follow the sequence of stable oscillatory regimes by varying D_1 at fixed $B = 5.5$.

A simpler situation occurs when the value of D_1 passes through D_1'' from the right to the left. Regardless of the number of cells in the array, two stable asymmetric periodic orbits always branch off P_H (a symmetry breaking bifurcation). Two asymmetric orbits (resulting from the reflection symmetry) have monotonic concentration profiles along the array, see Figs 6.40 (a)–(e). When the value of D_1 is decreased, the amplitude of the periodic orbits decreases and finally the orbits 'shrink' into two asymmetric steady states (see Fig. 6.41). For

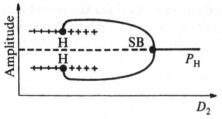

Fig. 6.41. Solution diagram of periodic orbits and stationary points, schematically: +++++ stable stationary points; + + + + unstable stationary points; ——— stable periodic orbits; – – – – – unstable periodic orbits; H – Hopf bifurcation; P_H – homogenous periodic orbit; SB – symmetry breaking bifurcation.

larger values of B the bifurcation at $D_1 = D_1''$ becomes subcritical and the branching pair of asymmetric orbits is unstable. This causes hysteresis and a jump to a new regime which can be chaotic.

A more complex sequence of bifurcations arises when D_1 passes through the lower critical value D_1' from the left to the right. A single periodic orbit branches off P_H via a period-doubling bifurcation; further development with increasing D_1 may also lead to chaos. A final disappearance of the orbits bifurcating at D_1' seems to be caused by the interaction of stable periodic and unstable steady states leading to a homoclinic or a heteroclinic loop (similar to the case of two coupled cells).

Chaotic attractors

Subsequent bifurcations leading to the formation of a chaotic invariant set correspond to two scenarios:

(a) sequence of period-doubling bifurcations,
(b) transition via a two-dimensional torus.
In general, the second type of scenario includes several possibilities; the two most important are: (i) formation of a three-dimensional torus, and following formation of a chaotic set or (ii) a synchronization on a two-dimensional torus, its destruction resulting from the bifurcations of resonant (i.e. periodic) solutions, and the subsequent formation of a chaotic set. In the cases discussed here the transition via destruction of the two-torus seems to occur. The chaotic set formed can be either attracting or nonattracting, depending on the values of parameters B and D_1. The transition between these two states may be accompanied either by intermittency or by transient chaotic oscillations. Both phenomena can often be observed in this system.

The scenario of period-doubling bifurcations, the subsequent formation of a chaotic attractor accompanied by intermittency and its destruction indicated by a transient chaos, is observed when $B = 5.9$ for all N. The region of existence of chaotic oscillations in the parametric space is directly connected with the regions of existence of asymmetric periodic orbits with monotonic profiles (see Fig. 6.40). The basic branch of periodic orbits on which the period-doubling occurs arise at the point D_1'', i.e. as a pair of asymmetric orbits. This bifurcation is subcritical (for $B \gtrsim 5.8$), see Fig. 6.42. The chaotic set exists for values of D_1

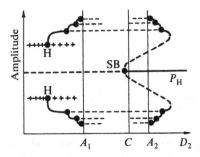

Fig. 6.42. Region of chaotic oscillations occuring via the period-doubling sequence, schematically. For legend see Fig. 6.41.

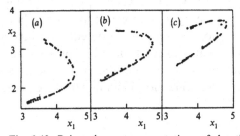

Fig. 6.43. Poincaré map representations of chaotic attractors from the parameter region shown in Fig. 6.42. $A = 2$; $B = 5.9$, $q = 0.1$: (a) $N = 3$, $D_1 = 2.814$; (b) $N = 4$, $D_1 = 4.77$; (c) $N = 5$, $D_1 = 7.27$.

between two accumulation points of the period-doubling bifurcations A_1, A_2. In the interval $A_1 < D_1 < C$ either a chaotic or a periodic attractor exists; at the point C a boundary crisis accompanied by a transient chaos occurs, and in the interval $C < D_1 < A_2$ the chaotic set is not attracting. The corresponding chaotic oscillations always have a strictly monotonic profile. The above scheme holds independently of the value of N, but the interval of the values of D_1 where the chaotic set exists is shifted with increasing N to higher values.

Poincaré maps of chaotic attractors close to the point C of a boundary crisis are for $N = 3, 4, 5$ shown in Fig. 6.43 (a)–(c). The Poincaré section Σ in the $2N$-dimensional state space of Eqs. (6.26) can be obtained by a generalization of the Poincaré section for two coupled cells. We set

$$\Sigma = \{(x_1, y_1, ..., x_N, y_N) \in \mathsf{R}^{2N}; \ \sum_{k=1}^{N} (x_k - y_k) + N(B/A - A) = 0\}$$

and use the projection into the (x_1, x_2) plane. Table 6.5 shows that for $N = 2$ the Lyapunov dimension of the attractor only slightly exceeds 2, i.e. the Poincaré map representation of the attractor is approximately located along a one-dimensional curve. This 'one-dimensional behaviour' is independent of N, as is evident from the comparison of Figs 6.30 (a) and 6.43 (a)–(c). It is probably caused by the fact that with increasing N the values of D_1 at which the chaotic attractor exists also increase and thus the relative intensity of dissipation in the system is preserved, i.e. the number of degrees of freedom increases with N but the increased dissipation causes contraction to a set of an approximately constant low dimension.

The second scenario, the transition via a torus, was observed for $B = 5.5$, i.e. in the region where no hysteresis between the homogeneous and nonhomogeneous regimes exists. Here, however, a simple transition to chaotic behaviour independent of N was not found. An example of a chaotic attractor resulting from a torus destruction is shown in Fig. 6.44 for $N = 5$.

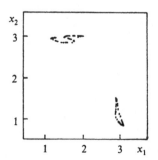

Fig. 6.44. Chaotic attractor after a torus destruction: $A = 2$, $B = 5.5$; $q = 0.1$; $D_1 = 0.014$; $N = 5$.

In cyclic arrays, the period-doubling transition to chaos which could be compared with that occurring in linear arrays was not found. However, an analogous dynamic behaviour is very likely to occur after proper adjustment of the values of A, B and q.

An analogue of chaotic behaviour arising via a torus destruction in the linear array with $N = 5$ exists for the cyclic array with the same number of cells. As in the linear array, the spatial structure is not preserved in a simple way with varying N.

REFERENCES

6.1   Alexander J. C. Spontaneous oscillations in two 2−component cells coupled by diffusion. Preprint, Maryland, Univ. of Maryland 1984.

6.2   Alligood K. T., Yorke E. D. and Yorke J. A. Why period−doubling cascades occur: Periodic orbit creation followed by stability shedding. *Physica* **28D** (1987) 197.

6.3   Aronson D. G., Doedel E. J. and Othmer H. G. An analytical and numerical study of the bifurcations in a system of linearly − coupled oscillators. *Physica* **25D** (1987) 20.

6.4   Babloyantz A. and Destexhe A. Is the normal heart a periodic oscillator ? *Biol. Cybern.* **58** (1988) 203.

6.5   Bailey J. E. Periodic operation of chemical reactors: A review. *Chem. Eng. Commun.* **1** (1973) 111.

6.6   Bamon R., Malta I. P., Pacifico M. J. and Takens F. Rotation intervals of endomorphisms of the circle. *Ergod. Th. & Dynam. Sys.* **4** (1984) 493.

6.7   Bar−Eli K. On the stability of coupled chemical oscillations. *Physica* **14D** (1985) 242.

6.8   Belair J. and Glass L. Universality and self−similarity in the bifurcations of circle map. *Physica* **16D** (1985) 143.

6.9   Blechman I. I. *Synchronization of Dynamical Systems*. Moscow, Nauka, 1971 (in Russian).

6.10   Blechman I. I. *Synchronization in Nature and Technics*. Moscow, Nauka, 1980 (in Russian).

6.11   Bryant C. and Jeffries C. Bifurcations of a forced magnetic oscillator near points of resonance. *Phys. Rev. Lett.* **53** (1984) 250; The dynamics of phase locking and points of resonance in a forced magnetic oscillator. *Physica* **25D** (1987) 196.

6.12   Buchholtz F., Freund A. and Schneider F. W. Periodic perturbation of the BZ−reaction in a CSTR : Chemical resonance, entrainment and quasi−periodic behaviour. In *Temporal Order*, ed. L. Rensing and N. I. Jaeger, Springer Series on Synergetics **29**, Berlin, Springer, 1984.

6.13   Dolník M., Finkeová J., Schreiber I. and Marek M. Dynamics of forced excitable and oscillatory chemical reaction systems. J. Phys. Chem. 93 (1989) 2 764.

6.14   Dolník M. and Marek M. Extinction of oscillations in forced and coupled reaction cells. *J. Phys. Chem.* **92** (1988) 2452.

6.15   Dolník M., Padušáková E. and Marek M. Periodic and aperiodic regimes in coupled reaction cells with pulse forcing. *J. Phys. Chem.* **91** (1987) 4407.

6.16   Dolník M., Schreiber I. and Marek M. Periodic and chaotic regimes in forced chemical oscillator. *Phys. Lett.* **100A** (1984) 316.

6.17   Dolník M., Schreiber I. and Marek M. Dynamic regimes in a periodically forced reaction cell with oscillatory chemical reaction. *Physica* **21D** (1986) 78.

6.18   Fiedler B. Global Hopf bifurcation of two parameter flows. *Arch. Rat. Mech. Anal.* **94** (1986) 59.

6.19   Field R. J. and Noyes R. M. Oscillations in chemical systems IV. Limit cycle behaviour in a model of a real chemical reaction. *J. Chem. Phys.* **60** (1974) 1877.

6.20   Glass L. and Winfree A. T. Discontinuities in phase resetting experiments. *Am. J. Physiol.* **246** (1984) R251.

6.21   Glass L., Graves C., Petrillo G. A. and Mackey M. C. Unstable dynamics of a periodically driven oscillator in the presence of noise. *J. Theor. Biol.* **86** (1980) 455.

6.22   Glass L., Guevara M. R., Shrier A. and Perez R. Bifurcation and chaos in a periodically stimulated cardiac cell. *Physica* **7D** (1983) 89.

6.23   Glass L., Guevara M. R., Belair J. and Shrier A. Global bifurcations of a periodically forced biological oscillator. *Phys. Rev.* **A29** (1984) 1348.

6.24   Gmitro J. I. and Scriven L. E. A physicochemical basis for pattern and rhythm. In *Intracellular Transport*, ed. K. B. Warren, New York, Academic Press, 1966.

6.25   Grebogi C., Ott E. and Yorke J. A. Basin boundary metamorphoses: Changes in accessible boundary orbits. *Physica* **24D** (1987) 243.

6.26   Guevara M. R. and Glass L. Phase locking, period doubling bifurcations and chaos in a mathematical model of a periodically driven oscillator : A theory for the entrainment of biological oscillators and the generation of cardiac disrhythmias. *J. Math. Biol.* **14** (1982) 1.

6.27   Guevara M. R., Glass L. and Shrier A. Phase locking, period doubling bifurcations and irregular dynamics in periodically stimulated cardiac cells. *Science* **214** (1981) 1350.

6.28   Hanin J. (Ed.) *Dynamics of Cholinergic Function.* Adv. in Behav. Biol. **30**, New York, Plenum Press, 1986.

6.29   Herman M. R. Mesure Lebesgue et nombre de rotation. In *Geometry and Topology*, Lecture Notes in Math. **579**, New York, Springer, 1977, p. 271.

6.30   Kaneko K. Collapse of tori and genesis of chaos in nonlinear nonequilibrium systems. *Prog. Theor. Phys.* **72** (1984) 1084.

6.31   Kawato M. Transient and steady state phase response curves of limit cycle oscillators. *J. Math. Biol.* **12** (1981) 13.

6.32   Kawato M. and Suzuki R. Two coupled neural oscillators as a model of the circadion pacemaker. *J. Theor. Biol.* **86** (1980) 547.

6.33   Keener J. P. and Glass L. Global bifurcations of a periodically forced nonlinear oscillator. *J. Math. Biol.* **21** (1984) 175.

6.34   Kevrekidis I. G., Aris R. and Schmidt L. D. The stirred tank forced. *Chem. Engng. Sci.* **41** (1986) 1549.

6.35   Klíč A. Period doubling bifurcations in a two–box model of the Brusselator. *Aplikace Matematiky* **28** (1983) 335; Bifurcations of the periodic solutions in symmetric systems. *Aplikace Matematiky* **31** (1986) 37.

6.36   Lefever R. and Prigogine I. Symmetry breaking instabilities in dissipative systems. *J. Chem. Phys.* **48** (1968) 263.

6.37   Marek M. Periodic and aperiodic regimes in forced chemical oscillations. In *Temporal Order*, ed. L. Rensing and N. I. Jaeger, Berlin, Springer, 1984, p. 105.

6.38   Marek M. Complex behaviour of coupled reaction–diffusion–convection cells. In *Spatial Inhomogeneities and Transient Behaviour in Chemical Kinetics*, Manchester, Manchester Univ. Press, 1988.

6.39   Marek M., Dolník M. and Schreiber I. Dynamic patterns in interacting chemical cells and effects of external periodic forcing. In *Self Organization by Nonlinear Irreversible Processes*, ed. W. Ebeling and H. Ulbricht, Berlin, Springer, 1986, p. 133.

6.40   Martinez H. M. Morphogenesis and chemical dissipative structures. A computer simulated case study. *J. Theor. Biol.* **36** (1972) 479.

6.41   Missiurewicz M. Twist sets for maps of the circle. *Ergod. Th. & Dynam. Sys.* **4** (1984) 391.

6.42   Nicolis G. and Prigogine I. *Self–Organization in Non–Equilibrium Systems.* New York, Wiley, 1977.

6.43   Noszticzius Z., Farkas H. and Schelly Z. A. Explodator : A new skeleton mechanism for the halate driven chemical oscillators. *J. Chem. Phys.* **80** (1984) 6062.

6.44 Perez—Pascual R. and Lomnitz—Adler J. Coupled relaxation oscillators and circle maps. *Physica* **30D** (1988) 61.

6.45 Raschman R., Schreiber I. and Marek M. Periodic and aperiodic regimes in linear and cyclic arrays of coupled reaction—diffusion cells. In Lect. in Appl. Math. **24**, Providence, A. M. S., 1986, p. 81.

6.46 Ruoff P. Phase—response relationships of the closed bromide—perturbed Belousov—Zhabotinsky reaction. Evidence of bromide control of the oscillating state without use of a bromide—detecting device. *J. Phys. Chem.* **88** (1984) 2851.

6.47 Schneider F. W. Periodic perturbations of chemical oscillators: Experiments. *Ann. Rev. Phys. Chem.* **36** (1985) 347.

6.48 Schreiber I. and Marek M. Strange attractors in coupled reaction—diffusion cells. *Physica* **5D** (1982) 258.

6.49 Schreiber I. and Marek M. Transition to chaos via two—torus in coupled reaction diffusion cells. *Phys. Lett.* **91A** (1982) 263.

6.50 Schreiber I., Holodniok M., Kubíček M. and Marek M. Periodic and aperiodic regimes in coupled dissipative chemical oscillators. *J. Stat. Phys.* **43** (1986) 489.

6.51 Schreiber I., Kubíček M. and Marek M. On coupled cells. In *New Approaches to Nonlinear Problems in Dynamics*, ed. by P. J. Holmes, Philadelphia, SIAM Publ., 1980, p. 496.

6.52 Schreiber I., Kubíček M. and Marek M. Impaired diffusion coupling — source of arrhythmia in cell systems. *Z. Naturforsch.* **39c** (1984) 1170.

6.53 Ševčíková H., Suchanová D. and Marek M. The patterns of phase resetting of oscillations in Zhabotinski reaction. *Sci. Papers of the Prague Inst. Chem. Technol.* **K17** (1982) 137.

6.54 Sporns O., Roth S. and Seeling F. F. Chaotic dynamics of two coupled biochemical oscillators. *Physica* **26D** (1987) 215.

6.55 Swift J. W. and Wiesenfeld K. Suppression of period doubling in symmetric systems. *Phys. Rev. Lett.* **52** (1984) 705.

6.56 Tyson J. and Kauffman S. Control of mitosis by a continuous biochemical oscillation: Synchronization, spatially inhomogeneous oscillations. *J. Math. Biol.* **1** (1975) 289.

6.57 Wang X.—J. and Nicolis G. Bifurcation phenomena in coupled oscillators: Normal form analysis and numerical simulations. *Physica* **26D** (1987) 140.

6.58 Winfree A. T. *The Geometry of Biological Time*. New York, Springer, 1980.

7

Chaos in distributed systems, perspectives

7.1 Introduction

Several strategies have been followed in theoretical investigations of spatio-temporal coherence and chaos. Spatially distributed nonlinear dynamical systems have been studied as cellular automata[7.3, 7.28, 7.94] coupled map lattices[7.4, 7.41, 7.57, 7.63, 7.76, 7.90] and as partial differential equations (PDEs) which, after a proper reduction[7.51, 7.5] or discretization, are studied by various numerical or analogue methods[7.14, 7.17, 7.22, 7.54].

Spatially extended dynamical systems may be classified according to continuity or discreteness in space, time and local state. Thus PDEs are continuous in space, time and local state, iterated functional equations are discrete in time and continuous in space and local state, chains of oscillators (discussed in Chapter 5) are discrete in space and continuous in time and local state, lattice dynamical systems are discrete in space and time but continuous in local state and, finally, cellular automata are discrete in all three aspects.

Computer time requirements for the study of evolution of spatio-temporal patterns increase from cellular automata to PDEs drastically. However, seemingly different models of different physical systems have revealed the same types of phenomena (stationary space-time patterns, similar routes to chaos, space-time intermittency). Relations between the results reached through different approaches are gradually being developed. It is expected that in some cases mappings from classes of coupled-map lattices and cellular automata to classes of PDEs will be established in the future.

Various examples of experimental observations of chaotic behaviour in distributed electronic, solid state, chemical and hydrodynamic systems have been discussed in Chapter 5. In this chapter we shall first briefly review most important approaches to a description of spatio-temporal patterns in cellular automata, coupled map lattices and most often studied systems described by nonlinear PDEs and then finish with remarks on perspective role of studies of chaotic motions.

7.2 Cellular automata

Cellular automata, which were introduced by von Neumann[7.80], are discrete dynamical systems consisting of a discrete lattice of cells and a rule which operates on the lattice. Each cell of the lattice has a finite number of states (for example zeros and ones) but the lattice may be infinite in extent. The rule (deterministic and translationally invariant) gives the state of a cell in terms of the states of cells in a finite neighbourhood about the cell at the previous time step[7.33, 7.80, 7.92]. Wolfram[7.94] presents two simple examples of cellular automata: (a) an automaton, where the rule is defined as $a_i^{(t+1)} = (a_{i-1}^{(t)} + a_{i+1}^{(t)}) \bmod 2$ and (b) an automaton, where the site values 0 or 1 are updated at each step according to the rule $a_i^{(t+1)} = a_{i-1}^{(t)} \oplus (a_i^{(t)} \vee a_{i+1}^{(t)})$ (here $\oplus$ denotes addition modulo 2, and $\vee$ Boolean disjunction). Patterns generated by such a cellular automaton's evolution from a simple initial state appears random.

Let us briefly illustrate the above definition on the class of one dimensional automata considered by Wilbur, Lipman and Shamma[7.91]. Any cell is required to have a state in the set $B = \{0, 1\}$ of binary digits. A rule is any mapping R: $B^3 \rightarrow B$. Hence there are 2^8 different rules corresponding to the decimal numbers 0 to 255. Wolfram[7.92] uses these numbers to represent the corresponding rules. Let us for any positive k denote a k-tuple any element of B^k, $(i_1, i_2, ..., i_k)$. For a given rule there is a function r defined on the set of all k-tuples, $k \geq 3$, mapping a k-tuple into a k-2 tuple. Let us define the function r,

$$r(i_1, i_2, ..., i_k) = (R(i_1, i_2, i_3), R(i_2, i_3, i_4), ..., R(i_{k-2}, i_{k-1}, i_k)), \qquad (7.1)$$

and a circularizing function c,

$$c(i_1, i_2, ..., i_k) = (i_k, i_1, i_2, ..., i_k, i_1). \qquad (7.2)$$

A cellular automaton $a(R, N)$ is defined by the rule R and a positive integer N as the composite mapping $r \, . \, c$ restricted to act on N-tuples. Any N-tuple is a possible state of the cellular automaton. A local pattern is a k-tuple with k much smaller than N.

It is difficult to determine the nature of the evolution of a cellular automaton directly from a description of the rule. A simply described rule operated iteratively from a simple initial state may give rise to very complex distributions at large time. A number of classifications of rules and attractors were introduced and discussed in Refs 7.33 and 7.94. In principle, the evolution may lead to a homogeneous state (for example, all sites have value 0) or to a set of stable or periodic structures or to a chaotic pattern. Monte Carlo simulation methods have often been used in investigation of statistical behaviour of cellular automata. Wilbur, Lipman and Shamma[7.91] have shown that cellular automaton evolution may be approximated by a Markov process. Gutowitz, Victor and Knight[7.49] have proposed a 'local structure theory', a finitely parametrized procedure for the determination of the statistical features of a cellular automaton's evolution; it is an analytic alternative to Monte Carlo methods. A lattice gas automaton was proposed for the modelling of developed turbulence[7.39] and it has been demonstrated that many statistical characteristics of fully developed turbulence may be described in this way[7.29, 7.30]. From the point of view of chaos the cellular automata are best suitable for studies of the propagation and distribution of the local chaos among a large number of sites.

7.3 Coupled map lattices

Coupled map lattices have been used in the 80s to study two aspects of chaos in distributed systems at the same time: the creation of local chaos and its propagation and distribution among a large number of degrees of freedom. The degrees of freedom are spatially distributed and exhibit translational symmetry. This makes the description of complex spatially extended systems simpler than that of general dynamical systems without the symmetry.

A coupled map lattice is a dynamical system with discrete time, discrete space and continuous state[7.21]. Let us consider a dynamical variable x_n^i where the superscript denotes the site's location and the subscript time. A simple logistic lattice may be defined as

$$x_{n+1}^i = rx_n^i(1 - x_n^i) + \varepsilon(x_n^{i+1} + x_n^{i-1}).$$ (7.3)

Here $f(x_n^i) = rx_n^i(1 - x_n^i)$, $x \in [0, 1]$ and $r \in [0, 4]$ describes the local dynamics and ε the coupling strength to neighbouring sites. If we consider the one-dimensional lattice of N sites then the vector $x_n = (x_n^0, x_n^1, ..., x_n^{N-1})$ characterizes the current state of the system and is called a pattern (or field). On the contrary to cellular automata, the lattice dynamical systems with continuous state variables have the capability of local information production. However, there are similarities in behaviour of both systems; for example, the binary space-time symbolic dynamics of the one-dimensional logistic lattice behaves like several cellular automata.

In the description of behaviour of coupled map lattices we shall follow the formulation accepted by Crutchfield and Kaneko[7.21]. Let us consider the following lattice dynamical system

$$x_{n+1}^i = f(x_n^i) + \varepsilon_0 \, g(x_n^i) + \varepsilon_R \, g(x_n^{i+1}) + \varepsilon_L \, g(x_n^{i-1}).$$ (7.4)

The most often studied examples of local dynamics are

(a) the logistic map $f(x) = rx(1 - x)$,
(b) the circle map $f(x) = \omega + x + k \sin(2\pi x)$,
(c) shift or tent maps (piecewise linear functions).

The Hénon map may be used in coupled lattices with two local variables. The coupling form may be local (coupling between the neighbouring sites) or nonlocal. The nearest neighbour coupling, where the dynamics of the site i is affected only by the interaction with the sites $i - 1$ and $i + 1$ is considered in (7.4). The vector $\varepsilon = (\varepsilon_0, \varepsilon_L, \varepsilon_R)$ is the coupling kernel and the function $g(x)$ describes the coupling dynamics; linear coupling corresponds to $g(x) = x$ and the 'future coupling' to $g(x) = f(x)$. In the coupling kernel the choice $\varepsilon_0 = 0$, $\varepsilon_R = \varepsilon_L$ corresponds to additive coupling, $-\varepsilon_0/2 = \varepsilon_R = \varepsilon_L$ to Laplacian coupling, and $-\varepsilon_0 = \varepsilon_L$, $\varepsilon_R = 0$ to uni-directional coupling (used in the modelling of the open flows), respectively. Fixed, periodic, free, noise or periodically driven boundary conditions may be used. For initial conditions often spatially periodic functions as $\sin(2\pi ki/N)$ or uniformly distributed random functions with a chosen mean and variance are used.

An extensive simulations of various types of coupled map lattices have been reported by Kaneko[7.57-7.63], Kapral and Oppo[7.65], Kapral[7.64] and Crutchfield and Kaneko[7.21]. Here we shall briefly mention several examples of observed chaotic behaviour and the ways of its characterization.

Crutchfield and Kaneko[7.21] have discussed examples of the coupled logistic lattice with future Laplacian, linear Laplacian and simple additive coupling. The last case is defined as

$$x_{n+1}^i = f(x_n^i) + (x_n^{i+1} + x_n^{i-1}) . \tag{7.5}$$

As discussed in Chapter 2, the single logistic map shows a period-doubling route to chaos when r is increased. In the logistic lattice the temporal period-doubling induces formation of spatial domain structures of phase coherent sites – domains. Domains are separated by walls or kinks. Kuznetsov and Pikovsky[7.71, 7.72] have studied spatial self-similarity, universality and scaling of spatial structures (kinks) arising via the period-doubling mechanism. When the pattern of chaotic states develops via the period-doubling in the lattice, some of the low-periodicity domain structures remain even if the patterns are chaotic. With a further increase of the nonlinearity (r) spatial structures with a characteristic wavelength appear in the patterns. As the local dynamics period-doubles with the change of r or ε, the spatial wavelengths also decrease in a regular manner. This period-doubling of domains introduces spatial self-similarity in the lattice pattern; the spatial structures dominate at higher values of r and as the nonlinearity is increased further, the period-doubling sequence loses its stability to a sequence exhibiting intermittency via pattern competition. Solution diagrams in the form of site histograms may be used for characterization of the subsequent pattern development. A dependence of the bifurcation behaviour on the system size (numbers of sites) N shows transition to chaos at some critical size $N_c = = N_c(r, \varepsilon)$.

A large class of coupled map lattices (for example model (7.5)) exhibits an anti-correlated zigzag instability in which the uniform field $x^i = const$ loses stability to a state of the form $x^i = const + (-1)^i \times$ amplitude. Kapral[7.64] has found that in two-dimensional lattices this pattern appears as a checkerboard pattern. From the zigzag structure the transition to chaos via torus occurs. The quasiperiodicity in time introduces spatial quasiperiodicity.

In the map lattices with the circle map and the Laplacian future coupling the so called soliton turbulence is observed. As solitons the particle-like structures which propagate, collide and annihilate are denoted. The transition to fully developed turbulence from a homogeneous or a periodic state occurs via the soliton turbulence. In strongly coupled circle maps also spatial Pomeau-Manneville intermittency has been observed. The spatio-temporal pattern of laminar clusters is characterized by the existence of two speeds: the initial burst propagation speed (describing the propagation of turbulent regions into a laminar region) and the laminar cluster propagation speed. Intermittency can also occur via pattern competition.

Asymmetric coupling may be used to model open flows. Open flow lattices may be described by the model

$$x_{n+1}^i = f(x_n^i) + \varepsilon[\alpha f(x_n^{i+1}) + (1 - \alpha) f(x_n^{i-1}) - f(x_n^i)] . \tag{7.6}$$

The degree of uni-directional coupling is determined by the value of α. When $\alpha = 0$ the coupling is uni-directional, from the left to the right. For example, a logistic open flow lattice with the additive future-coupling exhibits spatial period-doubling transition to chaos. The mechanism of the spatial period-doubling can be understood in terms of convective instability[7.27]. Transient spatial chaos may be also observed in a piecewise linear lattice[7.21].

The above reported types of behaviour can be interactively explored on a personal computer or a scientific work station. The coupled map lattice models are also very conveniently studied on parallel computers. However, an interactive type of work and an interactive graphics are essential for simulations. Fascinating structures of chaotic attractors with self-similar features and Julia-like boundaries can be observed already for the symmetric coupling of two one-dimensional logistic maps[7.76]. Qualitative characteristics (visualizations) of coupled map lattices include space-amplitude plots, spatial return maps and spacetime diagrams. Quantitative characteristics of pattern dynamics of spatio-temporal chaos in coupled map lattices include, among others, spatial power spectra, spatio-temporal power spectra and LEs[7.61].

The statistics of the spatial pattern can be represented by the spatial power spectra

$$S(k) = \langle\!\langle S(k, n)\rangle\!\rangle = \left\langle\!\!\left\langle \left|\frac{1}{N} \sum_{j=1}^{N} x_n^j \, e^{2\pi i k j}\right|^2 \right\rangle\!\!\right\rangle. \tag{7.7}$$

Here $\langle\!\langle \ \rangle\!\rangle$ denotes the long time average after transients have decayed. Kaneko[7.61, 7.62] has used the spatial power spectra $S(k)$ to analyze intermittency associated with the pattern competition in a logistic map lattice. He has shown the coexistence of a peak at k_p and a broadband noise at $k \approx 0$. The spatio-temporal characteristics were quantitized by a spatio-temporal power spectra $P(k, \omega)$

$$P(k, \omega) = \left\langle\!\!\left\langle \left|\frac{1}{M} \sum_{j=1}^{M} x_n^j \, e^{2\pi i k j + 2\pi i \omega n}\right|^2 \right\rangle\!\!\right\rangle. \tag{7.8}$$

Either the summation over all lattice points $j = 1, ..., N$, or a summation only over a restricted region (a window) $j = 1, 2, ..., M (M \leq N)$ may be used. The spectral strength of some spectral mode $P(k, \omega)$ changes with the window size M. If the oscillation has some finite spatial correlation length l, the spectral strength decreases with the window size for $M < l$ and approaches a constant value for $M > l$. In the neighbourhood of the pattern competition intermittency, the spatial correlation length can be very large.

Keeler and Farmer[7.66] have studied robust space-time intermittency in a one-dimensional lattice of coupled quadratic maps. They have found that spatial domains arise naturally and the motion of the domain walls causes spatially localized changes from chaotic to almost periodic behaviour. The almost periodic phases have eigenvalues close to one resulting in long-lived laminar bursts. They have also used frequency histograms, Lyapunov vectors and spatial power spectra to obtain quantitative characteristics of the observed spatio-temporal patterns.

Kaneko[7.61] has also introduced other ways of computing pattern distribution and pattern entropy, similar to techniques used in the phase transition studies[7.2]. Computation of LEs, Lyapunov vectors and co-moving LEs may be used for the quantification of the flow of disturbances in the open flow coupled maps lattices. For example Deissler and Kaneko[7.27, 7.61] have introduced the co-moving LE in the following way. First, they have made transformation to moving frame (velocity v) by choosing $i' = i - [vn]$ (here $[*]$ denotes the integer part of $*$). The co-moving LE is then calculated from the largest eigenvalue of the long time average of the product of Jacobi matrices in the moving frame. For example, for the open flow coupled logistic map lattice

$$x_{n+1}^i = (1 - \varepsilon) f(x_n^i) + \varepsilon f(x_n^{i-1}) \tag{7.9a}$$

the co-moving LE of the evolution from the homogeneous point solution $x^i = = x^* = 1/2r(\sqrt{(1 + 4r)} - 1)$ is determined as

$$\lambda(v) = \lambda_0 \log \left[\frac{1 - \varepsilon}{1 - v} \right] + \log \left[\frac{\varepsilon(1 - v)}{v(1 - \varepsilon)} \right]^v, \tag{7.9b}$$

where $\lambda_0 = \log |f'(x^*)|$. For the co-moving LE of a spatially homogeneous and temporally periodic or chaotic state the corresponding LE of the one-dimensional map orbit $\{f^n(x)\}$ is substituted for λ_0 in (7.9b).

We can observe from (7.9b) that even if the conventional LE is negative, the maximum of the co-moving LE $(\lambda(\varepsilon))$ can be positive; the homogeneous state is then convectively unstable. Deissler and Kaneko[7.27] have presented an example, where the conventional LE is negative $(\lambda(0) < 0)$ even if the pattern and the dynamics are chaotic; the co-moving LE is positive only within the velocity band $v_1 < v < v_2$.

7.4 Partial differential equations

Physical models of spatially distributed problems are usually written down either in the form of PDEs or continuous time lattice models. Through the method of Poincaré section, PDEs can be reduced to iterated functional mappings, continuous in space but discrete in time. If the functional map is further reduced to a discrete spatial lattice map we can transfer the results of the study of map lattices to those for PDEs. Thus Keeler and Farmer[7.66] studied lattice of quadratic maps corresponding to a driven reaction-diffusion equation.

There are several rigorous results showing that the attractors of many infinite-dimensional dynamical systems are of finite dimension and have a discrete spectrum of LEs[7.5, 7.9, 7.18, 7.20, 7.37, 7.42, 7.43, 7.88]. It has been established in most of the above quoted papers that in the studied systems compact attractors exist and a finite dimension space confine the attractors[7.19, 7.38, 7.55]. This was first established for autonomous dissipative parabolic equations (see a review in Ref. 7.88) and then the estimates of upper and lower bounds on the dimension of attractors were derived for other types of nonlinear distributed evolution equations. Information content of distributed dynamical systems has been studied by Grassberger[7.46] and methods of the algebraic approximation of attractors by Foias and Temam[7.35, 7.36]

Chaotic solutions have recently been found in many systems of nonlinear partial differential equations[7.34, 7.70, 7.83, 7.87, 7.88, 7.95-7.97]. In most cases numerical simulation of the corresponding PDEs have been used, in several cases centre manifold arguments have been applied and, for example, the presence of horseshoes have been demonstrated[7.52]. For example, Holmes and Marsden[7.53] have discussed the presence of chaos in a problem from magnetoelasticity. Guckenheimer[7.47] proved the presence of the Shilnikov mechanism near a codimension two bifurcation in a reaction-diffusion problem described by PDEs in one dimension. The same mechanism was identified in thermosolutal convection problems[7.6-7.8]. The presence of turbulence in the models of reaction-diffusion systems was studied by Kuramoto[7.68, 7.69] and an important question of pattern selection and propagation was studied by Ben-Jacob *et al*[7.10].

However, in studies of three-dimensional convection the mechanism of transition to chaotic solutions often remains unclear[7.22]. As was illustrated in Chapter 5, the period-doubling route to chaos was observed in several convection experiments, but mostly when the systems were constrained by boundaries so that only a few rolls were present (a few modes were excited). A cascade of period-doubling bifurcations leading to aperiodicity was observed in the numerical studies of two-dimensional double-diffusive convection[7.67]. However, geometrical constraints of two-dimensionality and symmetry about roll centres and imposed boundary conditions probably forced solutions of the partial differential equations to behave similarly to low-dimensional models.

The fluid-dynamical systems mentioned above and in Chapter 5 were mostly closed-flow systems, i.e. systems, where the flow 'closes back on itself' (no fluid enters or leaves the system under consideration). The open flow systems can behave in generically different ways from the closed ones. In the closed flow systems the boundary is fixed so that only certain eigenfunctions are reflected in evolving stationary or dynamic patterns.

In the open flow systems, the flow can be laminar at one location, transitional at another and turbulent at yet another downstream location. Hence it is necessary to consider temporal characteristics of the system at various spatial positions. Sreenivasan[7.86] has characterized chaotic attractors observed in various open flow systems (e.g. the wake behind a circular cylinder, an axisymmetric jet and a curved pipe) by values of the correlation dimension ranging from 2.6 to 6.3. He found large spatial variations of the dimension at two different spatial positions in the same flow with the same streamwise location and the same value of the Reynolds number. The dimension calculations at higher Reynolds numbers became uncertain.

The influence of external disturbances (noise) is also more difficult to ascertain in open fluid systems as the 'upstream influence' is most important. The effects of noise on open flow systems were studied by Deissler using the time-dependent generalized Ginzburg–Landau equation[7.23, 7.26]. Deissler and Kaneko[7.27] have introduced velocity – dependent LEs as a measure of chaos in open flow systems.

Deissler[7.25] have also demonstrated by solving numerically the corresponding Orr-Sommerfeld equation that the plane Poiseuille flow is absolutely stable but convectively unstable. The generation of spatio-temporal intermittency in the form of turbulent bursts, spots and slugs in open flows may then result from the amplification of external noise in convectively unstable systems, as was also confirmed by the numerical simulation of the generalized Ginzburg–Landau equation[7.26].

Most informations on chaotic behaviour was obtained for several model PDEs. Perhaps the largest progress has been achieved in condensed matter physics in the study of sine-Gordon-like systems[7.11–7.15, 7.31, 7.56]. The studies are centred on the driven, damped sine-Gordon equation[7.15]

$$\varphi_{tt} - \varphi_{xx} + \sin \varphi = \varepsilon(\Gamma \sin \omega t - \alpha \varphi_t), \tag{7.10}$$

e.g. with periodic boundary conditions of period L

$$\varphi(x + L, t) = \varphi(x, t),$$

$$\varphi(x, t = 0) = \varphi_{in}(x),$$

$$\dot{\varphi}(x, t = 0) = v_{in}(x). \tag{7.11}$$

Here α denotes the strength of the dissipation, Γ the amplitude of the driver, ω the frequency of the ac driver, L the spatial period and the initial data are defined by φ_{in}, v_{in}. Eilbeck et al.[7.31] have observed chaos in the inhomogeneously driven damped sine-Gordon equation; Bishop et al.[7.15] studied coherent spatial structures and temporal chaos in a perturbed sine-Gordon system and Imada[7.56] found that chaotic behaviour in the same system has been generated by a soliton–soliton interaction. The results of numerical studies of nonlinear dynamics in the driven damped sine-Gordon system in one and two spatial dimensions have been reviewed by Bishop and Lomdahl[7.12].

A numerical procedure for the nonlinear spectral transformation of spatially periodic fields (both temporally periodic and chaotic) applying a soliton wave train basis was developed and used to demonstrate that in this case also a spatially coherent low-dimensional chaos exists[7.82]. This spectral transformation in principle enables us to identify significant spatial modes and thus also helps to formulate appropriate lower-dimensional models.

Bishop et. al[7.14] have studied this system in detail in a near integrable situation $(0 < \varepsilon \ll 1)$ close to a nonlinear resonance region. They found a generic quasiperiodic route to intermittent chaos and used the above mentioned nonlinear spectral transform to identify and quantify spatio-temporal attractors in terms of a small number of soliton modes of the corresponding integrable system. They also used these analytic coordinates to identify homoclinic orbits as possible sources of chaotic behaviour.

Second often studied PDE is the nonlinear Schrödinger equation[7.42]

$$i\,\frac{\partial u}{\partial t} + \frac{\partial^2 u}{\partial x^2} + g(|u|^2)\,u + i\gamma u = f\,. \tag{7.12}$$

Dirichlet (7.13a), Neumann (7.13b) or periodic (7.13c) boundary conditions

$$u(0, t) = u(L, t) = 0\,, \tag{7.13a}$$

$$\frac{\partial u}{\partial x}(0, t) = \frac{\partial u}{\partial x}(L, t) = 0\,, \tag{7.13b}$$

$$u(x, t) = u(x + L, t) \tag{7.13c}$$

are used in the studies (f here denotes an external excitation). Ghidaglia[7.42] has shown analytically that the long time behaviour of solutions is described by an attractor which captures all the trajectories and has estimated the uniform LEs on this attractor. Blow and Doran[7.16] and Nozaki and Bekki[7.81] observed numerically and explained by physical arguments that low-dimensional chaotic attractors exist in this system. Aceves and coworkers[7.1] have used the mathematical model for the optically bistable ring cavity consisting essentially of a nonlinear Schrödinger type equation periodically forced and damped. They used, among other approaches, the nonlinear spectral transform idea, i.e. the projection of the perturbed field into a normal mode basis defined by an exactly integrable unperturbed problem.

The amplitude evolution of instability waves in a large variety of dissipative systems close to criticality may be described by the time-dependent Ginzburg-Landau equation

$$\frac{\partial u}{\partial t} = au - v_{\mathrm{g}}\,\frac{\partial u}{\partial x} + b\,\frac{\partial^2 u}{\partial x^2} - c|u|^2 u\,. \tag{7.14}$$

Here the dependent variable $u(x, t)$ is in general complex, a, b, and c are complex constants; v_{g} is the group velocity and $b_{\mathrm{r}} \geq 0$. This equation results from expanding the variables of the studied problem in powers of $R - R_{\mathrm{c}}$, where the control parameter R (the Reynolds number, the Taylor number or the Rayleigh number) is near the critical value R_{c}. Ghidaglia and Heron[7.43] have shown analytically that all the solutions to the Ginzburg–Landau PDE are captured by a finite-dimensional attractor and have derived lower and upper bounds on the dimension of the attractor. Moon, Huerre and Redekopp[7.77, 7.78] demonstrated numerically the presence of the chaotic attractor. Deissler[7.24, 7.26] has used generalized Ginzburg–Landau equation for the studies of the external noise in open flow systems.

A different model equation which has been studied in detail is the Kuramoto–Sivashinsky equation describing a small perturbation $u(x, t)$ of a metastable planar front or interface (flames, thin viscous fluid films flowing over inclined planes). The equation can be written in the form

$$u_t + v u_{xxxx} + u_{xx} + \tfrac{1}{2}(u_x)^2 = 0, \qquad (x, t) \in \mathrm{R}^1 \times \mathrm{R}_+ ,$$

$$u(x, 0) = u_0(x) , \qquad u(x + L, t) = u(x, t) . \tag{7.15}$$

Here the subscripts denote partial differentiation, v the positive fourth order viscosity and u_0 is L-periodic, where L is the size of a typical pattern scale.

Hyman, Nicolaenko and Zaleski[7.54] have presented a large number of numerical results illustrating evolution of attractors in the Kuramoto–Sivashinsky equation with the variation of L. For low values of L the trivial solution $u(x, t) \equiv 0$ is asymptotically stable; with the increase of L successive bifurcations to S^1 families of steady spatially periodic solutions with subsequently increasing number of maxima have occurred (only the first solution have been stable). Bifurcations from these solutions have given rise to travelling waves, and more complicated, pulse like chaotic solutions. The authors have suggested that these are connected with homoclinic or heteroclinic orbits to steady solutions. Quasiperiodic solutions in the form of modulated travelling waves have been also observed. Hyman, Nicolaenko and Zaleski have also presented detailed statistical characteristics of the observed chaotic solutions and discussed stringent requirements on the accuracy of numerical methods (spectral techniques) used for the simulations.

Armbruster, Guckenheimer and Holmes[7.5] have studied dynamical behaviour of (7.15) with $u_x(0, t) = u_x(L, t)$ in dependence on the length of the system. They reduced the PDE to a centre-unstable manifold for parameter values near that at which the second Fourier mode bifurcates. The local invariant manifold has been found to be a graph over the eigenspace of the first two modes; the approximation of the modes by a Taylor series including terms of order three has yielded a reduced vector field accurate to fourth order. Results of the analysis of this system have agreed qualitatively and quantitatively with detailed numerical simulations of the PDE reported by Hyman, Nicolaenko and Zaleski[7.54].

7.5 Perspectives

Interest in the studies of deterministic chaos and turbulence is increasing. It is connected both with a rapidly increasing number of concrete problems in different fields of natural sciences and engineering where chaos is observed and with the possibility of advancing our understanding of the mutual relations between dynamical and statistical laws governing the behaviour of various phenomena in nature. We have tried in this book to give at least partial answers to three most important questions in the study of chaotic behaviour:

(a) what are the sources of chaos in deterministic dissipative systems?
(b) what are the most common mechanisms for the transition to chaos when a characteristic parameter is varied?
(c) what means are suitable for the characterization of chaotic behaviour with respect to their applicability to a large number of different physical systems?

The use of algorithms for the reconstruction of orbits in a finite dimensional state space and for the computation of dimensions of chaotic attractors from the experimental time series holds great promise for the development of relatively simple models of the seemingly nondeterministic random behaviour with a number of applications particularly in natural and engineering sciences. However, it is necessary to bear in mind that when the fractal dimension becomes larger (reflecting the larger complexity of the data), it may be impossible to extract laws from the data, as it was documented by Guckenheimer and Buzyna[7.48] in the analysis of geostrophic turbulence. The finiteness of available computer systems also forms a practical limitation both to experimental data processing and to the numerical studies of mathematical models.

Construction of concrete models based on experimental data would profit from further studies of the effects of noise on the properties of chaotic attractors, from the development of the methodology for specific filtration of the data (the above mentioned nonlinear spectral transformation and methods discussed in Chapter 4 may serve as examples) and its use in an interactive experimentation.

Recently Matsumoto[7.74] have described three extremely simple electronic circuits in which chaotic phenomena can be observed. The circuits can be built even by high school students and one can both observe on an oscilloscope and even listen to chaotic oscillations. More advanced readers can confirm these chaotic phenomena by simulation on a personal computer and theoretically inclined ones can prove rigorously by means of the Shilnikov theorem that there is a homoclinic orbit and a horshoe near it.

Perhaps the most important applications of the studies of deterministic chaos appear in living systems. For example, biological clocks in vertebrates are considered to consist of a system of two chemical oscillators which mutually interact and create a basic rhythm in the organism. It was determined experimentally that this rhythm is regular (periodic) for most mammals, while cats have an aperiodic basic rhythm[7.79]; chaos in random neural networks has been studied by Sompolinsky and Crisanti[7.85]. Hence the application of the concepts of a deterministic chaos can help in an understanding of such systems.

REFERENCES

7.1 Aceves A., Adachihara H., Jones Ch., Lerman J. C., McLaughlin D. W., Moloney J. V. and Newell A. C. Chaos and coherent structures in partial differential equations. *Physica* **18D** (1986) 85.

7.2 Amit D. J. *Field Theory, Renormalization Group and Critical Phenomena.* New York, McGraw Hill, 1978.

7.3 Aoki K. and Mugibayashi N. Cellular automata and coupled chaos developed in a lattice chain of N equivalent switching elements. *Phys. Lett.* **A114** (1986) 425.

7.4 Aoki K. and Mugibayashi N. Onset of turbulent pattern in a coupled map lattice. Case for soliton−like behavior. *Phys. Lett.* **128A** (1988) 349.

7.5 Armbruster D., Guckenheimer J. and Holmes P. Kuramoto−Sivashinsky dynamics on the center−unstable manifold. Preprint, Ithaca, Cornell University, 1987.

7.6 Arneodo A. and Thual O. Direct numerical simulations of a triple convection problem versus normal form prediction. *Phys. Lett.* **109A** (1985) 367.

7.7 Arneodo A., Coullet P. H. and Spiegel E. A. The dynamics of triple convection. *Geophys. Astrophys. Fluid. Dyn.* **31** (1985) 1.

7.8 Arneodo A., Coullet P. H., Spiegel E. A. and Tresser C. Asymptotic chaos. *Physica* **14D** (1985) 327.

7.9 Babin A. V. and Vishik M. I. Attractors of partial differential equations and estimates of their dimension. *Russ. Math. Surv.* **38** (1983) 151.

7.10 Ben−Jacob E., Brand H., Dee G., Kramer L. and Langer J. S. Pattern propagation in nonlinear dissipative systems. *Physica* **14D** (1986) 348.

7.11 Bennet B., Bishop A. R. and Trullinger S. E. Coherence and chaos in the driven damped sine−Gordon. *Z. Phys.* **B47** (1982) 265.

7.12 Bishop A. R. and Lomdahl P. S. Nonlinear dynamics in driven damped sine−Gordon systems. *Physica* **18D** (1986) 54.

7.13 Bishop A. R., Fesser K., Lomdahl P. S., Kerr W. C., Williams M. B. and Trullinger S. E. Coherent spatial structure versus time chaos in a perturbed sine−Gordon system. *Phys. Rev. Lett.* **50** (1983) 1095.

7.14 Bishop A. R., Forest M. G., McLaughlin D. W. and Overman II E. A. A quasi−periodic route to chaos in a near−integrable PDE. *Physica* **23D** (1986) 293.

7.15 Bishop A. R., Krumhansl J. A. and Trullinger S. E. Solitons in condensed matter: a paradigm. *Physica* **1D** (1980) 1.

7.16 Blow K. J. and Doran N. J. Global and local chaos in the pumped nonlinear Schrodinger equation. *Phys. Rev. Lett.* **52** (1984) 526.

7.17 Canuto C., Hussaini M. Y., Quarteroni A. and Zang T. A. *Spectral Methods in Fluid Dynamics.* Springer Series in Computational Physics, New York, Springer, 1988.

7.18 Constantin P., Foias C. and Temam R. Attractors representing turbulent flows. *Memoirs of A.M.S.* **53** (1985) 314.

7.19 Constantin P., Foias C. and Temam R. On the dimension of the attractors in two−dimensional turbulence. *Physica* **30D** (1988) 284.

7.20   Constantin P., Foias C., Manley O. and Temam R. Determining modes and fractal dimension of turbulent flows. *J. Fluid Mech.* **150** (1985) 427.

7.21   Crutchfield J. P. and Kaneko K. Phenomenology of spatio—temporal chaos. In *Directions in Chaos*, ed. Hao Bai—lin, Singapore, World Scientific, 1987.

7.22   Curry J. H., Herring J. R., Loncaric J. and Orszag S. A. Order and disorder in two and three—dimensional Bénard convection. *J. Fluid. Mech.* **147** (1984) 1.

7.23   Deissler R. J. Noise—sustained structure, intermittency and the Ginzburg—Landau equation. *J. Stat. Phys.* **40** (1985) 371.

7.24   Deissler R. J. Spatially growing waves, intermittency and convective chaos in an open—flow system. *Physica* **25D** (1986) 448.

7.25   Deissler R. J. The convective nature of instability in plane Poiseuille flow. *Phys. Fluids* **30** (1987) 2303.

7.26   Deissler R. J. Turbulent bursts, spots and slugs in a generalized Ginzburg—Landau equation. *Phys. Lett.* **120A** (1987) 334.

7.27   Deissler R. J. and Kaneko K. Velocity dependent Lyapunov exponents as a measure of chaos for open flow systems. *Phys. Lett.* **119A** (1987) 397.

7.28   Dress A. W. M., Gerhardt M. and Schuster H. Cellular automata simulating the evolution of structure through the synchronization of oscillators. In *From Chemical to Biological Organization*, ed. by Markus M., Muller S. C., Nicolis G., Springer Series on Synergetics, Springer, New York, 1988.

7.29   Duong—Van M. Fully developed turbulence via Feigenbaum's period—doubling bifurcations. Preprint, Berkeley, Univ. of California at Berkeley, 1987.

7.30   Duong—Van M., Feit M. D., Keller P. and Pound M. The nature of turbulence in a triangular lattice gas automaton. *Physica* **23D** (1986) 448.

7.31   Eilbeck J. C., Lomdahl P. S. and Newell A. C. Chaos in the inhomogenously driven sine—Gordon equation. *Phys. Lett.* **87A** (1981) 1.

7.32   Farmer J. D. Chaotic attractors of an infinite—dimensional dynamical system. *Physica* **4D** (1982) 366.

7.33   Farmer J. D., Toffoli T. and Wolfram S. (Eds.) *Cellular Automata*. Amsterdam, North Holland, 1984.

7.34   Firth W. J. Optical memory and spatial chaos. *Phys. Rev. Lett.* **61** (1988) 329.

7.35   Foias C. and Temam R. Approximation algebrique des attracteurs. I. Le cas de la dimension finie. II. Le cas de la dimension infinie. *C. R. Acad. Sci. Paris* **307** (1988) 5, 67.

7.36   Foias C. and Temam R. The algebraic approximation of attractors. *Physica* **32D** (1988) 163.

7.37   Foias C., Nicolaenko B., Sell G. R. and Temam R. Inertial manifolds for dissipative partial differential equations. *C.R.A.S Paris I* **301** (1985) 139.

7.38   Foias C., Sell G. and Temam R. Inertial manifolds for nonlinear evolutionary equations. *J. Diff. Eq.* **73** (1988) 309.

7.39   Frisch U., Hasslacher B. and Pomeau Y. A lattice gas automaton for the Navier—Stokes equations. *Phys. Rev. Lett.* **41** (1986) 1505.

7.40   Geist K. and Lauterborn W. Chaos upon soliton decay in a perturbed periodic Toda chain. *Physica* **23D** (1986) 374.

7.41   Geist K. and Lauterborn W. The nonlinear dynamics of the damped and driven Toda chain. I. Energy and bifurcation diagrams. *Physica* **31D** (1988) 103.

7.42   Ghidaglia J. M. Comportement de dimension finie pour les equations de Schrödinger faiblement amorties. *Ann. de l'IHP, Analyse Non Lineaire* (1988).

7.43   Ghidaglia J. M. and Heron B. Dimension of the attractors associated to the Ginzburg—Landau partial differential equation. Preprint, Orsay, Universite de Paris—Sud, 1987.

7.44   Gledzer E. B., Dolzhansky F. V. and Obukhov A. M. *Systems of Hydrodynamical Type and Their Applications*. Moscow, Nauka, 1981 (in Russian).

7.45 Grassberger P. Chaos and diffusion in deterministic cellular automata. *Physica* **10D** (1984) 52.

7.46 Grassberger P. Information content and predictability of lumped and distributed dynamical systems. Preprint, Wuppertal, University of Wuppertal, 1987.

7.47 Guckenheimer J. On a codimension two bifurcation. In *Dynamical Systems and Turbulence*, ed. D. A. Rand and L. S. Young, Lecture Notes in Math. **898**, Heidelberg, Springer, 1981, p. 99.

7.48 Guckenheimer J. and Buzyna G. Dimension measurements for geostrophic turbulence. *Phys. Rev. Lett.* **51** (1983) 1438.

7.49 Gutowitz H. A., Victor J. D. and Knight B. W. Local structure theory for cellular automata. *Physica* **28D** (1987) 18.

7.50 Hart J. E. A model for the transition to baroclinic chaos. *Physica* **20D** (1986) 350.

7.51 Henry D. *Geometric Theory of Semilinear Parabolic Equations.* Lecture Notes in Mathematics **840**, New York, Springer, 1981.

7.52 Holmes P. Chaotic motions in a weakly nonlinear model for surface waves. *J. Fluid. Mech.* **162** (1986) 365.

7.53 Holmes P. and Marsden J. A partial differential equation with infinitely many periodic orbits: Chaotic oscillations of a forced beam. *Arch. Rat. Mech. Anal.* **76** (1981) 135.

7.54 Hyman J. M., Nicolaenko B. and Zaleski S. Order and complexity in the Kuramoto – Sivashinsky model of weakly turbulent interfaces. *Physica* **23D** (1986) 265.

7.55 Iljashenko Yu. S. On the dimensionality of attractor of K – systems in an infinite – dimensional space. *Reports of Moscow State University* **24** (1983) 52 (in Russian).

7.56 Imada M. Chaos caused by the soliton – soliton interaction. *J. Phys. Soc. Japan* **52** (1983) 1946.

7.57 Kaneko K. Spatiotemporal intermittency in coupled map lattices. *Prog. Theor. Phys.* **74** (1985) 1033.

7.58 Kaneko K. Turbulence in coupled map lattices. *Physica* **18D** (1986) 475.

7.59 Kaneko K. Lyapunov analysis and information flow in coupled map lattices. *Physica* **23D** (1986) 436.

7.60 Kaneko K. *Collapse of Tori and Genesis of Chaos in Dissipative Systems.* Singapore, World Scientific, 1986.

7.61 Kaneko K. Pattern dynamics in spatiotemporal chaos. Preprint, Tokyo, Institute of Physics, 1987.

7.62 Kaneko K. Pattern competition intermittency and selective flicker noise in spatiotemporal chaos. *Phys. Lett.* **125A** (1987) 25.

7.63 Kaneko K. Phenomenology and characterization of spatiotemporal chaos. In *Dynamical Systems and Singular Phenomena*, ed. G. Ikegami, Singapore, World Scientific, 1987.

7.64 Kapral R. Pattern formation in two – dimensional arrays of coupled, discrete – time oscillators. *Phys. Rev.* **31A** (1985) 3868.

7.65 Kapral R. and Oppo G. – L. Competition between stable states in spatially – distributed systems. *Physica* **23D** (1986) 455.

7.66 Keeler J. D. and Farmer J. D. Robust space – time intermittency and $1/f$ noise. *Physica* **23D** (1986) 413.

7.67 Knobloch E., Moore D. R. and Toomre T. Transitions to chaos in two – dimensional double diffusive convection. *J. Fluid. Mech.* **166** (1986) 409.

7.68 Kuramoto Y. (Ed.) *Chemical Oscillations, Waves and Turbulence.* Heidelberg, Springer, 1984.

7.69 Kuramoto Y. Phase dynamics of weakly unstable periodic structures. *Prog. Theor. Phys.* **71** (1984) 1182.

7.70 Kuramoto Y. and Yamada T. Turbulent state in chemical reactions. *Progr. Theor. Phys. Japan* **56** (1976) 679.

7.71 Kuznetsov S. P. and Pikovsky A. S. Universality of period doubling bifurcations in one – dimensional dissipative media. *Rep. of Inst. of Higher Education, sect. Radiophysics*, **28** (1985) 308 (in Russian).

7.72 Kuznetsov S. P. and Pikovsky A. S. Universality and scaling of period doubling bifurcations in a dissipative distributed medium. *Physica* **19D** (1986) 3841.

7.73 Mandelbrot B. B. On the geometry of homogeneous turbulence, with stress on the fractal dimension of the iso—surface of scalars. *J. Fluid Mech.* **72** (1975) 401.

7.74 Matsumoto T. Chaos in electronic circuits. *Proc. of the IEEE* **75** (1987) 1033.

7.75 McLaughlin D. W., Moloney J. V. and Newell A. C. Solitary waves as fixed points of infinite dimensional maps in an optical bistable ring cavity. *Phys. Rev. Lett.* **51** (1983) 75.

7.76 Metzler W., Beau W., Frees W. and Ueberla A. Symmetry and self—similarity with coupled logistic maps. *Z. Naturforsch.* **42a** (1987) 310.

7.77 Moon H. T., Huerre P. and Redekopp L. G. Three—frequency motion and chaos in the Ginzburg—Landau equation. *Phys. Rev. Lett.* **49** (1982) 458.

7.78 Moon H. T., Huerre P. and Redekopp L. G. Transitions to chaos in the Ginzburg—Landau equation. *Physica* **7D** (1983) 135.

7.79 Moore—Ede M. C., Sulzman F. M. and Fuller Ch. A. *The Clocks That Time Us: Physiology of the Circadian Timing System.* Cambridge, Ma., Harward University Press, 1982.

7.80 von Neumann J. In *Theory of Self—Reproducing Automata*, ed. A. W. Burks, Urbana, University of Illinois Press, 1966.

7.81 Nozaki K. and Bekki N. Low—dimensional chaos in a driven damped nonlinear Schrödinger equation. *Physica* **21D** (1986) 381.

7.82 Overmann II E. A., McLaughlin D. W. and Bishop A. R. Coherence and chaos in the driven damped sine—Gordon equation: Measurement of the soliton spectrum. *Physica* **19D** (1986) 1.

7.83 Pikovskii A. S. Chaotic autowaves. *Lett. to JTP* **11** (1985) 672 (in Russian).

7.84 Rabinovich M. I. Stochastic self—oscillations and turbulence. *Sov. Phys. Usp.* **21** (1978) 443.

7.85 Sompolinsky H. and Crisanti A. Chaos in random neural networks. *Phys. Rev. Lett.* **61** (1988) 259.

7.86 Sreenivasan K. R. Chaos in open flow systems. In *Dimensions and Entropies in Chaotic Systems*, ed. G. Mayer—Kress, Berlin, Springer, 1986.

7.87 Temam R. Dynamical systems, turbulence and the numerical solution of the Navier—Stokes equations. In Proc. of the Conf. *Numerical Methods in Fluid Dynamics*, Williamsburg, June 1988.

7.88 Temam R. *Infinite—Dimensional Dynamical Systems in Mechanics and Physics.* Applied Mathematical Sciences **68**, New York, Springer, 1988.

7.89 Toda M. *Theory of Nonlinear Lattices.* Heidelberg, Springer, 1981.

7.90 Tongue B. H. On obtaining global nonlinear system characteristics through interpolated cell mapping. *Physica* **28D** (1987) 401.

7.91 Wilbur W. J., Lipman D. J. and Shamma S. A. On the prediction of local patterns in cellular automata. *Physica* **19D** (1986) 397.

7.92 Wolfram S. Statistical mechanics of cellular automata. *Rev. Mod. Phys.* **55** (1983) 601.

7.93 Wolfram S. Origins of randomness in physical systems. *Phys. Rev. Lett.* **55** (1985) 449.

7.94 Wolfram S. (Ed.) *Theory and Applications of Cellular Automata.* Singapore, World Scientific, 1986.

7.95 Yamada T. and Kuramoto Y. A reduced model showing chemical turbulence. *Progr. Theor. Phys. Japan* **56** (1976) 681.

7.96 Yamazaki H., Oono Y. and Hirakawa K. Experimental study on chemical turbulence. *J. Phys. Soc. Japan* **44** (1978) 335.

7.97 Yoshimura K. and Watanabe S. Behavior of nonlinear evolution equation with first order dispersion. *J. Phys. Soc. Japan* **51** (1982) 3028.

7.98 Zaslavskii G. M., Sagdeev P. E., Usikov D. A. and Chernikov A. A. Minimal chaos, stochastic spider's web and structures with a symmetry of a quasicrystal type. *Uspekhi fyz. nauk* **156** (1988) 193 (in Russian).

APPENDIX A

Normal forms and their bifurcation diagrams

Consider the following system of differential equations

$$\dot{x} = f(x, \alpha),$$
(A.1)

where $x = (x_1, ..., x_n)$ is an n-vector of *state space variables* and $\alpha = (\alpha_1, ..., \alpha_d)$ is a d-vector of *parameters* (n, d being reasonably small). Suppose that $f(x, \alpha)$ is sufficiently smooth with respect to its arguments and the system (A.1) does not possess any special properties (e.g. symmetries).

A qualitative analysis of the system (A.1) at some fixed value of α consists in finding its *phase portrait*, that is a topological structure of a partition of the state space into trajectories. A *structuraly stable* system does not change its phase portrait under small perturbations of parameters while the phase portrait of a *structuraly unstable* system undergoes a qualitative change – a *bifurcation*. Points in the parameter space corresponding to structuraly unstable systems are called bifurcation values.

Partition of a neighbourhood of a given bifurcation value α_c into connected sets, each coresponding to the same phase portrait in some region of the state space, is called a *bifurcation diagram*.

Every structuraly unstable system is characterized by fulfilling k *bifurcation conditions* (in the form of equalities) together with a certain number of *non-degeneracy conditions* (in the form of nonequalities). Bifurcation conditions define a manifold in parameter space of codimension k, $0 < k \le d$, (i.e. its dimension is $d - k$) and nondegeneracy conditions define regions on these manifolds. Every point in a particular region corresponds to bifurcation of codimension k. When some nondegeneracy condition is violated, i.e. on the boundary of a codimension k region, a bifurcation of codimension $k+1$ occurs.

To investigate a given codimension k bifurcation, it is sufficient to consider any k-parametric system containing this bifurcation and satisfying non-degeneracy conditions. The bifurcation diagram obtained in this way has a universal character and fully describes the given bifurcation in a concrete system.

More specifically, a *centre manifold theorem* applies and states that given a stationary point x_0 with a centre manifold of dimension n_c at a bifurcation value α_c of codimension k, the centre manifold can be smoothly parametrized by α in the neighbourhood of α_c so that all qualitative changes in dynamics occur on this parametrized (unfolded) centre manifold. A similar theorem can be applied to periodic orbits (using the Poincaré mapping).

The motion on the unfolded manifold can be systematically simplified to a (polynomial) system of n_c differential equations with k parameters

$$\dot{\mathbf{u}} = v(\mathbf{u}, \boldsymbol{\beta}), \qquad \mathbf{u} = (u_1, ..., u_{nc}), \qquad \boldsymbol{\beta} = (\beta_1, ..., \beta_k), \tag{A.2}$$

called a *normal form*. This is the simplest (but not unique) model system possessing the studied bifurcation. At $\boldsymbol{\beta} = \mathbf{0}$, k bifurcation conditions and necessary nondegeneracy conditions are satisfied by (A.2). Any system (A.1) can be put, by a nonlinear coordinate transformation, into a form which is identical to (A.2) up to terms of a certain order in a neighbourhood of α_c. Terms of higher order do not change the topological structure of the bifurcation diagram. Thus the qualitative bifurcation diagram of (A.1) can be obtained from (A.2).

The outlined problems are treated in detail in a number of papers and books; the interested reader can find some of them in the References to this Appendix and to Chapter 3.

In the following, we illustrate the topic on several *local bifurcations* of stationary points for one and two-dimensional systems of ODEs (i.e. $n_c = 1, 2$). If $n_c > 1$ then *global bifurcations* involving *periodic, homoclinic* and *heteroclinic* orbits may emerge in the vicinity of the studied codimension k stationary point. Thus global bifurcations occur naturally when studying local ones. If $n_c > 2$, one can expect chaotic behaviour of normal forms which makes the classification of bifurcations extremely difficult.

A.1 Dynamical systems on a line

Consider the scalar differential equation

$$\dot{x} = f(x, \alpha), \tag{A.3}$$

where $x \in R$ and $\alpha \in R^d$. Let $x_0 = x_0(\alpha)$ be a stationary point of (A.3) for some fixed α and let $\lambda(\alpha) = \partial f(x_0, \alpha)/\partial x$ be the eigenvalue determining the stability of x_0. A bifurcation of x_0 of codimension one is given by the bifurcation condition

$$\lambda(\alpha) = 0 \tag{A.4}$$

and by a nondegeneracy condition

$$a(\alpha) = \frac{1}{2} \frac{\partial^2 f(x_0, \boldsymbol{\alpha})}{\partial x^2} \neq 0 , \qquad (A.5)$$

i.e. x_0 is a doubly degenerate stationary point of (A.3).

The normal form for this case depends on one parameter β_1 (codimension one bifurcation):

$$\dot{u} = \beta_1 + au^2 , \qquad a \neq 0 . \qquad (A.6)$$

The system (A.6) as well as the original system (A.3) undergoes a limit point bifurcation, see Fig. A.1.

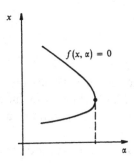

Fig. A.1. Two stationary points merge at the limit point bifurcation.

Consider now a more degenerate situation where (A.5) turns into equality, that is, we have simultaneously

$$\lambda(\alpha) = 0 , \qquad a(\alpha) = 0 , \qquad (A.7)$$

and we add to these two bifurcation conditions a nondegeneracy condition

$$b(\alpha) = \frac{1}{6} \frac{\partial^3 f(x_0, \boldsymbol{\alpha})}{\partial x^3} \neq 0 , \qquad (A.8)$$

i.e., x_0 is a triply degenerate stationary point of (A.3).

A normal form for this situation depends on two parameters $(\beta_1, \beta_2) = \boldsymbol{\beta}$:

$$\dot{u} = v(u, \boldsymbol{\beta}) = \beta_1 + \beta_2 u + bu^3 , \qquad b \neq 0 . \qquad (A.9)$$

This corresponds to a codimension two bifurcation at $\beta_1 = \beta_2 = 0$. Bifurcation diagram for the system (A.9) is shown in Fig. A.2. Two codimension one curves S_1, S_2 merge at a codimension two point A. The regions 1 and 2 correspond to a structurally stable situation. The dynamics on the line u corresponding to individual pieces of the partitioned parameter space is shown via the graphs of $v(u)$ in Fig. A.2. There are three stationary points in the region 2 (two stable and

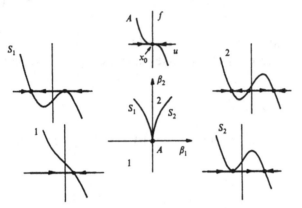

Fig. A.2. Bifurcation diagram for the 'cusp' bifurcation.

one unstable) and one stable stationary point in the region 1. Points on S_1, S_2 correspond to limit point bifurcation described by (A.6) which is included in (A.9) in a natural way. However, unlike (A.6), the system (A.9) contains a cusp point A of codimension two.

Let us proceed further in the construction of a more degenerate codimension three bifurcation. Bifurcation conditions are now

$$\lambda(\alpha) = 0\,, \qquad a(\alpha) = 0\,, \qquad b(\alpha) = 0 \qquad\qquad\qquad \text{(A.10)}$$

and the nondegeneracy condition is

$$c(\alpha) = \frac{1}{24}\frac{\partial^4 f(x_0,\,\alpha)}{\partial x^4} \neq 0\,, \qquad\qquad\qquad\qquad \text{(A.11)}$$

i.e. x_0 is a quadruply degenerate stationary point of (A.3). The normal form that describes this bifurcation is

$$\dot{u} = v(u,\,\beta) = \beta_1 + \beta_2 u + \beta_3 u^2 + cu^4\,, \qquad c \neq 0\,. \qquad\qquad \text{(A.12)}$$

A (three-dimensional) bifurcation diagram for (A.12) is shown in Fig. A.3. Structurally stable (three-dimensional) pieces of the parameter space correspond to either 2 or 4 (hyperbolic) stationary points which pairwise merge at the codimension 1 surface S having a shape of a '*swallow tail*'. Bifurcation points on S correspond to (A.6), while bifurcation points on codimension two curves A_1, A_2 correspond to (A.9); the most degenerate codimension 3 point x is described by (A.12) at $\beta_1 = \beta_2 = \beta_3 = 0\,.$

A convenient two-dimensional representation of Fig. A.3 can be constructed as follows: Consider a ball of a small radius in the parameter space centred at $\beta = 0$. The intersection of the bifurcation surfaces and lines with the surface of the ball contains most of the information of Fig. A.3. Separating the surface of the ball into two half surfaces and projecting into a plane we obtain Fig. A.4

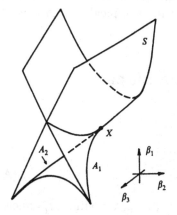

Fig. A.3. Bifurcation diagram for the 'swallow tail' bifurcation.

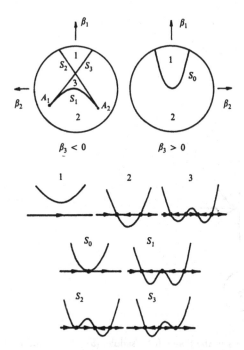

Fig. A.4. Projected bifurcation diagram from Fig. A.3.

which is supplied by the graphs of $v(u)$ in individual regions of the partitioned parameter space.

The one-dimensional system (A.3) can be seen as a system that governs the dynamics on an unfolded centre manifold of dimension $n_c = 1$ in the neighbourhood of the bifurcation value α_c and the normal forms (A.6), (A.9) and (A.12) qualitatively describe some typical situations which may occur when a stationary point of (A.3) bifurcates. Next we examine typical bifurcations which may occur when $n_c = 2$.

A.2 Dynamical systems on a plane

Consider the two-dimensional system

$$\dot{x}_1 = f_1(x_1, x_2, \alpha) ,$$

$$\dot{x}_2 = f_2(x_1, x_2, \alpha) , \qquad\qquad\qquad (A.13)$$

depending on the d-dimensional parameter $\alpha = (\alpha_1, ..., \alpha_d)$ with f_1, f_2 sufficiently smooth. Let $x_0(\alpha)$ be a stationary point of (A.13) for some fixed α. The local stability of x_0 is given by the eigenvalues $\lambda_1(\alpha), \lambda_2(\alpha)$ of the Jacobi matrix

$$A(\alpha) = \left\{ \frac{\partial f_i(x_0, \alpha)}{\partial x_j} \right\} .$$

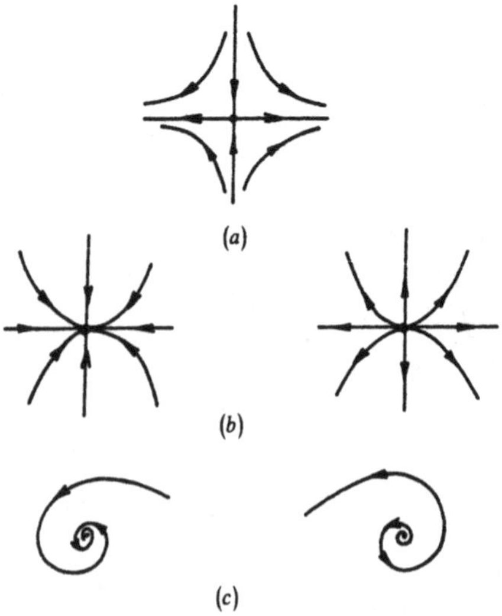

Fig. A.5. Hyperbolic stationary points in the plane: (a) – saddle; (b) – stable and unstable node; (c) – stable and unstable focus.

The phase portrait in the vicinity of x_0 does not qualitatively change if the real parts of λ_1, λ_2 are different from zero. Thus x_0 is *hyperbolic* and three topological types of stationary points exist: *saddle* $(\lambda_1 > 0, \lambda_2 < 0)$, *stable node* or *focus* $(\text{Re}(\lambda_{1,2}) < 0)$ and *unstable node* or *focus* $(\text{Re}(\lambda_{1,2}) > 0)$, see Fig. A.5. Trajectories which are asymptotic to a saddle when $t \to \pm\infty$ are called *separatrices*.

A stationary point x_0 is called *multiply degenerate* if $\det A = \lambda_1\lambda_2 = 0$ and *neutral* if $\text{tr } A = \lambda_1 + \lambda_2 = 0$. There exist two possible codimension 1 bifurcations. The first one is characterized by the bifurcation condition

$$\lambda_1(\alpha) = 0 , \tag{A.14}$$

i.e. $\det A(\alpha) = 0$. The normal form describing this bifurcation is

$$\dot{u}_1 = \beta_1 + au_1^2 ,$$

$$\dot{u}_2 = \lambda_2 u_2 , \tag{A.15}$$

and nondegeneracy conditions are

$$\lambda_2(\alpha) \neq 0 , \qquad a(\alpha) \neq 0 . \tag{A.16}$$

The condition $a \neq 0$ is analogous to (A.5). In fact, the dynamics of u_1 and u_2 according to (A.15) are separated and only the first equation governs the dynamics on an unfolded centre manifold and determines the bifurcation behaviour. Two stationary points – a *saddle* and a *node* – merge at the point of bifurcation which is called *limit point* or *saddle-node bifurcation*, see Fig. A.6.

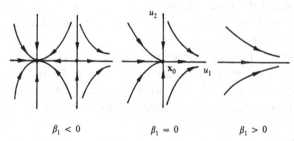

$$\beta_1 < 0 \qquad\qquad \beta_1 = 0 \qquad\qquad \beta_1 > 0$$

Fig. A.6. Changes in the phase portrait corresponding to the limit point bifurcation.

The second codimension 1 bifurcation is defined by a pair of purely imaginary eigenvalues,

$$\text{Re}(\lambda_{1,2}) = 0 , \qquad \det A > 0 . \tag{A.17}$$

A normal form for this bifurcation is conveniently written in a complex variable $z = u_1 + iu_2$,

$$\dot{z} = (\beta_1 + i\omega) z + l_1 z |z|^2 , \tag{A.18}$$

where $l_1(\alpha)$ is a real number and $\omega(\alpha) = \mathrm{Im}\,(\lambda_1(\alpha))$. The nondegeneracy conditions are

$$\omega(\alpha) \neq 0 , \qquad l_1(\alpha) \neq 0 . \tag{A.19}$$

This bifurcation does not have an analogue in one-dimensional systems. The changes of the phase portrait corresponding to this bifurcation are shown in Fig. A.7. The stationary point $z = 0$ changes its stability from a stable focus

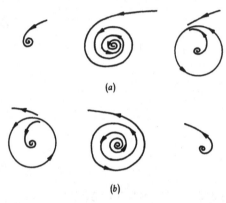

(a)

(b)

Fig. A.7. Changes in the phase portrait corresponding to the Hopf bifurcation.

for $\beta_1 < 0$ to an unstable focus for $\beta_1 > 0$. The topological structure on both sides of the bifurcation value $\beta_1 = 0$ depends on l_1. If $l_1 < 0$ then a *stable* periodic trajectory of a small diameter exists for $\beta_1 > 0$ while an *unstable* periodic trajectory exists for $\beta_1 < 0$ if $l_1 > 0$. The bifurcation described by (A.18) and (A.19) is called *Hopf bifurcation*.

When moving on a manifold in the parameter space of (A.13) corresponding to a codimension 1 bifurcation, a nondegeneracy condition may be violated and a codimension 2 bifurcation occurs. This may arise in several distinct ways.

The first bifurcation of this kind is characterized by the conditions

$$\lambda_1(\alpha) = 0 , \qquad a(\alpha) = 0 . \tag{A.20}$$

The normal form and the nondegeneracy conditions are

$$\dot{u} = \beta_1 + \beta_2 u_1 + b u_1^3 ,$$

$$\dot{u}_2 = \lambda_2 u_2 , \tag{A.21}$$

and

$$b \neq 0 , \qquad \lambda_2 \neq 0 . \tag{A.22}$$

This situation is quite analogous to (A.9) as the equations (A.21) are separated. The bifurcation diagram, see Fig. A.8, is the same as in Fig. A.2, but the state space is now two-dimensional.

The second codimension 2 bifurcation (which is not found in one-dimensional systems) is characterized by bifurcation conditions

$$\text{Re}\left(\lambda_{1,2}(\alpha)\right) = 0, \qquad l_1(\alpha) = 0. \tag{A.23}$$

The normal form together with nondegeneracy conditions reads

$$z = (\beta_1 + i\omega) z + \beta_2 z |z|^2 + l_2 z |z|^4, \qquad \omega \neq 0, \quad l_2 \neq 0. \tag{A.24}$$

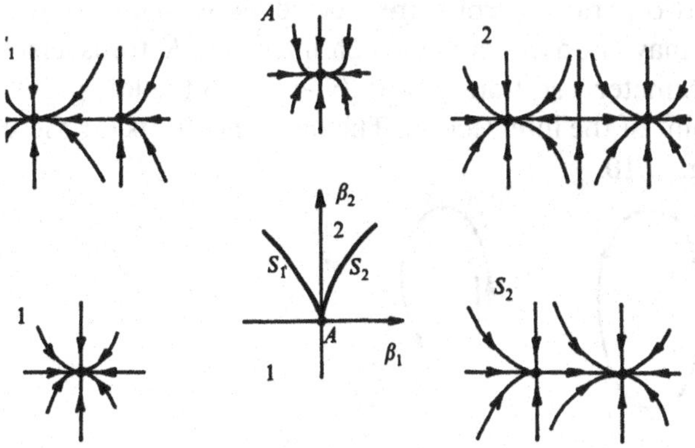

Fig. A.8. Bifurcation diagram for the 'cusp' bifurcation.

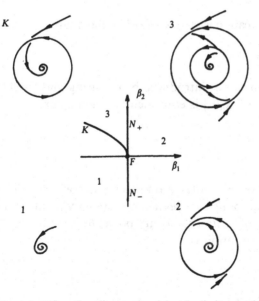

Fig. A.9. Bifurcation diagram involving the limit point bifurcation of periodic trajectories.

Let $l_2 > 0$. The bifurcation diagram (see Fig. A.9) contains a line N corresponding to Hopf bifurcation $(\mathrm{Tr}A = 0,\ \det A > 0)$. For $\beta_1 < 0$ the stationary point of (A.24) is a stable focus and for $\beta_1 > 0$ an unstable focus. Let us follow the phase portrait of (A.24) as parameters move around a circle of a small radius centred at the point $F = (0, 0)$. When crossing N from the left to the right a stable periodic trajectory is born from the stationary point. When crossing N_+ from the right to the left an unstable periodic trajectory arises from the stationary point. There are two periodic trajectories in the region 3, one is stable and the other is unstable. Both cycles merge and disappear at the curve K which separates the regions 3 and 1. Thus we observe a *nonlocal* bifurcation of codimension 1 at which two periodic orbits are created simultaneously (i.e. limit point bifurcation). We may choose a curve which intersects K transversely and parametrize it by a parameter γ so that $\gamma > 0$ $(\gamma < 0)$ to the left (right) from K and $\gamma = 0$ at the point of the intersection. The phase portraits for these three cases are shown in Fig. A.10.

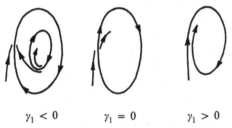

$$\gamma_1 < 0 \qquad\qquad \gamma_1 = 0 \qquad\qquad \gamma_1 > 0$$

Fig. A.10. The emergence of a pair of periodic trajectories in the state space.

The last codimension 2 bifurcation is characterized by the bifurcation conditions

$$\lambda_1(\alpha) = 0\,, \qquad \lambda_2(\alpha) = 0\,. \tag{A.25}$$

The nondegeneracy condition can be introduced by requiring that the linearization matrix A of (A.13) around $x_0(\alpha)$ is equivalent to the matrix

$$\begin{bmatrix} 0 & 1 \\ 0 & 0 \end{bmatrix}.$$

These conditions define a manifold in the parameter space which is located in the intersection of the manifold of *doubly degenerate* stationary points (defined by $\det A = 0$) with the manifold of *neutral* stationary points (defined by $\mathrm{Tr}\,A = 0$).

The normal form for this situation is

$$\dot{u}_1 = u_2\,,$$

$$\dot{u}_2 = \beta_1 + \beta_2 u_1 + a u_1^2 + d u_1 u_2\,, \qquad a \neq 0\,,\ d \neq 0\,. \tag{A.26}$$

Bifurcation diagram is shown in Fig. A.11. The boundary S of the region 1 corresponds to limit point bifurcation: the curve N corresponds to Hopf bifurcation. Let us follow the phase portrait of (A.26) as the point (β_1, β_2) moves

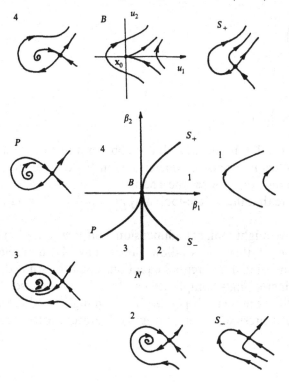

Fig. A.11. Bifurcation diagram involving the saddle-loop bifurcation.

clockwise around a circle centred at $B = (0, 0)$. There are no stationary points or cycles in the region 1. When crossing S_-, two stationary points (a saddle and a stable node) emerge via the limit point bifurcation. Further on the node turns into a focus which loses its stability via Hopf bifurcation at the point on N. The created stable periodic trajectory disappears on the curve P as a *saddle-loop*. The

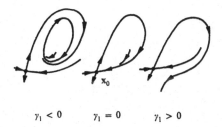

$\gamma_1 < 0 \qquad \gamma_1 = 0 \qquad \gamma_1 > 0$

Fig. A.12. Changes in the phase potrait for the saddle-loop bifurcation.

period of the cycle tends to infinity at this *nonlocal* bifurcation point. Finally, when crossing S_+ from the left to the right both stationary points merge and disappear.

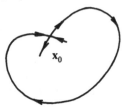

Fig. A.13. A 'large' saddle-loop.

We have observed here another nonlocal codimension 1 bifurcation – *saddle-loop bifurcation*. On introducing a new parameter γ as in the case of the limit point bifurcation of periodic trajectories, the change of the phase portrait is shown in Fig. A.12. We note that the saddle-loop may also look as shown in Fig. A.13.

Proceeding in this way, we might look at codimension 3 bifurcations of planar differential equations, one of them – 'swallow tail' – have been discussed previously. Then three-dimensional differential equations could be studied, etc. The bifurcations become increasingly complicated and are beyond the scope of this Appendix. There is still a number of open problems in higher codimension bifurcations. We refer the interested reader to original literature listed below.

REFERENCES

A.1 Andronov A. A., Leontovich E. A., Gordon I. I. and Maier A. G. *Bifurcation Theory of Dynamical Systems on a Plane*. Moscow, Nauka, 1967 (in Russian).

A.2 Arnold V. I. Lectures on bifurcations in versal families. *Russ. Math. Surv.* **27** (1972) 54.

A.3 Arnold V. I. *Geometrical Methods in the Theory of Ordinary Differential Equations*. New York, Springer, 1982 (Russian original, Moscow, Nauka 1978).

A.4 Arnold V. I. *Singularity Theory : Selected Papers*. Cambridge, Cambridge University Press, 1981.

A.5 Bazykin A. D., Kuznetsov Yu. A. and Khibnik A. U. Bifurcation diagrams of dynamical systems on a plane. Pushchino, USSR Academy of Sciences, 1985 (in Russian).

A.6 Bogdanov R. I. Versal deformations of a singular point on the plane in the case of zero eigenvalues. *Funct. Anal. and Appl.* **9** (1975) 144.

A.7 Chow S. N. and Hale J. K. *Methods of Bifurcation Theory*. New York, Springer, 1982.

A.8 Gambaudo J. M., Glendinning P. and Tresser C. The gluing bifurcation: I. Symbolic dynamics of the closed curves. *Nonlinearity* **1** (1988) 203.

A.9 Golubitsky M. and Schaeffer D. *Singularities and Groups in Bifurcation Theory*. Vol. I. New York, Springer, 1985.

A.10 Golubitsky M., Schaeffer D. and Stewart I. *Singularities and Groups in Bifurcation Theory*. Vol. II. New York, Springer, 1988.

A.11 Guckenheimer J. On a codimension two bifurcation. In *Dynamical Systems*, Progress in Mathematics **8**, Boston, Birkhauser, 1981, p. 99.

A.12 Guckenheimer J. Multiple bifurcation problems for chemical reactors. *Physica* **20D** (1986) 1.

A.13 Guckenheimer J. and Dangelmayr G. On a four parameter family of planar vector fields. *Arch. Rat. Mech. Anal.* **97** (1987) 321.

A.14 Guckenheimer J. and Holmes P. J. *Nonlinear Oscillations, Dynamical Systems and Bifurcations of Vector Fields*. New York, Springer, 1983, 1986.

A.15 Holmes P. J. Center manifolds, normal forms and bifurcations of vector fields with application to coupling between periodic and steady motions. *Physica* **2D** (1981) 449.

A.16 Iooss G. and Joseph D. D. *Elementary Stability and Bifurcation Theory*. New York, Springer, 1981.

A.17 Khazin L. G. and Shnol E. E. *Stability of Critical Equilibria*. Pushchino, USSR Academy of Sciences, 1985 (in Russian).

A.18 Medved' M. On a codimension three bifurcation. *Czechoslovak Math. J.* **33** (1984) 3.

A.19 Swift J. W. Hopf bifurcation with the symmetry of the square. *Nonlinearity* **1** (1988) 333.

A.20 Takens F. Singularities of vector fields. *Publ. Math. IHES* **43** (1974) 47.

APPENDIX B

CONT – a program for construction of solution and bifurcation diagrams

This is the simplest self-consistent version of a continuation program developed and used in our research group. The program is able to find a starting point in a state space and to continue this point as one or two parameters are varied. The description of the method of continuation used in the program as well as references are contained in Chapter 4. Hints for running the program are found in comment lines to the main program. The subroutine PRSTR defining the equations and the Jacobi matrix must be supplied by the user. As a test model, two coupled Brusselators discussed in detail in Chapter 6 are defined in the sample subroutine PRSTR. Also included are input data for obtaining solution diagrams in Figs 6.14 and 6.17

CONT package is available through Internet upon request. For information send and e-mail message to marek@ vscht.cs or schrig@ vscht.cs.

```
C
C*********************************************************************************
C
C    CONT - PROGRAM FOR CONSTRUCTION OF PARAMETER DEPENDENCE OF
C           STATIONARY OR PERIODIC SOLUTIONS OF ORDINARY DIFFERENTIAL
C           OR DIFFERENCE EQUATIONS.
C
C
C       1) CONTINUATION OF STATIONARY AND PERIODIC SOLUTIONS
C          OF AUTONOMOUS ORDINARY DIFFERENTIAL EQUATIONS - AODES
C       2) CONTINUATION OF PERIODIC SOLUTIONS OF DIFFERENCE
C          EQUATIONS (= ITERATED MAPPINGS) - DIFES
C       3) CONTINUATION OF PERIODIC SOLUTIONS OF PERIODICALLY
C          FORCED ORDINARY DIFFERENTIAL EQUATIONS - FODES
C       4) CONTINUATION OF LOCAL BIFURCATION POINTS ASSOCIATED
C          WITH THE ABOVE TYPES OF SOLUTIONS (LIMIT POINTS - LP, HOPF
C          BIFURCATION POINTS - HBP, PERIOD DOUBLING POINTS - PDP AND
C          TORUS BIFURCATION POINTS - TBP ARE INCLUDED IN THIS VERSION)
C
C*********************************************************************************
C
C INPUT DATA ARE SPECIFIED ON UNIT WITH LOGICAL NUMBER LR AND ARE READ
C IN SUBPROGRAMS MAIN AND INIT.
C
C FOR THE SAKE OF SIMPLICITY THE STRUCTURE OF DATA IN INPUT FILE IS THE
C SAME FOR ALL TYPES OF CONTINUATIONS. THIS IMPLIES THAT IN CERTAIN
C CASES INPUT VALUES OF SOME CONSTANTS AND VARIABLES ARE FORMALLY
C SUPPLIED BUT UNUSED DURING THE COMPUTATIONS. SUCH ENTRIES ARE PLACED
C IN < > BELOW AND THEIR INPUT VALUES MAY BE ARBITRARY. ENTRIES WITH
C SPECIFIED DEFAULT VALUES ARE PLACED IN [ ]. SOMETIMES BOTH CASES CAN
C OCCUR SIMULTANEOUSLY WHICH IS INDICATED BY [< >].
C
C TO START CONTINUATION THE USER MUST SUPPLY DATA AS DESCRIBED BELOW.
C THE STARTING POINT ON THE SOLUTION CURVE (OR ITS ESTIMATE) IS DEFINED
C BY ASSIGNING INITIAL VALUES TO X(),ALPHA,BETA,<ARG>,<PER>. THE OTHER
C INPUT ENTRIES SPECIFY THE ACTUAL CONTINUATION PROBLEM AND METHODS OF
C COMPUTATION.
C
C NAML   : LENGTH OF ALPHANUMERICAL STRING USED AS THE NAME OF THE
C          ANALYZED MODEL
C
C NAME() : ARRAY SPECIFYING THE MODEL NAME
C
C NREAD  : SPECIFICATION OF THE INPUT DATA
C          NREAD  = 1 READ INPUT DATA WITH UNSPECIFIED DEFAULT VALUES AS
C                     FOLLOWS:
C                     NAML,(NAME(I),I=1,NAML)                 (I5/(70A1))
C                     NREAD,ITYP,ICON,<ISHT>,NDIM,NPAR,<IPER>      (7I5)
C                     PAR(I),I=1,NPAR                          (4F15.7)
C                     X(I),I=1,NDIM                            (4F15.7)
C                     XLOW(I),I=1,NDIM                         (4F15.7)
C                     XUPP(I),I=1,NDIM                         (4F15.7)
```

```
C                        ALPHA,BETA,<ARG>,<PER>                    (4F15.7)
C                        ALFLOW,BETLOW,<ARGLOW>,<PERLOW>           (4F15.7)
C                        ALFUPP,BETUPP,<ARGUPP>,<PERUPP>           (4F15.7)
C                        INITAL,NFIX,KDIR,INTG,MAXOUT,NPRNT,<ISRCH>  (7I5)
C            NREAD  = 2 THE SAME AS FOR NREAD = 1 AND IN ADDITION REPLACE
C                        DEFAULT INPUT VALUES OF SCALARS BY READING THEM AS
C                        FOLLOWS:
C                        H1,HMIN1,HMAX1,<ACCUR1>                    (4E15.7)
C                        <H2>,<HMIN2>,<HMAX2>,<ACCUR2>             (4E15.7)
C                        EPS,<HX>,EMACH                            (4E15.7)
C                        ITIN,NCORR,NCRAD                             (7I5)
C            NREAD  = 3 THE SAME AS FOR NREAD = 2 AND IN ADDITION REPLACE
C                        DEFAULT INPUT VALUE OF X0() BY READING IT AS
C                        FOLLOWS:
C                        <X0(I),I=1,NDIM>                          (4F15.7)
C
C ITYP   : TYPE OF CONTINUATION PROBLEM
C            ITYP   = 1 STATIONARY SOLUTIONS OF AODES
C            ITYP   = 2 PERIODIC SOLUTIONS OF AODES
C            ITYP   = 3 PERIODIC SOLUTIONS OF DIFES
C            ITYP   = 4 PERIODIC SOLUTIONS OF FODES WITH FORCING PERIOD
C                        AS FIXED PARAMETER STORED IN PER
C            ITYP   = 5 PERIODIC SOLUTIONS OF FODES WITH FORCING PERIOD
C                        AS VARIABLE PARAMETER STORED IN ALPHA
C            ITYP   = 6 PERIODIC SOLUTIONS OF FODES WITH FORCING PERIOD
C                        AS VARIABLE PARAMETER STORED IN BETA
C
C ICON   : TYPE OF CONTINUED SOLUTION
C            ICON   = 1 CONTINUATION WITH RESPECT TO ALPHA FOR FIXED BETA
C            ICON   = 2 CONTINUATION WITH RESPECT TO BETA FOR FIXED ALPHA
C            ICON   = 3 CONTINUATION OF THE LIMIT POINTS (BOTH ON
C                        STATIONARY AND PERIODIC SOLUTIONS)
C            ICON   = 4 CONTINUATION OF THE PERIOD-DOUBLING POINTS
C            ICON   = 5 CONTINUATION OF THE HOPF BIFURCATION POINTS (ON
C                        STATIONARY SOLUTIONS) OR TORUS BIFURCATION POINTS
C                        (ON PERIODIC SOLUTIONS)
C
C ALPHA,BETA : FORMAL NAMES OF TWO CHOSEN VARIABLE PARAMETERS
C                  USED FOR CONTINUATION
C
C <ARG> : EITHER IMAGINARY PART (FOR HBP) OR ARGUMENT MODULO 1 (FOR TBP)
C          OF THE CRITICAL EIGENVALUE
C
C <PER> : EITHER THE PERIOD OF AODES OR
C          THE FORCING PERIOD OF FODES IF IT IS USED AS FIXED PARAMETER
C
C ALFLOW,BETLOW,<ARGLOW>,<PERLOW> : LOWER BOUNDS FOR VARIABLE PARAMETERS
C ALFUPP,BETUPP,<ARGUPP>,<PERUPP> : UPPER BOUNDS FOR VARIABLE PARAMETERS
C
C <IPER>: (DISCRETE) PERIOD OF THE PERIODIC SOLUTION OF DIFES OR THE
C          (DISCRETE) PERIOD OF THE PERIODIC ORBIT OF  THE POINCARE MAP
C          ASSOCIATED WITH FODES (THE PERIOD OF THE CONTINUOUS TIME
```

```
C             SYSTEM IS IPER*FORCING PERIOD IN THE LATTER CASE)
C
C NDIM   : NUMBER OF MODEL EQUATIONS (= THE DIMENSION OF THE STATE
C             SPACE)
C
C NPAR   : NUMBER OF FIXED PARAMETERS
C
C PAR()  : ARRAY CONTAINING NPAR FIXED PARAMETERS (IN ADDITION ONE
C             POSSIBLE FIXED PARAMETER - FORCING PERIOD OF FODES IS STORED
C             IN PER)
C
C <ISHT>: SPECIFIES THE METHOD OF SOLUTION OF THE BOUNDARY VALUE PROBLEM
C             FOR PERIODIC SOLUTIONS OF AODES, DIFES OR FODES
C             ISHT  = 1 SINGLE SHOOTING METHOD
C             ISHT  > 1 MULTIPLE SHOOTING METHOD AT ISHT POINTS
C                       EQUIDISTANTLY DISTRIBUTED WITHIN THE PERIOD OF THE
C                       SOLUTION (APPLIES ALSO FOR DIFES WHERE ISHT=IPER IS
C                       SET AUTOMATICALLY)
C
C X()    : ARRAY OF VARIABLES CONTAINING STATE SPACE POINT (OR ISHT STATE
C             SPACE POINTS IF MULTIPLE SHOOTING IS USED) PLUS A RELEVANT
C             SUBSET OF VARIABLE PARAMETERS ALPHA,BETA,ARG,PER
C
C XLOW() : LOWER BOUNDS FOR X()
C
C XUPP() : UPPER BOUNDS FOR X()
C
C ONLY FIRST NDIM ENTRIES OF X(),XLOW(),XUPP(), ARE SPECIFIED
C IN INPUT DATA, THE OTHER ARE FILLED OUT AUTOMATICALLY.
C
C INITAL: THE SEARCH FOR STARTING POINT AND THE FOLLOWING CONTINUATION
C             INITAL = 0 CONTINUATION WITH KNOWN STARTING POINT
C             INITAL = 1 SEARCH FOR STARTING POINT FROM AN ESTIMATE FOLLOWED
C                        BY CONTINUATION
C             INITAL = 2 SEARCH FOR STARTING POINT ONLY
C
C NFIX   : INDEX OF VARIABLE FROM X() ARRAY (INCLUDING THE RELEVANT
C             SUBSET OF VARIABLE PARAMETERS ALPHA,BETA,ARG,PER) WHICH IS
C             FIXED IN THE COURSE OF THE SEARCH FOR STARTING POINT
C
C KDIR   : DIRECTION OF CONTINUATION AT THE STARTING POINT
C             KDIR  =  1 INCREASE OF X(NFIX)
C             KDIR  = -1 DECREASE OF X(NFIX)
C             KDIR  =  2 INCREASE OF X(NFIX) FOLLOWED AFTER COMPLETING
C                        THE CONTINUATION ALONG THIS DIRECTION BY DECREASE
C                        OF X(NFIX)
C             KDIR  = -2 DECREASE OF X(NFIX) FOLLOWED AFTER COMPLETING
C                        THE CONTINUATION ALONG THIS DIRECTION BY INCREASE
C                        OF X(NFIX)
```

```
C
C INTG   : SPECIFIES THE METHOD FOR THE INTEGRATION OF CONTINUATION
C           EQUATIONS
C           INTG   = 1 ADAMS-BASHFORTH METHODS (UP TO 4-TH ORDER)
C           INTG   = 2 RUNGE-KUTTA-MERSON METHOD
C
C MAXOUT: MAXIMUM NUMBER OF POINTS ON THE CONTINUATION CURVE
C
C NPRNT  : SPECIFICATION OF OUTPUT RESULTS
C           NPRNT  = 1 WRITE X(), RE(), RI(), NCR, ARCL, H1, NDIR() AND
C                      SQUAR ON UNIT WITH LOGICAL NUMBER LW
C           NPRNT  = 2 WRITE X() AND RE(), RI() ON UNIT LW
C           NPRNT  = 3 WRITE X() ON UNIT LW
C
C<ISRCH>: SPECIFIES IF A NEW STATE SPACE POINT SATISFYING THE ADDITIONAL
C           EQUATION FOR FIXING A POINT ON THE PERIODIC SOLUTION OF AODES
C           SHOULD BE FOUND
C           ISRCH  = 0 NO SEARCH
C           ISRCH  = 1 SEARCH
C
C [H1,HMIN1,HMAX1]  : INITIAL,MINIMAL AND MAXIMAL STEPS FOR THE
C           CONTINUATION EQUATIONS SOLVED BY PREDICTOR (EITHER ADAMS-
C           -BASHFORTH METHODS OR RUNGE-KUTTA-MERSON METHOD) + CORRECTOR
C           (NEWTON METHOD)
C
C [<ACCUR1>] : ERROR TOLERANCE FOR RUNGE-KUTTA-MERSON METHOD USED FOR
C           CONTINUATION EQUATIONS
C           ACCUR1 > 0 RELATIVE ERROR CONTROL
C           ACCUR1 < 0 ABSOLUTE ERROR CONTROL
C
C RUNGE-KUTTA-MERSON METHOD IS USED FOR THE INTEGRATION OF MODEL ODES +
C VARIATIONAL ODES.
C
C [<H2>,<HMIN2>,<HMAX2>,<ACCUR2>] : INITIAL,MINIMAL AND MAXIMAL STEPS
C           AND ERROR TOLERANCE FOR RUNGE-KUTTA-MERSON METHOD USED FOR
C           MODEL + VARIATIONAL EQUATIONS
C           ACCUR1 > 0 RELATIVE ERROR CONTROL
C           ACCUR1 < 0 ABSOLUTE ERROR CONTROL
C
C [EPS] : ERROR TOLERANCE FOR NEWTON METHOD
C           EPS    > 0 RELATIVE ERROR CONTROL
C           EPS    < 0 ABSOLUTE ERROR CONTROL
C
C [<HX>]: RELATIVE STEP FOR DETERMINATION OF FIRST DERIVATIVES
C           BY NUMERICAL DIFFERENCING (HX IS USED IN CONTINUATION OF
C           BIFURCATION POINTS ,I.E. ICON > 2)
C
C [EMACH]: MACHINE ACCURACY - THE SMALLEST POSITIVE NUMBER SUCH THAT
C           MACHINE DISTINGUISHES BETWEEN NUMBERS 1.0 AND 1.0+EMACH
C
C [ITIN]: MAXIMAL NUMBER OF INITIAL NEWTON ITERATIONS
C
```

```
C [NCORR]: MAXIMAL NUMBER OF CORRECTING NEWTON ITERATIONS
C
C [NCRAD]: ADDITIONAL NEWTON CORRECTION
C          NCRAD  = 0 NO ADDITIONAL NEWTON CORRECTION
C          NCRAD  = 1 ADDITIONAL NEWTON CORRECTION
C
C [<X0()>]: ARRAY OF NDIM NUMBERS FOR DETERMINATION OF
C          CHARACTERISTIC POLYNOMIAL
C
C *************************************************************************
C
C OUTPUT DATA ARE SPECIFIED ON UNIT WITH LOGICAL NUMBER LW AND ARE
C WRITTEN IN SUBPROGRAMS INIT AND DERPAR. ONLY LIMITED INFORMATION ABOUT
C THE CONTINUATION CURVE AND STABILITY IS GIVEN IN THE PROGRAM. FURTHER
C SPECIFICATIONS ON OUTPUT DATA ARE LEFT TO USER.
C
C X()    : CURRENT POINT ON CONTINUATION CURVE
C
C RE(),RI() : ARRAYS OF REAL AND IMAGINARY PARTS OF EIGENVALUES
C            CHARACTERIZING THE LINEAR STABILITY OF THE STATIONARY OR
C            PERIODIC SOLUTION
C
C NCR    : NUMBER OF NEWTON CORRECTIONS OF THE CURRENT POINT
C
C ARCL   : CURRENT LENGTH OF THE CONTINUATION CURVE
C
C H1     : INTEGRATION STEP FOR CONTINUATION EQUATIONS SUGGESTED FOR THE
C          COMPUTATION OF THE NEW POINT
C
C SQUAR  : EUCLIDEAN NORM OF RESIDUAL VECTOR
C
C NDIR(): ARRAY SPECIFYING THE DIRECTION OF THE TANGENT VECTOR TO
C          THE CONTINUATION CURVE AT THE CURRENT POINT X()
C          NDIR(I) =  1 IF X(I) INCREASES WITH INCREASING ARCL
C          NDIR(I) = -1 IF X(I) DECREASES WITH INCREASING ARCL
C
C *************************************************************************
C
C LIST OF USED SUBPROGRAMS: MAIN  ,INIT   ,DERPAR,SPTST ,PNTRS ,INITAS,
C                           INITMS,GAUSE  ,ADAMS ,MERS1 ,PRST1 ,PRST3 ,
C                           FCTN  ,FCT    ,FCSTS ,FCSSH ,FCMSH ,FCDER ,
C                           FCAODE,CHRULE,FCAD  ,COEF  ,DETER ,GAUSD ,
C                           INTEG ,ODEDRV,INSECT,SURF  ,GSURF ,MERS2 ,
C                           RHS   ,PRST2 ,PRSTR .
C
C THE MODEL EQUATIONS AND THE JACOBI MATRIX OF RIGHT HAND SIDES SHOULD
C BE SPECIFIED IN THE SUBPROGRAM PRSTR, WHICH MUST HAVE THE FOLLOWING
C FORM:
```

```
C
C       SUBROUTINE PRSTR(NDIM,NVAR,N,T,X,F,G)
C       IMPLICIT REAL*8(A-H,O-Z)
C       DIMENSION X(NDIM),F(NDIM),G(NDIM,NVAR)
C       COMMON/FIXP/PAR(20)
C       COMMON/VARP/ALPHA,BETA,ARG,PER
C
C       F(1) = ...
C       F(2) = ...
C          .
C          .
C          .
C       IF(N.EQ.NDIM) RETURN
C
C       G(1,1) = ...
C       G(2,1) = ...
C          .
C          .
C          .
C
C       RETURN
C       END
C
C WHERE
C
C T    :  TIME (EXPLICITLY OCCURS ONLY FOR FODES)
C X()  :  ARRAY OF NDIM STATE SPACE VARIABLES
C F()  :  ARRAY OF NDIM RIGHT HAND SIDES DEPENDING ON X;ALPHA,BETA,PAR()
C          (IN ADDITION F DEPENDS EXPLICITLY ON T AND EVENTUALLY ON PER
C          FOR FODES, SEE ABOVE)
C G(,) :  NDIM BY NVAR=NDIM+2 MATRIX OF FIRST DERIVATIVES,
C          G = [DF/DX,DF/DALPHA,DF/DBETA]
C
C TO RUN THE PROGRAM CONT THE USER MUST SUPPLY SUBPROGRAM RG FROM
C EISPACK-LINPACK LIBRARY AND COMPILE/LINK IT WITH CONT. SUBPROGRAM RG
C COMPUTES THE EIGENVALUES (AND EIGENVECTORS) OF A REAL GENERAL MATRIX.
C
C ***********************************************************************
C
C DIMENSIONS OF ARRAYS THROUGHOUT THE PROGRAM ARE DEFINED AUTOMATICALLY
C BY SETTING POINTERS INTO WORK ARRAYS XW(),IW() WHICH ARE DECLARED IN
C SUBPROGRAM MAIN. ONLY PAR() MUST BE DECLARED SEPARATELY.
C
C THE DIMENSION OF XW() IS SET TO 10000 IN THIS VERSION AND SHOULD BE AT
C LEAST (17*NDIM+74)*NDIM+108 OR (3*NDIM*ISHT+38)*NDIM*ISHT+
C +(14*NDIM+36)*NDIM+108 IF MULTIPLE SHOOTING IS USED
C
C THE DIMENSION OF IW() IS SET TO 1000 IN THIS VERSION AND SHOULD BE AT
C LEAST 6*NDIM+20 OR (5*ISHT+1)*NDIM+20 IF MULTIPLE SHOOTING IS USED
C
C THE ARRAY NAME() IS EQUIVALENT TO IW(), HENCE NAML MUST BE SMALLER
C THAN THE DIMENSION OF IW()
```

```
C
C THE ARRAY PAR() IS DECLARED IN COMMON FIELD PAR IN SUBPROGRAMS INIT
C AND PRSTR. ITS DIMENSION IS SET TO 20 IN THIS VERSION.
C
C ***********************************************************************
C
C MAIN PROGRAM
C
C SUBPROGRAMS CALLED: PNTRS,INIT,DERPAR
C FORTRAN LIBRARY FUNCTIONS:  NONE
C
C ----------------------------------------------------------------------
      IMPLICIT REAL*8(A-H,O-Z)
      DIMENSION XW(10000),IW(1000),NAME(1)
      EQUIVALENCE (IW(1),NAME(1))
      COMMON/PRNT/LR,LW
      COMMON/TYPE/ITYP,ICON,ISHT,NS,IPER
      COMMON/PNTR1/NXLOW,NXUPP,NPREF,NXSP,NDELTA,NDXDT,NF,NG,NXOLD,
     & NDXOLD,NRES1
      COMMON/PNTR2/NRE,NRI,NB,NEIGBH,NFV1,NRES2
      COMMON/PNTR4/NA,NX0,NXA,NFA,NGA,NGB,NGC,NRES4
      COMMON/PNTR6/NNDIR1,NMARK,NIRR,NIRK,NIVR
C ----------------------------------------------------------------------
C
C SET INPUT AND OUTPUT LOGICAL DEVICE NUMBER
C
      LR=1
      LW=2
C
C NAME OF THE MODEL
C
C ++++++++++++++++++++++++++++++++++++++++++++++++++++++++++++++++++++++++
      READ(LR,1000) NAML,(NAME(I),I=1,NAML)
C ++++++++++++++++++++++++++++++++++++++++++++++++++++++++++++++++++++++++
C
C SPECIFICATION OF THE PROBLEM
C
C ++++++++++++++++++++++++++++++++++++++++++++++++++++++++++++++++++++++++
      READ(LR,1100)NREAD,ITYP,ICON,ISHT,NDIM,NPAR,IPER
C ++++++++++++++++++++++++++++++++++++++++++++++++++++++++++++++++++++++++
C
1000  FORMAT(I5/(70A1))
1100  FORMAT(7I5)
      NS=NDIM
      IF(ITYP.EQ.3.AND.ISHT.GT.1)ISHT=IPER
      IF(ITYP.EQ.1)ISHT=0
      IF(ITYP.EQ.2)IPER=1
      IF(ITYP.GT.1.AND.ISHT.GT.1)NS=NDIM*ISHT
      NVAR=NDIM+2
      NMERS=NDIM*(NVAR+1)
      N=NS
      IF(ICON.EQ.3.OR.ICON.EQ.4)N=NS+1
```

```
        IF(ICON.EQ.5)N=NS+2
        IF(ITYP.EQ.2)N=N+1
        N1=N+1
C
C SET POINTERS
C
        CALL PNTRS(NDIM,NVAR,NS+3,NS+4,NMERS)
C
C INPUT DATA AND INITIALIZATION
C
        CALL INIT(NREAD,NDIM,NVAR,NPAR,N,N1,XW(1),XW(NXLOW),XW(NXUPP),
       & XW(NPREF),EPS,INITAL,ITIN,MAXOUT,NCORR,NCRAD,NPRNT,KDIR,NFIX,
       & INTG,XW(NX0),XW,NAML,NAME)
C
C EXECUTE THE CONTINUATION
C
        CALL DERPAR(NDIM,NVAR,N,N1,XW(1),XW(NXLOW),XW(NXUPP),XW(NPREF),
       & EPS,INITAL,ITIN,MAXOUT,NCORR,NCRAD,NPRNT,KDIR,NFIX,INTG,XW(NXSP),
       & XW(NDELTA),XW(NDXDT),XW(NF),XW(NG),XW(NXOLD),XW(NDXOLD),
       & XW(NRE),XW(NRI),XW(NB),XW,IW(1),IW(NMARK),IW)
C
        STOP
        END
C
C ****************************************************************************
C
        SUBROUTINE INIT(NREAD,NDIM,NVAR,NPAR,N,N1,X,XLOW,XUPP,PREF,EPS,
       & INITAL,ITIN,MAXOUT,NCORR,NCRAD,NPRNT,KDIR,NFIX,INTG,X0,XW,NAML,
       & NAME)
C
C ****************************************************************************
C
C READ AND WRITE INPUT DATA AND PREPARE DATA FOR CONTINUATION SUBPROGRAM
C
C SUBPROGRAMS CALLED: INITAS,INITMS
C FORTRAN LIBRARY FUNCTIONS:  DFLOAT,DABS,DSIGN
C
C --------------------------------------------------------------------------
        IMPLICIT REAL*8(A-H,O-Z)
        DIMENSION X(N1),XLOW(N1),XUPP(N1),PREF(N1),X0(NDIM),XW(1),
       & NAME(NAML)
        COMMON/PRNT/LR,LW
        COMMON/TYPE/ITYP,ICON,ISHT,NS,IPER
        COMMON/FIXP/PAR(20)
        COMMON/VARP/ALPHA,BETA,ARG,PER
        COMMON/MER1/H1,HMIN1,HMAX1,ACCUR1
        COMMON/MER2/H2,HMIN2,HMAX2,ACCUR2
        COMMON/NMDR/HX
        COMMON/TOLS/EMACH,EMIN,EACC
        COMMON/PNTR4/NA,NX0,NXA,NFA,NGA,NGB,NGC,NRES4
C --------------------------------------------------------------------------
C
```

```
C READ INPUT DATA
C
C FIXED PARAMETERS
C
C +++++++++++++++++++++++++++++++++++++++++++++++++++++++++++++++++++++++++++
      READ(LR,1000)(PAR(I),I=1,NPAR)
C +++++++++++++++++++++++++++++++++++++++++++++++++++++++++++++++++++++++++++
C
C ESTIMATE OF STARTING POINT IN STATE SPACE (OF DIMENSION NDIM),
C LOWER AND UPPER BOUNDS FOR X()
C
C +++++++++++++++++++++++++++++++++++++++++++++++++++++++++++++++++++++++++++
      READ(LR,1000)(X(I),I=1,NDIM)
      READ(LR,1000)(XLOW(I),I=1,NDIM)
      READ(LR,1000)(XUPP(I),I=1,NDIM)
C +++++++++++++++++++++++++++++++++++++++++++++++++++++++++++++++++++++++++++
C
C VARIABLE PARAMETERS AND THEIR LOWER AND UPPER BOUNDS
C
C +++++++++++++++++++++++++++++++++++++++++++++++++++++++++++++++++++++++++++
      READ(LR,1000)ALPHA,BETA,ARG,PER
      READ(LR,1000)ALFLOW,BETLOW,ARGLOW,PERLOW
      READ(LR,1000)ALFUPP,BETUPP,ARGUPP,PERUPP
C +++++++++++++++++++++++++++++++++++++++++++++++++++++++++++++++++++++++++++
C
C CONTROL PARAMETERS FOR CONTINUATION
C
C +++++++++++++++++++++++++++++++++++++++++++++++++++++++++++++++++++++++++++
      READ(LR,1100)INITAL,NFIX,KDIR,INTG,MAXOUT,NPRNT,ISRCH
C +++++++++++++++++++++++++++++++++++++++++++++++++++++++++++++++++++++++++++
C
C RELEVANT SUBSET OF VARIABLE PARAMETERS ALPHA,BETA,ARG,PER AND THEIR
C LOWER AND UPPER BOUNDS INCLUDED IN X
C
      IF(ICON.EQ.2) GO TO 10
      X(NS+1)=ALPHA
      XLOW(NS+1)=ALFLOW
      XUPP(NS+1)=ALFUPP
10    IF(ICON.NE.2) GO TO 20
      X(NS+1)=BETA
      XLOW(NS+1)=BETLOW
      XUPP(NS+1)=BETUPP
20    IF(ICON.LT.3) GO TO 30
      X(NS+2)=BETA
      XLOW(NS+2)=BETLOW
      XUPP(NS+2)=BETUPP
30    IF(ICON.NE.5) GO TO 40
      X(NS+3)=ARG
      XLOW(NS+3)=ARGLOW
      XUPP(NS+3)=ARGUPP
40    IF(ITYP.NE.2) GO TO 50
      X(N+1)=PER
```

```
          XLOW(N+1)=PERLOW
          XUPP(N+1)=PERUPP
50        CONTINUE
C
C DEFAULT VALUES FOR SOME CONSTANTS (WHICH CAN EVENTUALLY BE SPECIFIED
C IN THE INPUT FILE)
C
          H1=0.1
          HMIN1=1.0E-6
          HMAX1=1.0
          ACCUR1=1.0E-4
C
          H2=0.01
          HMIN2=1.0E-12
          HMAX2=1.0
          ACCUR2=1.0E-7
C
          EPS=1.0E-6
          HX=1.0E-12
          EMACH=1.0E-15
C
          ITIN=20
          NCORR=5
          NCRAD=1
C
          IF(NREAD.LT.2) GO TO 60
C +-+-+-+-+-+-+-+-+-+-+-+-+-+-+-+-+-+-+-+-+-+-+-+-+-+-+-+-+-+-+-+-+-+-
          READ(LR,1200)H1,HMIN1,HMAX1,ACCUR1
          READ(LR,1200)H2,HMIN2,HMAX2,ACCUR2
          READ(LR,1200)EPS,HX,EMACH
          READ(LR,1100)ITIN,NCORR,NCRAD
C +-+-+-+-+-+-+-+-+-+-+-+-+-+-+-+-+-+-+-+-+-+-+-+-+-+-+-+-+-+-+-+-+-+-
60        CONTINUE
C
C DEFAULT VALUE FOR X0() (WHICH CAN EVENTUALLY BE SPECIFIED IN THE INPUT
C FILE)
C
          DO 70 I=1,NDIM
            X0(I)=DFLOAT((-1)**I*I)
70        CONTINUE
C
          IF(NREAD.LT.3) GO TO 80
C +-+-+-+-+-+-+-+-+-+-+-+-+-+-+-+-+-+-+-+-+-+-+-+-+-+-+-+-+-+-+-+-+-+-
          READ(LR,1000)(X0(I),I=1,NDIM)
C +-+-+-+-+-+-+-+-+-+-+-+-+-+-+-+-+-+-+-+-+-+-+-+-+-+-+-+-+-+-+-+-+-+-
80        CONTINUE
1000  FORMAT(4F15.7)
1100  FORMAT(7I5)
1200  FORMAT(4E15.7)
C
C CHECK THE INPUT DATA
C
```

```
      EMIN=EMACH**0.7
      EACC=EMIN
      IF(DABS(ACCUR2).LT.EMIN)ACCUR2=DSIGN(EMIN,ACCUR2)
      IF(ITYP.NE.1.AND.ITYP.NE.3)EACC=DABS(ACCUR2)
      IF(DABS(EPS).LT.EACC)EPS=DSIGN(EACC,EPS)
      IF(DABS(ACCUR1).LT.EACC)ACCUR1=DSIGN(EACC,ACCUR1)
      IF(HX.LT.EMIN)HX=EMIN
      EACC=EACC**0.7
      IF(ITYP.NE.1.AND.ITYP.NE.3.AND.HX.LT.EACC) HX=EACC
      IF(NFIX.GT.N1) NFIX=NS+1
C
C INITIALIZE PREFERENCES FOR GAUSSIAN ELIMINATION WITH PIVOTING
C
      DO 90 I=1,N1
         PREF(I)=1.0
90    CONTINUE
C
C PRELIMINARY SEARCH FOR A POINT ON PERIODIC ORBIT OF AODES SATISFIYNG
C THE ADDITIONAL (SO CALLED ANCHOR) EQUATION THAT DEFINES THE POINCARE
C SURFACE (IF DESIRED)
C
      IF(ITYP.EQ.2.AND.ISRCH.EQ.1)CALL INITAS(NDIM,NVAR,XW(1),XW,1,NEND)
C
C EMPTY ENTRIES ARE FILLED IF MULTIPLE SHOOTING IS USED
C
      IF(ITYP.GT.1.AND.ISHT.GT.1)  CALL INITMS(NDIM,NVAR,N1,X,XLOW,XUPP,
     &                          PREF,XW(NXA),XW(NFA),XW(NGA),XW,1)
C
C PRINT INPUT DATA
C
      IF(ITYP.EQ.1)WRITE(LW,2000)
      IF(ITYP.EQ.2)WRITE(LW,2100)
      IF(ITYP.EQ.3)WRITE(LW,2200)
      IF(ITYP.GE.4)WRITE(LW,2300)
      WRITE(LW,2400)  (NAME(I),I=1,NAML)
      IF(ICON.EQ.1)WRITE(LW,2500)
      IF(ICON.EQ.2)WRITE(LW,2600)
      IF(ICON.EQ.3)WRITE(LW,2700)
      IF(ICON.EQ.4)WRITE(LW,2800)
      IF(ICON.EQ.5.AND.ITYP.EQ.1)WRITE(LW,2900)
      IF(ICON.EQ.5.AND.ITYP.GT.1)WRITE(LW,3000)
      IF(ISHT.EQ.1.AND.ITYP.GT.1)WRITE(LW,3100)
      IF(ISHT.GT.1.AND.ITYP.GT.1)WRITE(LW,3200)
      WRITE(LW,3300)(PAR(I),I=1,NPAR)
      WRITE(LW,3400)ALPHA
      WRITE(LW,3500)BETA
      IF(ICON.EQ.5)WRITE(LW,3600)ARG
      IF(ITYP.EQ.2)WRITE(LW,3700)PER
      IF(ITYP.EQ.4)WRITE(LW,3800)PER
      IF(ITYP.GE.3)WRITE(LW,3900)IPER
      WRITE(LW,4000)(X(I),I=1,N1)
      WRITE(LW,4100)(XLOW(I),I=1,N1)
```

```
            WRITE(LW,4200)(XUPP(I),I=1,N1)
            IF(INTG.EQ.1)WRITE(LW,4300)H1,HMIN1,HMAX1
            IF(INTG.EQ.2)WRITE(LW,4400)H1,HMIN1,HMAX1,ACCUR1
            IF(ITYP.EQ.2.OR.ITYP.GT.3)WRITE(LW,4500)H2,HMIN2,HMAX2,ACCUR2
            IF(ICON.EQ.1.OR.ICON.EQ.2)WRITE(LW,4600)EPS,EMACH
            IF(ICON.GE.3)WRITE(LW,4700)EPS,HX,EMACH
            WRITE(LW,4800)INITAL,ITIN,NCORR,NCRAD,NFIX,KDIR,INTG,MAXOUT,NPRNT,
           &ISRCH
      C
      2000  FORMAT('       *********************************************'/
           &       '       * STATIONARY SOLUTIONS OF AUTONOMOUS ODES *'/
           &       '       *********************************************')
      2100  FORMAT('       *********************************************'/
           &       '       * PERIODIC SOLUTIONS OF AUTONOMOUS ODES *'/
           &       '       *********************************************')
      2200  FORMAT('       ************************************************'/
           &       '       * PERIODIC SOLUTIONS OF DIFFERENCE EQUATIONS *'/
           &       '       ************************************************')
      2300  FORMAT('       *************************************************'/
           &       '       * PERIODIC SOLUTIONS OF PERIODICALLY FORCED ODES *'/
           &       '       *************************************************')
      2400  FORMAT(/(' ',70A1))
      2500  FORMAT(// CONTINUATION OF THE SOLUTION WITH RESPECT TO ALPHA FOR',
           & ' FIXED BETA')
      2600  FORMAT(// CONTINUATION OF THE SOLUTION WITH RESPECT TO BETA FOR',
           & ' FIXED ALPHA')
      2700  FORMAT(// CONTINUATION OF THE LIMIT POINT')
      2800  FORMAT(// CONTINUATION OF THE PERIOD-DOUBLING POINT')
      2900  FORMAT(// CONTINUATION OF THE HOPF BIFURCATION POINT')
      3000  FORMAT(// CONTINUATION OF THE TORUS BIFURCATION POINT')
      3100  FORMAT(// SIMPLE SHOOTING')
      3200  FORMAT(// MULTIPLE SHOOTING')
      3300  FORMAT(// FIXED PARAMETERS - PAR()'/(1X,5E14.6))
      3400  FORMAT(// VARIABLE PARAMETER WITH FORMAL NAME ALPHA =',E14.6)
      3500  FORMAT(// VARIABLE PARAMETER WITH FORMAL NAME  BETA =',E14.6)
      3600  FORMAT(// VARIABLE PARAMETER ARG =',E14.6)
      3700  FORMAT(// VARIABLE PARAMETER PER =',E14.6)
      3800  FORMAT(// FIXED PARAMETER PER =',E14.6)
      3900  FORMAT(// DISCRETE PERIOD IPER =',I5)
      4000  FORMAT(// ESTIMATE OF STARTING POINT - X()'/(1X,5E14.6))
      4100  FORMAT(// LOWER BOUNDS - XLOW()'/(1X,5E14.6))
      4200  FORMAT(// UPPER BOUNDS - XUPP()'/(1X,5E14.6))
      4300  FORMAT(// INITIAL,MINIMAL AND MAXIMAL STEPS FOR CONTINUATION',
           & ' METHOD - H1,HMIN1,HMAX1'/3E14.6)
      4400  FORMAT(// INITIAL,MINIMAL,MAXIMAL STEPS AND TOLERANCE FOR',
           & ' INTEGRATION'/' OF CONTINUATION EQUATIONS - H1,HMIN1,HMAX1,',
           & 'ACCUR1'/4E14.6)
      4500  FORMAT(// INITIAL,MINIMAL,MAXIMAL STEPS AND TOLERANCE FOR',
           & ' INTEGRATION'/' OF MODEL + VARIATIONAL EQUATIONS - H2,HMIN2,',
           & 'HMAX2,ACCUR2'/4E14.6)
      4600  FORMAT(// EPS,EMACH',2E14.6)
      4700  FORMAT(// EPS,HX,EMACH',3E14.6)
```

```
4800 FORMAT(/'     INITAL, ITIN, NCORR, NCRAD,   NFIX,   KDIR,   INTG,',
     & 'MAXOUT, NPRNT, ISRCH'/1X,10I7)
C
      RETURN
      END
C
C ********************************************************************
C
      SUBROUTINE DERPAR(NDIM,NVAR,N,N1,X,XLOW,XUPP,PREF,EPS,INITAL,ITIN,
     &                  MAXOUT,NCORR,NCRAD,NPRNT,KDIR,NFIX,INTG,XSP,
     &                  DELTA,DXDT,F,G,XOLD,DXOLD,RE,RI,B,XW,NDIR,MARK,
     &                  IW)
C
C ********************************************************************
C
C CONTINUATION SUBPROGRAM BASED ON ARCLENGTH PARAMETRIZATION
C
C SUBPROGRAMS CALLED: FCTN,GAUSE,SPTST,RG,PRST3,ODEDRV,INITAS,INITMS,
C                     ADAMS,MERS1
C FORTRAN LIBRARY FUNCTIONS:  DSQRT,DABS,IABS
C
C --------------------------------------------------------------------
      IMPLICIT REAL*8(A-H,O-Z)
      DIMENSION B(NDIM,NDIM),RE(NDIM),RI(NDIM),X(N1),XLOW(N1),XUPP(N1),
     & PREF(N1),DXDT(N1),XSP(N1),F(N1),G(N,N1),DELTA(N1),XOLD(N1),
     & DXOLD(N1),MARK(N1),NDIR(N1),XW(1),IW(1)
      COMMON/PRNT/LR,LW
      COMMON/TYPE/ITYP,ICON,ISHT,NS,IPER
      COMMON/VARP/ALPHA,BETA,ARG,PER
      COMMON/MER1/H1,HMIN1,HMAX1,ACCUR1
      COMMON/MER2/H2,HMIN2,HMAX2,ACCUR2
      COMMON/TOLS/EMACH,EMIN,EACC
      COMMON/PNTR2/NRE,NRI,NB,NEIGBH,NFV1,NRES2
      COMMON/PNTR3/NFHY,NGHX,NXWR1,NFWR1,NYK1,NYK21,NYK31,NRES3
      COMMON/PNTR4/NA,NX0,NXA,NFA,NGA,NGB,NGC,NRES4
      COMMON/PNTR5/NXINT,NFINT,NXWR2,NFWR2,NYK2,NYK22,NYK32
      COMMON/PNTR6/NNDIR1,NMARK,NIRR,NIRK,NIVR
      DATA INDIC,INDSP/'****','    '/
      DATA IUNST,ISTAB,INEUT/' UNS',' STB',' NTR'/
C --------------------------------------------------------------------
C
      IDIR=1
      PRFX=PREF(NFIX)
      PREF(NFIX)=0.0
      IF (INITAL.EQ.0) GO TO 50
C
C INITIAL NEWTON ITERATIONS
C
      DO 30 L=1,ITIN
        CALL FCTN(NDIM,NVAR,N,N1,X,F,G,B,XW(NEIGBH),XW(NFHY),XW(NGHX),
     &            XW,IW,IERR)
        IF (IERR.NE.0) GO TO 520
```

```
              SQUAR=0.0
              DO 10 I=1,N
                 SQUAR=SQUAR+F(I)**2
10            CONTINUE
              LL=L-1
              SQUAR=DSQRT(SQUAR)
              WRITE (LW,1700) LL,(X(I),I=1,N1),SQUAR
              CALL GAUSE(N,N1,G,F,IERR,PREF,DELTA,K,XW(NFHY),XW(NGHX),
     &                   IW(NIRR),IW(NIRK))
              IF (IERR.NE.0) GO TO 530
              P=0.0
              W=1.0
              DO 20 I=1,N1
                 X(I)=X(I)-F(I)
                 IF(EPS.GT.0.0)  W=1.0/(EACC+DABS(X(I)))
                 P=P+W*DABS(F(I))
20            CONTINUE
C
C CHECK IF THE FIXED POINT ON THE PERIODIC ORBIT OF AODES IS NOT CLOSE
C TO STATIONARY POINT
C
              IF(ITYP.EQ.2) CALL SPTST(NDIM,NVAR,N1,X,XW,EPS,ISPT)
              IF(ITYP.EQ.2.AND.ISPT.NE.0) GO TO 550
              IF (P.LE.DABS(EPS)) GO TO 40
30            CONTINUE
C
C OVERFLOW IN INITIAL NEWTON METHOD
C
              WRITE (LW,1800) ITIN
              RETURN
C
C INITIAL NEWTON METHOD HAS CONVERGED
C
40            CONTINUE
              WRITE (LW,1900)(X(I),I=1,N1)
C
C COMPUTE EIGENVALUES USING THE LINPACK PACKAGE
C
              CALL RG(NDIM,NDIM,B,RE,RI,0,XW(NEIGBH),IW(NIVR),XW(NFV1),IERR)
              IF(IERR.NE.0) WRITE(LW,2000)
              IF(IERR.EQ.0) WRITE(LW,2100)(RE(I),RI(I),I=1,NDIM)
50        CONTINUE
C
C CHECK WHETHER THE STARTING POINT LIES WITHIN PRESCRIBED BOUNDS
C
              DO 60 I=1,N1
                 IF (X(I).LT.XLOW(I).OR.X(I).GT.XUPP(I)) GO TO 540
60            CONTINUE
              IF (INITAL.EQ.2) RETURN
C
C CONTINUATION OF THE SOLUTION CURVE FROM THE STARTING POINT
C
```

```
            WRITE (LW,2200)
            IF(NPRNT.NE.3)WRITE(LW,2300)
70          CONTINUE
            KOUT=0
            NFIXB=0
            MADMS=0
            IWRN=O
            ISWTCH=0
            IBACK=0
            NUEVL1=1
            NUEVL2=-1
            ARCL=0.0
C
C SUCCESSIVE COMPUTATION OF AT MOST MAXOUT POINTS ON CONTINUATION CURVE
C
            NOUT=1
80          CONTINUE
C
            IF(NOUT.EQ.KOUT+1.AND.NOUT.GT.1)  X(NS+1)=XSP(NS+1)
            IF(NFIXB.GT.0)  X(NFIXB)=XBOUND
            NC=0
90          CONTINUE
C
C ONE ITERATION IN NEWTON METHOD
C
            CALL FCTN(NDIM,NVAR,N,N1,X,F,G,B,XW(NEIGBH),XW(NFHY),XW(NGHX),
      &               XW,IW,IERR)
            IF (IERR.NE.0) GO TO 520
            SQUAR=0.0
            DO 100 I=1,N
              SQUAR=SQUAR+F(I)**2
100         CONTINUE
            SQUAR=DSQRT(SQUAR)
            CALL GAUSE(N,N1,G,F,IERR,PREF,DELTA,K,XW(NFHY),XW(NGHX),
      &               IW(NIRR),IW(NIRK))
            IF (IERR.NE.0) GO TO 530
            IF (NCRAD.EQ.1) SQUAR=-SQUAR
            P=0.0
            W=1.0
            DO 110 I=1,N1
              IF(EPS.GT.0.0) W=1.0/(EACC+DABS(X(I)))
              P=P+W*DABS(F(I))
110         CONTINUE
C
C CHECK IF THE FIXED POINT ON THE PERIODIC ORBIT OF AODES IS NOT CLOSE
C TO STATIONARY POINT
C
            IF(ITYP.EQ.2) CALL SPTST(NDIM,NVAR,N1,X,XW,EPS,ISPT)
            IF(ITYP.EQ.2.AND.ISPT.NE.0) GO TO 550
C
            IF (P.LE.DABS(EPS)) GO TO 150
            DO 120 I=1,N1
```

```
                X(I)=X(I)-F(I)
120         CONTINUE
            NC=NC+1
            IF(NC.LT.NCORR) GO TO 90
C
C NEWTON METHOD DOES NOT CONVERGE
C
            IF(NOUT.EQ.1) GO TO 560
            IF(NOUT.EQ.KOUT+1) GO TO 280
130         IF(NFIXB.EQ.0) GO TO 135
              PREF(NFIXB)=PRFB
              NFIXB=0
135         CONTINUE
C
C RERUN FROM PREVIOUS POINT WITH A SMALLER STEP FOR PREDICTOR
C
            DO 140 I=1,N1
              II=NNDIR1+I-1
              NDIR(I)=IW(II)
              X(I)=XOLD(I)
              DXDT(I)=DXOLD(I)
140         CONTINUE
            H1=(ARCL-ARCOLD)/3.0
            IF(H1.LT.DABS(HMIN1)) GO TO 560
            ARCL=ARCOLD
            K=KOLD
            MADMS=0
            NOUT=NOUT-1
            GO TO 460
150         CONTINUE
C
C NEWTON METHOD HAS CONVERGED
C
            IF (NCRAD.EQ.0) GO TO 170
C
C ADDITIONAL NEWTON CORRECTION
C
            DO 160 I=1,N1
              X(I)=X(I)-F(I)
160         CONTINUE
C
170         CONTINUE
            IF(NOUT.GT.1)GOTO 190
C
C DETERMINE THE INITIAL DIRECTION ALONG THE CURVE AND STORE THE FIRST
C POINT
C
            PREF(NFIX)=PRFX
            IF(KDIR.GE.0)NDIR(K)=1
            IF(KDIR.LT.0)NDIR(K)=-1
            IF(KDIR.GT.0.AND.IDIR.EQ.2)NDIR(K)=-1
            IF(KDIR.LT.0.AND.IDIR.EQ.2)NDIR(K)=1
```

```
                CALL PRST3 (N1,DELTA,NDIR,K,DXDT)
                NDIRSP=NDIR(NS+1)
                H1SP=H1
                DO 180 I=1,N1
                  XSP(I)=X(I)
180         CONTINUE
190      CONTINUE
C
C EVALUATE THE NORMALIZED TANGENT VECTOR TO CONTINUATION CURVE
C
         CALL PRST3 (N1,DELTA,NDIR,K,DXDT)
C
         IF(NOUT.EQ.1) GO TO 230
           IF(NOUT.EQ.KOUT+1) GO TO 280
C
C CHECK THE CONTINUITY OF THE NEW POINT
C
         DIST=0.0
         DO 200 I=1,N1
           DIST=DIST+(X(I)-XOLD(I))**2
200      CONTINUE
         DIST=DSQRT(DIST)
         IF(DIST.GT.(ARCL-ARCOLD)) GO TO 130
C
         IF(NFIXB.GT.0) NFIXB=-1
         IF(NFIXB.EQ.-1) GO TO 320
C
         IF(INTG.NE.1.OR.NC.GT.1) GO TO 210
C
C SET NEW STEP FOR ADAMS-BASHFORTH METHOD
C
           IF(H1.EQ.DABS(HMAX1)) GO TO 210
           H1=(3.0-DFLOAT(NC))*H1
           IF(H1.GT.DABS(HMAX1)) H1=DABS(HMAX1)
           MADMS=0
210      CONTINUE
C
C CHECK THE DIRECTION ALONG THE CONTINUATION CURVE
C
           IDRCTN=1
           DO 220 I=1,N1
             II=NNDIR1+I-1
             IF(ISWTCH.EQ.1.AND.I.LE.NS) GO TO 220
             IF(NDIR(I).EQ.-IW(II)) GO TO 220
             IDRCTN=0
220        CONTINUE
           IF(IDRCTN.NE.1) GO TO 230
             NDIR(K)=-NDIR(K)
             CALL PRST3 (N1,DELTA,NDIR,K,DXDT)
230      CONTINUE
C
C CHECK WHETHER THE NEXT POINT WILL BE WITHIN PRESCRIBED BOUNDS
```

```
C
            DO 240 I=1,N1
            IF (H1*DXDT(I).LE.0.8*(XLOW(I)-X(I))) GO TO 250
            IF (H1*DXDT(I).GE.0.8*(XUPP(I)-X(I))) GO TO 260
240         CONTINUE
            GO TO 280
250         XBOUND=XLOW(I)
            GO TO 270
260         XBOUND=XUPP(I)
270         NFIXB=I
            MADMS=0
            PRFB=PREF(NFIXB)
            PREF(NFIXB)=0.0
            H1=(XBOUND-X(NFIXB))/DXDT(NFIXB)
            IF(H1.LE.0.0) NFIXB=-1
            GO TO 320
280         IF (NOUT.LE.4) GO TO 320
C
C CLOSED CURVE TEST
C
            P=0.0
            W=1.0
            DO 290 I=1,N1
            IF(ISWTCH.EQ.1.AND.I.LE.NS)GO TO 290
            IF(EPS.GT.0.0) W=1.0/(EACC+DABS(X(I)))
            P=P+W*DABS(X(I)-XSP(I))
290         CONTINUE
            IF (P.LE.DABS(EPS)) GO TO 320
            IF(NOUT.NE.KOUT+1) GO TO 310
            DO 300 I=1,N1
            II=NNDIR1+I-1
            NDIR(I)=IW(II)
            X(I)=XOLD(I)
            DXDT(I)=DXOLD(I)
300         CONTINUE
            PREF(NS+1)=PRFC
            H1=H1OLD
            K=KOLD
            ARCL=ARCOLD
            MADMS=0
            IBACK=0
            NOUT=NOUT-1
            KOUT=KOUT-1
            GO TO 460
310         CONTINUE
            IF(H1*DABS(DXDT(NS+1)).LE.0.8*DABS(X(NS+1)-XSP(NS+1)))GOTO 320
            IF((XSP(NS+1)-X(NS+1))*DFLOAT(NDIR(NS+1)).LE.0.0) GO TO 320
            IF(H1*DABS(DXDT(NS+1)).LE.0.9*DABS(X(NS+1)-XSP(NS+1)))GOTO 320
C
C CLOSE CURVE IS EXPECTED
C
            IF(NDIR(NS+1).NE.NDIRSP) IBACK=1
```

```
              H1OLD=H1
              H1=DABS(X(NS+1)-XSP(NS+1))/DABS(DXDT(NS+1))
              KOUT=NOUT
              PRFC=PREF(NS+1)
              PREF(NS+1)=0.0
              MADMS=0
320           CONTINUE
C
C PRINT OUTPUT RESULTS
C
              DO 330 I=1,N1
                MARK(I)=INDSP
330           CONTINUE
              MARK(K)=INDIC
              WRITE(LW,2400)NOUT,(X(I),MARK(I),I=1,N1)
              IF(NPRNT.EQ.3)GOTO 360
C
C COMPUTE EIGENVALUES USING THE LINPACK PACKAGE
C
              CALL RG(NDIM,NDIM,B,RE,RI,0,XW(NEIGBH),IW(NIVR),XW(NFV1),IERR)
              IF(IERR.NE.0) GO TO 350
C
                NUEVL=0
                ESTB=DABS(EPS)**0.5
                DO 340 I=1,NDIM
                  IF(ITYP.EQ.1)STABCN=RE(I)
                  IF(ITYP.GT.1)STABCN=DLOG(DSQRT(RE(I)**2+RI(I)**2+EMACH))
                  MARK(I)=IUNST
                  IF(STABCN.LT.0.0.AND.DABS(STABCN).GT.ESTB)MARK(I)=ISTAB
                  IF(STABCN.GT.0.0.AND.DABS(STABCN).GT.ESTB) NUEVL=NUEVL+1
                  IF(DABS(STABCN).LE.ESTB) MARK(I)=INEUT
340             CONTINUE
                WRITE(LW,2500)(RE(I),RI(I),MARK(I),I=1,NDIM)
                IF(NUEVL.EQ.NUEVL2) GO TO 360
                IF(NUEVL1.EQ.0) WRITE(LW,2600) NUEVL2,NUEVL
                NUEVL2=NUEVL
                NUEVL1=0
                GO TO 360
350           WRITE(LW,2000)
              NUEVL1=1
              NUEVL2=-1
360           MARK(K)=INDIC
              IF(NPRNT.EQ.1)WRITE(LW,2700)NC,ARCL,H1,SQUAR,
     &                                    (NDIR(I),MARK(I),I=1,N1)
              IF(NFIXB.EQ.-1) GO TO 540
              IF(NOUT.EQ.KOUT+1.AND.IBACK.EQ.0.AND.NOUT.GT.1) GO TO 580
              IF(NOUT.EQ.KOUT+1.AND.IBACK.EQ.1.AND.NOUT.GT.1) GO TO 570
C
              IF(ITYP.NE.2) GO TO 430
C
C CHECK IF THE FIXED POINT ON THE PERIODIC ORBIT OF AODES DOES NOT
C ESCAPE FROM THE POINCARE SURFACE DEFINED BY THE ANCHOR EQUATION
```

```
C
            DO 370 I=1,NDIM
              II=NXINT+I-1
              XW(II)=X(I)
370         CONTINUE
            CALL ODEDRV(NDIM,NVAR,NDIM,0.0D0,0.05D0*PER,H2,HMIN2,HMAX2,
         &            XW(NXINT),XW(NFINT),ACCUR2,NDIM,IMSG,1,XW)
            IF(IMSG.GT.3) GO TO 390
            DO 380 I=1,NDIM
              II=NXINT+I-1
              XW(II)=X(I)
380         CONTINUE
            CALL ODEDRV(NDIM,NVAR,NDIM,0.0D0,0.95D0*PER,H2,HMIN2,HMAX2,
         &            XW(NXINT),XW(NFINT),ACCUR2,NDIM,IMSG,0,XW)
            CALL ODEDRV(NDIM,NVAR,NDIM,0.0D0,0.04D0*PER,H2,HMIN2,HMAX2,
         &            XW(NXINT),XW(NFINT),ACCUR2,NDIM,IMSG,1,XW)
            IF(IMSG.GT.3) GO TO 390
C
C THE INTERSECTION WITH THE POINCARE SURFACE WAS NOT FOUND
C CONTINUE WITH THE OLD POINT
C
            GO TO 430
C
390         CONTINUE
C
C CHOOSE A NEW POINT
C
            CALL INITAS(NDIM,NVAR,X,XW,0,NEND)
            IF(NEND.EQ.2) GO TO 420
            IF(ISHT.GT.1) CALL INITMS(NDIM,NVAR,N1,X,XLOW,
         &     XUPP,PREF,XW(NXA),XW(NFA),XW(NGA),XW,0)
            CALL FCTN(NDIM,NVAR,N,N1,X,F,G,B,XW(NEIGBH),XW(NFHY),
         &            XW(NGHX),XW,IW,IERR)
            IF (IERR.NE.0) GO TO 520
            DO 400 I=1,NDIM
              II=NRES3+I-1
              XW(II)=PREF(I)
              PREF(I)=1000.0
400         CONTINUE
            CALL GAUSE(N,N1,G,F,IERR,PREF,DELTA,K,XW(NFHY),XW(NGHX),
         &            IW(NIRR),IW(NIRK))
            DO 410 I=1,NDIM
              II=NRES3+I-1
              PREF(I)=XW(II)
410         CONTINUE
            IF (IERR.NE.0) GO TO 530
            CALL PRST3(N1,DELTA,NDIR,K,DXDT)
            ISWTCH=1
            WRITE(LW,2800)
420         CONTINUE
430       CONTINUE
C
```

```
C SET PREFERENCES FOR GAUSSIAN ELIMINATION WITH PIVOTING
C
        DO 440 I=1,N1
          IF(PREF(I).EQ.0.0.OR.(I.GT.NDIM.AND.I.LE.NS)) GO TO 440
          PREF(I)=1.0/(DABS(DXDT(I))+1.0D-4)
440     CONTINUE
C
C STORE NECESSARY INFORMATION FOR EVENTUAL RESTART FROM THE CURRENT
C POINT
C
        DO 450 I=1,N1
          II=NNDIR1+I-1
          IW(II)=NDIR(I)
          XOLD(I)=X(I)
          DXOLD(I)=DXDT(I)
450     CONTINUE
        KOLD=K
        ARCOLD=ARCL
C
C ONE STEP OF PREDICTOR
C
460     GO TO (470,480),INTG
470     CALL ADAMS(N1,DXDT,MADMS,ARCL,H1,X,XW(NXWR1))
        GO TO 490
480     CALL MERS1(NDIM,NVAR,N,N1,ARCL,X,DXDT,H1,HMIN1,HMAX1,ACCUR1,
     &    1.0D0,N1,IWRN,IERR,XW(NXWR1),XW(NFWR1),XW(NYK1),XW(NYK21),
     &    XW(NYK31),XW,NDIR,IW(NNDIR1),IW)
        IF(IERR.LT.0) GO TO 520
        IF(IERR.GT.0) GO TO 530
C
C GO BACK TO NEWTON METHOD TO CORRECT THE PREDICTED NEW POINT IF MAXIMUM
C NUMBER OF POINTS IS NOT REACHED
C
490     NOUT=NOUT+1
        IF(NOUT.LE.MAXOUT) GO TO 80
C
C MAXIMUM NUMBER OF POINTS IS REACHED, EITHER RETURN OR CONTINUE FROM
C THE FIRST POINT IN THE OTHER DIRECTION ALONG THE CURVE
C
500     IDIR=IDIR+1
        IF(NOUT.EQ.1.AND.NFIXB.NE.-1) RETURN
        IF(IDIR.GT.IABS(KDIR)) RETURN
        DO 510 I=1,N1
          X(I)=XSP(I)
          IF(I.GT.NDIM.AND.I.LE.NS) GO TO 510
          PREF(I)=1.0
510     CONTINUE
        H1=H1SP
        PREF(NFIX)=0.0
        WRITE(LW,2900)
        GOTO 70
C
```

```
520    NXND=NX0+NDIM-1
       WRITE(LW,1000) (XW(I),I=NX0,NXND)
       GO TO 500
530    WRITE(LW,1100) (X(I),I=1,N1)
       GO TO 500
540    WRITE(LW,1200)
       GO TO 500
550    WRITE(LW,1300)
       GO TO 500
560    WRITE (LW,1400) NCORR,P
       GO TO 500
570    WRITE(LW,1500)
       GO TO 500
580    WRITE(LW,1600)
C
1000   FORMAT(/' DERPAR ERROR  SINGULAR MATRIX IN EVALUATION OF',
      & ' COEFFICIENTS'/' OF CHARACTERISTIC POLYNOMIAL'/
      & ' X0() =',(7X,4E16.8))
1100   FORMAT(/' DERPAR ERROR  SINGULAR JACOBIAN MATRIX FOR X() ='/
      & (1X,4E16.8))
1200   FORMAT(/' DERPAR OUT OF BOUNDS')
1300   FORMAT(/' DERPAR PERIODIC ORBIT OF AODES IS CLOSE TO STATIONARY ',
      & 'POINT'/' HOPF BIFURCATION IS EXPECTED')
1400   FORMAT(/' DERPAR ERROR  NUMBER OF NEWTON CORRECTIONS =',I5,' IS',
      & ' NOT SUFFICIENT'/' ERROR OF X() =',E14.6/)
1500   FORMAT(/' DERPAR ERROR  RETURN TO STARTING POINT ALONG ALREADY ',
      & 'COMPUTED BRANCH')
1600   FORMAT(/' DERPAR CLOSED CURVE MAY BE EXPECTED')
1700   FORMAT(/' DERPAR',I3,'.INITIAL NEWTON ITERATION - X(),SQUAR'/
      & (3X,4(E16.8)))
1800   FORMAT(/' DERPAR ERROR OVERFLOW',I5,2X,'INITIAL ITERATIONS')
1900   FORMAT(/' DERPAR AFTER INITIAL NEWTON ITERATIONS'/(3X,4E16.8))
2000   FORMAT(/' DERPAR ERROR IN THE COMPUTATION OF EIGENVALUES')
2100   FORMAT(/' EIGENVALUES   ',2(2E14.6,2X)/(15X,2(2E14.6,2X)))
2200   FORMAT(///'   *****************'/
      &         '   * DERPAR RESULTS *'/
      &         '   *****************'//
      &         ' VARIABLE CHOSEN AS INDEPENDENT IS MARKED BY *')
2300   FORMAT(   ' STABLE EIGENVALUE IS MARKED BY S'/
      &         ' UNSTABLE EIGENVALUE IS MARKED BY U'/
      &         ' EIGENVALUE CLOSE TO NEUTRAL ONE IS MARKED BY N')
2400   FORMAT(/' DERPAR POINT NO.',I4/' X() =',4(E16.8,A1)/
      & (6X,4(E16.8,A1)))
2500   FORMAT(' EIGENVALUES   ',2(2E14.6,A2)/(15X,2(2E14.6,A2)))
2600   FORMAT(/' DERPAR THE DIMENSION OF UNSTABLE MANIFOLD HAS CHANGED ',
      & 'FROM ',I2,' TO ',I2/' A BIFURCATION IS EXPECTED'/)
2700   FORMAT(' NCR =',I2,' ARCL =',E14.6,' H1 =',E14.6,' SQUAR ='
      & E14.6/' NDIR() =',14(I4,A1)/(1X,14(I4,A1)))
2800   FORMAT(/' DERPAR NEW POINT ON THE POINCARE SURFACE WAS CHOSEN')
2900   FORMAT(/' DERPAR RESTART FROM THE STARTING POINT IN THE OTHER ',
      & 'DIRECTION')
C
```

```
      RETURN
      END
C
C     **********************************************************************
C
      SUBROUTINE SPTST(NDIM,NVAR,N1,X,XW,EPS,ISPT)
C
C     **********************************************************************
C
C STATIONARY POINT TEST
C
C SUBPROGRAMS CALLED: ODEDRV
C FORTRAN LIBRARY FUNCTIONS: DABS
C
C     ------------------------------------------------------------------
      IMPLICIT REAL*8(A-H,O-Z)
      DIMENSION X(N1),XW(1)
      COMMON/VARP/ALPHA,BETA,ARG,PER
      COMMON/MER1/H1,HMIN1,HMAX1,ACCUR1
      COMMON/MER2/H2,HMIN2,HMAX2,ACCUR2
      COMMON/PNTR5/NXINT,NFINT,NXWR2,NFWR2,NYK2,NYK22,NYK32
C     ------------------------------------------------------------------
C
      ISPT=0
      DO 10 I=1,NDIM
        II=NXINT+I-1
        XW(II)=X(I)
10    CONTINUE
      CALL ODEDRV(NDIM,NVAR,NDIM,0.0D0,PER,H2,HMIN2,HMAX2,
     &            XW(NXINT),XW(NFINT),ACCUR2,NDIM,IMSG,2,XW)
      II=NXINT+NDIM
      SPT=DABS((XW(II)-PER)/PER)
      IF(SPT.LT.DABS(EPS)**0.333) ISPT=1
C
      RETURN
      END
C
C     **********************************************************************
C
      SUBROUTINE PNTRS(NDIM,NVAR,NMAX,N1MAX,NMERS)
C
C     **********************************************************************
C
C   SET POINTERS INTO WORK ARRAYS XW AND IW
C
C SUBPROGRAMS CALLED: NONE
C FORTRAN LIBRARY FUNCTIONS: NONE
C
C     ------------------------------------------------------------------
      COMMON/PNTR1/NXLOW,NXUPP,NPREF,NXSP,NDELTA,NDXDT,NF,NG,NXOLD,
     & NDXOLD,NRES1
      COMMON/PNTR2/NRE,NRI,NB,NEIGBH,NFV1,NRES2
```

```
      COMMON/PNTR3/NFHY,NGHX,NXWR1,NFWR1,NYK1,NYK21,NYK31,NRES3
      COMMON/PNTR4/NA,NX0,NXA,NFA,NGA,NGB,NGC,NRES4
      COMMON/PNTR5/NXINT,NFINT,NXWR2,NFWR2,NYK2,NYK22,NYK32
      COMMON/PNTR6/NNDIR1,NMARK,NIRR,NIRK,NIVR
C ------------------------------------------------------------------------
C
      NXLOW=1+N1MAX
      NXUPP=NXLOW+N1MAX
      NPREF=NXUPP+N1MAX
      NXSP=NPREF+N1MAX
      NDELTA=NXSP+N1MAX
      NDXDT=NDELTA+N1MAX
      NF=NDXDT+N1MAX
      NG=NF+N1MAX
      NXOLD=NG+NMAX*N1MAX
      NDXOLD=NXOLD+N1MAX
      NRES1=NDXOLD+N1MAX
C
      NRE=NRES1+NMAX*N1MAX
      NRI=NRE+NDIM
      NB=NRI+NDIM
      NEIGBH=NB+NDIM**2
      NFV1=NEIGBH+NDIM**2
      NRES2=NFV1+NDIM
C
      NFHY=NRES2+NDIM*NDIM
      NGHX=NFHY+N1MAX
      NXWR1=NGHX+NMAX*N1MAX
      NFWR1=NXWR1+N1MAX
      NYK1=NFWR1+N1MAX
      NYK21=NYK1+N1MAX
      NYK31=NYK21+N1MAX
      NRES3=NYK31+N1MAX
C
      NA=NRES3+NMAX*N1MAX
      NX0=NA+NDIM
      NXA=NX0+NDIM
      NFA=NXA+NDIM
      NGA=NFA+NDIM
      NGB=NGA+NDIM*NVAR
      NGC=NGB+NVAR**2
      NRES4=NGC+NDIM*NVAR
C
      NXINT=NRES4+NDIM*NDIM
      NFINT=NXINT+NMERS
      NXWR2=NFINT+NMERS
      NFWR2=NXWR2+NMERS
      NYK2=NFWR2+NMERS
      NYK22=NYK2+NMERS
      NYK32=NYK22+NMERS
C
      NNDIR1=1+N1MAX
```

```
      NMARK=NNDIR1+N1MAX
      NIRR=NMARK+N1MAX
      NIRK=NIRR+N1MAX
      NIVR=NIRK+NDIM
C
      RETURN
      END
C
C ******************************************************************************
C
      SUBROUTINE INITAS(NDIM,NVAR,X,XW,NINIT,NEND)
C
C ******************************************************************************
C
C FIND A FIXED PHASE POINT OF THE PERIODIC ORBIT OF AODES AS AN
C INTERSECTION POINT OF THE ORBIT WITH A POINCARE SURFACE DEFINED
C BY THE ANCHOR EQUATION X * F(X) = 0
C
C SUBPROGRAMS CALLED: ODEDRV
C FORTRAN LIBRARY FUNCTIONS: DABS
C
C ------------------------------------------------------------------------------
      IMPLICIT REAL*8(A-H,O-Z)
      DIMENSION X(NDIM),XW(1)
      COMMON/TYPE/ITYP,ICON,ISHT,NS,IPER
      COMMON/MER2/H2,HMIN2,HMAX2,ACCUR2
      COMMON/VARP/ALPHA,BETA,ARG,PER
      COMMON/PNTR2/NRE,NRI,NB,NEIGBH,NFV1,NRES2
      COMMON/PNTR4/NA,NX0,NXA,NFA,NGA,NGB,NGC,NRES4
      COMMON/PNTR5/NXINT,NFINT,NXWR2,NFWR2,NYK2,NYK22,NYK32
C ------------------------------------------------------------------------------
C
      NEND=0
      BGNT=0.0
      ENDT=0.98*PER
      DO 10 J=1,NDIM
        JJ=NXINT+J-1
        XW(JJ)=X(J)
        JJ=NEIGBH+J
        XW(JJ)=X(J)
10      CONTINUE
      XW(NEIGBH)=BGNT
      I=1
      IMSG=-1
20    CALL ODEDRV(NDIM,NVAR,NDIM,BGNT,ENDT,H2,HMIN2,HMAX2,XW(NXINT),
     &            XW(NFINT),ACCUR2,NDIM,IMSG,1,XW)
      IF(NINIT.EQ.1.AND.IMSG.EQ.2) RETURN
      IF(IMSG.GT.3) GO TO 30
C
C THE INTERSECTION WITH THE POINCARE SURFACE WAS NOT FOUND
C
      NEND=1
```

```
        IF(I.GT.1) GO TO 30
C
C X() IS NOT CHANGED
C
        NEND=2
        RETURN
C
C
30      CONTINUE
C
C THE INTERSECTION POINT OR THE END POINT WAS REACHED
C STORE THE TIME AND POINT
C
        DO 40 J=1,NDIM
          JJ=NEIGBH+J+I*(NDIM+1)
          IF(JJ.GE.NA) GO TO 50
          JF=NXINT+J-1
          XW(JJ)=XW(JF)
40      CONTINUE
        JJ=NEIGBH+I*(NDIM+1)
        XW(JJ)=ENDT
        IF(NEND.EQ.1) GO TO 60
        BGNT=ENDT
        ENDT=0.98*PER
        I=I+1
        GO TO 20
C
50      CONTINUE
C
C OVERFLOW OF THE BUFFER FOR STORAGE OF INTERSECTION POINTS
C
60      IF(I.GT.3) GO TO 90
          IF(NINIT.NE.1) GO TO 80
C
C THERE IS ONE OR TWO INTERSECTION POINTS, THE FIRST ONE IS TAKEN
C
          DO 70 J=1,NDIM
            JJ=NEIGBH+J+NDIM+1
            X(J)=XW(JJ)
70        CONTINUE
          RETURN
80        CONTINUE
C
C X() IS NOT CHANGED
C
        NEND=2
        RETURN
C
90      CONTINUE
C
C THERE ARE MORE THAN TWO INTERSECTION POINTS, CHOOSE ONE OF THEM
C
```

```
      PLMIN=DABS(PER)
      DO 100 J=1,I
        J1=NEIGBH+(J-1)*(NDIM+1)
        J2=J1+NDIM+1
        PL=DABS(XW(J2)-XW(J1))
        IF(PL.GT.PLMIN) GO TO 100
        JPLM=J1
        PLMIN=PL
100   CONTINUE
      PMIN=1.0E5
      II=I-1
      DO 110 J=1,II
        J1=NEIGBH+(J-1)*(NDIM+1)
        J2=J1+NDIM+1
        J3=J2+NDIM+1
        IF(J1.EQ.JPLM.OR.J2.EQ.JPLM) GO TO 110
        P=DABS((XW(J2)-XW(J1))/(XW(J3)-XW(J2)))
        IF(DABS(P-1.0).GT.PMIN) GO TO 110
        JPM=J2
        PMIN=DABS(P-1.0)
110   CONTINUE
C
C COPY THE CHOSEN POINT FROM WORK ARRAY INTO X
C
      DO 120 J=1,NDIM
        JJ=JPM+J
        X(J)=XW(JJ)
120   CONTINUE
C
      RETURN
      END
C
C *********************************************************************
C
      SUBROUTINE INITMS(NDIM,NVAR,N1,X,XLOW,XUPP,PREF,XA,FA,GA,XW,ICN)
C
C *********************************************************************
C
C   FILL EMPTY ENTRIES FOR MULTIPLE SHOOTING METHOD
C
C SUBPROGRAMS CALLED: INTEG
C FORTRAN LIBRARY FUNCTIONS: NONE
C
C -------------------------------------------------------------------
      IMPLICIT REAL*8(A-H,O-Z)
      DIMENSION X(N1),XLOW(N1),XUPP(N1),PREF(N1),XA(NDIM),
     & FA(NDIM),GA(NDIM,NVAR),XW(1)
      COMMON/TYPE/ITYP,ICON,ISHT,NS,IPER
      COMMON/PNTR5/NXINT,NFINT,NXWR2,NFWR2,NYK2,NYK22,NYK32
C -------------------------------------------------------------------
C
      DO 10 I=1,NDIM
```

```
          XA(I)=X(I)
10        CONTINUE
          DO 30 I=2,ISHT
            CALL INTEG(NDIM,NVAR,I-1,XA,FA,GA,XW(NXINT),XW(NFINT),XW)
            DO 20 K=1,NDIM
              XA(K)=FA(K)
              JJ=(I-1)*NDIM+K
              X(JJ)=FA(K)
              IF(ICN.EQ.0) GO TO 20
              XLOW(JJ)=XLOW(K)
              XUPP(JJ)=XUPP(K)
              PREF(JJ)=1000.0
20          CONTINUE
30        CONTINUE
C
          RETURN
          END
C
C     ********************************************************************
C
          SUBROUTINE GAUSE(N,N1,A,B,IERR,PREF,DELTA,K,Y,X,IRR,IRK)
C
C     ********************************************************************
C
C SOLUTION OF N LINEAR EQUATIONS FOR N1=N+1 UNKNOWNS
C BASED ON GAUSSIAN ELIMINATION WITH PIVOTING.
C
C N       : NUMBER OF EQUATIONS
C A(,)    : N X N1 MATRIX OF SYSTEM
C B()     : RIGHT-HAND SIDES AND AFTER RETURN RESULTS
C IERR    : IF (IERR.NE.0) AFTER RETURN THEN A WAS NUMERICALLY SINGULAR
C PREF(I) : PREFERENCE NUMBER FOR X(I) TO BE INDEPENDENT VARIABLE,
C           THE LOWER IS PREF(I) THE HIGHER IS PREFERENCE OF X(I)
C DELTA(I): COEFFICIENTS IN EXPLICIT DEPENDENCES OBTAINED IN FORM:
C           X(I)=B(I)+DELTA(I)*X(K) , I.NE.K.
C K       : RESULTING INDEX OF INDEPENDENT VARIABLE
C
C SUBPROGRAMS CALLED: NONE
C FORTRAN LIBRARY FUNCTIONS: DABS
C
C ----------------------------------------------------------------------
          IMPLICIT REAL*8(A-H,O-Z)
          DIMENSION A(N,N1),B(N1),PREF(N1),DELTA(N1),Y(N1),X(N1),IRR(N1),
     &              IRK(N1)
          COMMON/TOLS/EMACH,EMIN,EACC
C ----------------------------------------------------------------------
C
          ID=1
          IERR=0
          DO 10 I=1,N1
            IRK(I)=0
            IRR(I)=0
```

```
10      CONTINUE
20      IR=1
        IS=1
        AMAX=0.0
        DO 60 I=1,N
          IF (IRR(I)) 60,30,60
30        DO 50 J=1,N1
            P=PREF(J)*DABS(A(I,J))
            IF (P-AMAX) 50,50,40
40          IR=I
            IS=J
            AMAX=P
50        CONTINUE
60      CONTINUE
        IF (AMAX.GT.EMACH) GO TO 70
          IERR=1
          RETURN
C
70      IRR(IR)=IS
        DO 90 I=1,N
          IF (I.EQ.IR.OR.A(I,IS).EQ.0.0) GO TO 90
          P=A(I,IS)/A(IR,IS)
          DO 80 J=1,N1
            A(I,J)=A(I,J)-P*A(IR,J)
80        CONTINUE
          A(I,IS)=0.0
          B(I)=B(I)-P*B(IR)
90      CONTINUE
        ID=ID+1
        IF (ID.LE.N) GO TO 20
        DO 100 I=1,N
          IR=IRR(I)
          X(IR)=B(I)/A(I,IR)
          IRK(IR)=1
100     CONTINUE
        DO 110 K=1,N1
          IF(IRK(K).EQ.0) GO TO 120
110     CONTINUE
120     DO 130 I=1,N
          IR=IRR(I)
          Y(IR)=-A(I,K)/A(I,IR)
130     CONTINUE
        DO 140 I=1,N1
          B(I)=X(I)
          DELTA(I)=Y(I)
140     CONTINUE
        B(K)=0.0
        DELTA(K)=1.0
C
        RETURN
        END
C
```

```
C  ***********************************************************************
C
       SUBROUTINE ADAMS(N1,D,MADMS,T,H,X,DER)
C
C  ***********************************************************************
C
C  ADAMS-BASHFORTH METHODS FOR INTEGRATION OF CONTINUATION EQUATIONS
C
C  SUBPROGRAMS CALLED: NONE
C  FORTRAN LIBRARY FUNCTIONS: NONE
C
C  ---------------------------------------------------------------------
       IMPLICIT REAL*8 (A-H,O-Z)
       DIMENSION DER(4,N1),X(N1),D(N1)
C  ---------------------------------------------------------------------
C
       DO 20 I=1,3
         K=4-I
         DO 10 J=1,N1
           DER(K+1,J)=DER(K,J)
10       CONTINUE
20     CONTINUE
       MADMS=MADMS+1
       IF (MADMS.GT.4) MADMS=4
       DO 70 I=1,N1
         DER(1,I)=D(I)
         GO TO (30,40,50,60),MADMS
30       X(I)=X(I)+H*DER(1,I)
         GO TO 70
40       X(I)=X(I)+0.5*H*(3.0*DER(1,I)-DER(2,I))
         GO TO 70
50       X(I)=X(I)+H*(23.0*DER(1,I)-16.0*DER(2,I)+5.0*DER(3,I))/12.0
         GO TO 70
60       X(I)=X(I)+H*(55.0*DER(1,I)-59.0*DER(2,I)+37.0*DER(3,I)-
     &           9.0*DER(4,I))/24.0
70     CONTINUE
       T=T+H
C
       RETURN
       END
C
C***********************************************************************
C
       SUBROUTINE MERS1(NDIM,NVAR,NN,N,T,X,F,H,HMIN,HMAX,ACCUR,DRCTN,
     & KACC,IWRN,IERR,XWORK,FWORK,YK,YK2,YK3,XW,NDIR,NDIR1,IW)
C
C***********************************************************************
C
C  RUNGE-KUTTA-MERSON METHOD FOR INTEGRATION OF CONTINUATION EQUATIONS
C
C  SUBPROGRAMS CALLED: PRST1
C  FORTRAN LIBRARY FUNCTIONS: DABS
```

```
C
C------------------------------------------------------------------------
      IMPLICIT DOUBLE PRECISION (A-H,O-Z)
      DIMENSION X(N),F(N),XWORK(N),FWORK(N),YK(N),YK2(N),YK3(N),XW(1),
     &          NDIR(N),NDIR1(N),IW(1)
      COMMON/TYPE/ITYP,ICON,ISHT,NS,IPER
      COMMON/TOLS/EMACH,EMIN,EACC
      COMMON/PRNT/LR,LW
C------------------------------------------------------------------------
C
   10 DO 20 I=1,N
      YK(I)=H*F(I)/3.0
      XWORK(I)=X(I)+YK(I)
   20 CONTINUE
      CALL PRST1(NDIM,NVAR,NN,N,XWORK,FWORK,XW,IW,IERR)
      IF (IERR.NE.0) RETURN
      DO   30 I=1,N
      YK2(I)=H*FWORK(I)/3.0
      XWORK(I)=X(I)+0.5D0*(YK(I)+YK2(I))
   30 CONTINUE
      CALL PRST1(NDIM,NVAR,NN,N,XWORK,FWORK,XW,IW,IERR)
      IF (IERR.NE.0) RETURN
      DO   40 I=1,N
      YK2(I)=H*FWORK(I)/3.0
      XWORK(I)=X(I)+0.375D0*YK(I)+1.125D0*YK2(I)
   40 CONTINUE
      CALL PRST1(NDIM,NVAR,NN,N,XWORK,FWORK,XW,IW,IERR)
      IF (IERR.NE.0) RETURN
      DO   50 I=1,N
      YK3(I)=H*FWORK(I)/3.0
      XWORK(I)=X(I)+1.5D0*YK(I)+6.0*YK3(I)-4.5D0*YK2(I)
   50 CONTINUE
      CALL PRST1(NDIM,NVAR,NN,N,XWORK,FWORK,XW,IW,IERR)
      IF (IERR.NE.0) RETURN
C
      E=EACC*DABS(ACCUR)
      DO   60 I=1,N
      IF (I.GT.KACC) GO TO 60
      IF (ISHT.GT.1.AND.(I.GT.NDIM.AND.I.LE.NS)) GO TO 60
      P=DABS(0.2D0*(YK(I)+4.0*YK3(I)-4.5D0*YK2(I)-
     &   0.5D0*FWORK(I)*H/3.0))
      IF(ACCUR.GT.0.0)P=P/(DABS(XWORK(I))+EACC)
      IF (P.GT.E)  E=P
   60 CONTINUE
      EREL = E / DABS(ACCUR)
      IF (E-DABS(ACCUR)) 90,70,70
   70 IF (IWRN.EQ.1) GO TO 90
C
      ISTOP=0
      HTEMP=H
      H=0.8*H*EREL**(-0.2)
      IF(DABS(H).LT.DABS(HTEMP)/3.0) H=HTEMP/3.0
```

```fortran
      IF (DABS(H).GE.DABS(HMIN)) GO TO 10
        IWRN=1
        WRITE (LW,1000) H,HMIN,T
        H=DABS(HMIN)*DRCTN
        DO 80 I=1,N
          NDIR(I)=NDIR1(I)
   80   CONTINUE
      GO TO 10
C
   90 DO 100 I=1,N
        X(I)=X(I)+0.5D0*(YK3(I)*4.0+YK(I)+H*FWORK(I)/3.0)
  100 CONTINUE
      T=T+H
      HTEMP=H
      H=0.8*H*EREL**(-0.2)
      IF(DABS(H).GT.DABS(HTEMP)*3.0) H=HTEMP*3.0
      IF (DABS(H).GT.DABS(HMAX)) H=DABS(HMAX)*DRCTN
      IF (DABS(H).GE.DABS(HMIN)) GO TO 110
        H=DABS(HMIN)*DRCTN
        RETURN
  110 IF (IWRN.EQ.0) RETURN
      IF (ISTOP.EQ.1) RETURN
      IWRN=0
      WRITE (LW,1100) H,T
C
 1000 FORMAT (' MERS1 H=',E12.3,' IS TOO SMALL, REPLACED BY H=',E12.3/
     &  ' AT T=',E14.6)
 1100 FORMAT (' MERS1 H=',E12.3,' ALREADY SATISFIES DESIRED ACCURACY'/
     &  ' AT T=',E14.6)
C
      RETURN
      END
C
C ********************************************************************
C
      SUBROUTINE PRST1(NDIM,NVAR,N,N1,X,F,XW,IW,IERR)
C
C ********************************************************************
C
C RIGHT HAND SIDES OF CONTINUATION EQUATIONS
C
C SUBPROGRAMS CALLED: FCTN,GAUSE,PRST3
C FORTRAN LIBRARY FUNCTIONS: NONE
C
C ------------------------------------------------------------------
      IMPLICIT REAL*8(A-H,O-Z)
      DIMENSION X(N1),F(N1),XW(1),IW(1)
      COMMON/PNTR1/NXLOW,NXUPP,NPREF,NXSP,NDELTA,NDXDT,NF,NG,NXOLD,
     & NDXOLD,NRES1
      COMMON/PNTR2/NRE,NRI,NB,NEIGBH,NFV1,NRES2
      COMMON/PNTR3/NFHY,NGHX,NXWR1,NFWR1,NYK1,NYK21,NYK31,NRES3
      COMMON/PNTR6/NNDIR1,NMARK,NIRR,NIRK,NIVR
```

```
C    -------------------------------------------------------------------
C
      CALL FCTN(NDIM,NVAR,N,N1,X,XW(NF),XW(NG),XW(NB),XW(NEIGBH),
     &          XW(NFHY),XW(NGHX),XW,IW,IERR)
      IF(IERR.NE.0) GO TO 10
      CALL GAUSE(N,N1,XW(NG),XW(NF),IERR,XW(NPREF),XW(NDELTA),K,
     &          XW(NFHY),XW(NGHX),IW(NIRR),IW(NIRK))
      IF(IERR.NE.0) GO TO 10
      CALL PRST3(N1,XW(NDELTA),IW(1),K,F)
C
10    RETURN
      END
C
C    ********************************************************************
C
      SUBROUTINE PRST3(N1,DELTA,NDIR,K,DXDT)
C
C    ********************************************************************
C
C NORMALIZED TANGENT VECTOR TO CONTINUATION CURVE
C
C SUBPROGRAMS CALLED: NONE
C FORTRAN LIBRARY FUNCTIONS: DSQRT,DFLOAT
C
C    -------------------------------------------------------------------
      IMPLICIT REAL*8(A-H,O-Z)
      DIMENSION DXDT(N1),DELTA(N1),NDIR(N1)
C    -------------------------------------------------------------------
C
      DXK2=1.0
      DO 10 I=1,N1
        DXK2=DXK2+DELTA(I)**2
10    CONTINUE
      DXDT(K)=1.0/DSQRT(DXK2)*DFLOAT(NDIR(K))
      DO 20 I=1,N1
        NDIR(I)=1
        DXDT(I)=DELTA(I)*DXDT(K)
        IF(DXDT(I).LT.0.0)NDIR(I)=-1
20    CONTINUE
C
      RETURN
      END
C
C    ********************************************************************
C
      SUBROUTINE FCTN(NDIM,NVAR,N,N1,X,F,G,B,BH,FH,GH,XW,IW,IERR)
C
C    ********************************************************************
C
C RIGHT HAND SIDES AND JACOBI MATRIX FOR NEWTON METHOD
C
C SUBPROGRAMS CALLED: FCT
```

```
C FORTRAN LIBRARY FUNCTIONS: DABS
C
C ------------------------------------------------------------------------
      IMPLICIT REAL*8(A-H,O-Z)
      DIMENSION X(N1),F(N1),G(N,N1),FH(N1),GH(N,N1),XW(1),IW(1),
     & BH(NDIM,NDIM),B(NDIM,NDIM)
      COMMON/NMDR/HX
      COMMON/TYPE/ITYP,ICON,ISHT,NS,IPER
      COMMON/TOLS/EMACH,EMIN,EACC
C ------------------------------------------------------------------------
C
      CALL FCT(NDIM,NVAR,N,N1,0,X,F,G,B,XW,IW,IERR)
      IF(IERR.LT.0) RETURN
      IF(ICON.EQ.1.OR.ICON.EQ.2)RETURN
      IF(ISHT.GT.1) GO TO 20
        DO 10 I=1,NS
          HDER=HX*DABS(X(I))
          IF(DABS(HDER).LT.EMIN) HDER=EMIN
          X(I)=X(I)+HDER
          CALL FCT(NDIM,NVAR,N,N1,0,X,FH,GH,BH,XW,IW,IERR)
          IF(IERR.LT.0) RETURN
          G(NS+1,I)=(FH(NS+1)-F(NS+1))/HDER
          IF(ICON.EQ.5)G(NS+2,I)=(FH(NS+2)-F(NS+2))/HDER
          X(I)=X(I)-HDER
10      CONTINUE
        GO TO 50
C
20    DO 30 J=1,NS
      DO 30 I=1,NS
        GH(I,J)=G(I,J)
30    CONTINUE
      DO 40 IST=1,ISHT
      DO 40 J=1,NDIM
        I=(IST-1)*NDIM+J
        HDER=HX*DABS(X(I))
        IF(DABS(HDER).LT.EMIN) HDER=EMIN
        X(I)=X(I)+HDER
        CALL FCT(NDIM,NVAR,N,N1,IST,X,FH,GH,BH,XW,IW,IERR)
        IF(IERR.LT.0) RETURN
        G(NS+1,I)=(FH(NS+1)-F(NS+1))/HDER
        IF(ICON.EQ.5)G(NS+2,I)=(FH(NS+2)-F(NS+2))/HDER
        X(I)=X(I)-HDER
40    CONTINUE
C
50    NS1=NS+1
      DO 60 I=NS1,N1
        HDER=HX*DABS(X(I))
        IF(DABS(HDER).LT.EMIN) HDER=EMIN
        X(I)=X(I)+HDER
        CALL FCT(NDIM,NVAR,N,N1,0,X,FH,GH,BH,XW,IW,IERR)
        IF(IERR.LT.0) RETURN
        G(NS+1,I)=(FH(NS+1)-F(NS+1))/HDER
```

```
        IF(ICON.EQ.5)G(NS+2,I)=(FH(NS+2)-F(NS+2))/HDER
        X(I)=X(I)-HDER
60      CONTINUE
C
        RETURN
        END
C
C ***********************************************************************
C
        SUBROUTINE FCT(NDIM,NVAR,N,N1,IST,X,F,G,B,XW,IW,IERR)
C
C ***********************************************************************
C
C RIGHT HAND SIDES AND JACOBI MATRIX FOR NEWTON METHOD
C
C SUBPROGRAMS CALLED: FCSTS,FCSSH,FCMSH,FCDER,FCAD,FCAODE
C FORTRAN LIBRARY FUNCTIONS: NONE
C
C ----------------------------------------------------------------------
        IMPLICIT REAL*8(A-H,O-Z)
        DIMENSION X(N1),F(N1),G(N,N1),XW(1),IW(1),B(NDIM,NDIM)
        COMMON/VARP/ALPHA,BETA,ARG,PER
        COMMON/TYPE/ITYP,ICON,ISHT,NS,IPER
        COMMON/PNTR4/NA,NX0,NXA,NFA,NGA,NGB,NGC,NRES4
C ----------------------------------------------------------------------
C
        IF(ICON.EQ.1.OR.ICON.GE.3)ALPHA=X(NS+1)
        IF(ICON.EQ.2)BETA=X(NS+1)
        IF(ICON.GE.3)BETA=X(NS+2)
        IF(ICON.EQ.5)ARG=X(NS+3)
        IF(ITYP.EQ.2)PER=X(N+1)
C
        IF(ITYP.EQ.1)CALL FCSTS(NDIM,NVAR,N,N1,X,F,G,B,XW(NXA),XW(NFA),
     &  XW(NGA),XW)
        IF(ISHT.EQ.1)CALL FCSSH(NDIM,NVAR,N,N1,X,F,G,B,XW(NXA),XW(NFA),
     &  XW(NGA),XW(NGB),XW(NGC),XW)
        IF(ISHT.GT.1.AND.IST.EQ.0)CALL FCMSH(NDIM,NVAR,N,N1,X,F,G,B,
     &  XW(NXA),XW(NFA),XW(NGA),XW(NGB),XW(NGC),XW)
        IF(ISHT.GT.1.AND.IST.GT.0)CALL FCDER(NDIM,NVAR,N,N1,IST,X,G,B,
     &  XW(NXA),XW(NFA),XW(NGA),XW(NGB),XW(NGC),XW)
C
        IF(ICON.EQ.1.OR.ICON.EQ.2) GO TO 10
        CALL FCAD(NDIM,N1,F,B,XW(NA),XW,IW,IERR)
        IF(IERR.LT.0) GO TO 40
10      CONTINUE
        IF(IST.GT.0) GO TO 40
        DO 20 I=1,NS
          IF(ICON.EQ.2)G(I,NS+1)=G(I,NS+2)
          IF(ICON.EQ.5)G(I,NS+3)=0.0
20      CONTINUE
        IF(ITYP.EQ.2)CALL FCAODE(NDIM,NVAR,N,N1,X,F,G,XW(NXA),XW(NFA),
     &  XW(NGA))
```

```
            IF(ITYP.EQ.1)GOTO 40
            DO 30 I=1,NS
              F(I)=F(I)-X(I)
              G(I,I)=G(I,I)-1.0
30          CONTINUE
40      CONTINUE
C
        RETURN
        END
C
C ***********************************************************************
C
        SUBROUTINE FCSTS(NDIM,NVAR,N,N1,X,F,G,B,XA,FA,GA,XW)
C
C ***********************************************************************
C
C STATIONARY SOLUTIONS
C
C SUBPROGRAMS CALLED: INTEG
C FORTRAN LIBRARY FUNCTIONS: NONE
C
C ----------------------------------------------------------------------
        IMPLICIT REAL*8(A-H,O-Z)
        DIMENSION X(N1),F(N1),G(N,N1),XA(NDIM),FA(NDIM),GA(NDIM,NVAR),
       & B(NDIM,NDIM),XW(1)
        COMMON/PNTR5/NXINT,NFINT,NXWR2,NFWR2,NYK2,NYK22,NYK32
C ----------------------------------------------------------------------
C
        DO 10 I=1,NDIM
          XA(I)=X(I)
10      CONTINUE
        CALL INTEG(NDIM,NVAR,0,XA,FA,GA,XW(NXINT),XW(NFINT),XW)
        DO 40 I=1,NDIM
          DO 20 J=1,NDIM
            B(I,J)=GA(I,J)
20        CONTINUE
          DO 30 J=1,NVAR
            G(I,J)=GA(I,J)
30        CONTINUE
          F(I)=FA(I)
40      CONTINUE
        RETURN
        END
C
C ***********************************************************************
C
        SUBROUTINE FCSSH(NDIM,NVAR,N,N1,X,F,G,B,XA,FA,GA,GB,GC,XW)
C
C ***********************************************************************
C
C SIMPLE SHOOTING METHOD FOR PERIODIC SOLUTIONS
C SUBPROGRAMS CALLED: INTEG,CHRULE
```

```
C FORTRAN LIBRARY FUNCTIONS: NONE
C
C -------------------------------------------------------------------------
      IMPLICIT REAL*8 (A-H,O-Z)
      DIMENSION X(N1),F(N1),G(N,N1),XA(NDIM),FA(NDIM),GA(NDIM,NVAR),
     & GB(NVAR,NVAR),GC(NDIM,NVAR),B(NDIM,NDIM),XW(1)
      COMMON/TYPE/ITYP,ICON,ISHT,NS,IPER
      COMMON/PNTR5/NXINT,NFINT,NXWR2,NFWR2,NYK2,NYK22,NYK32
C -------------------------------------------------------------------------
C
      DO 20 I=1,NVAR
        DO 10 J=1,NVAR
          GB(J,I)=0.0
10      CONTINUE
        GB(I,I)=1.0
20    CONTINUE
      DO 30 I=1,NDIM
        XA(I)=X(I)
30    CONTINUE
      DO 60 K=1,IPER
        CALL INTEG(NDIM,NVAR,K,XA,FA,GA,XW(NXINT),XW(NFINT),XW)
        DO 40 I=1,NDIM
          XA(I)=FA(I)
40      CONTINUE
        CALL CHRULE(NDIM,NVAR,NVAR,NVAR,GA,GB,GC)
        DO 50 I=1,NVAR
        DO 50 J=1,NDIM
          GB(J,I)=GC(J,I)
50      CONTINUE
60    CONTINUE
      DO 90 I=1,NDIM
        DO 70 J=1,NDIM
          B(I,J)=GB(I,J)
70      CONTINUE
        DO 80 J=1,NVAR
          G(I,J)=GB(I,J)
80      CONTINUE
        F(I)=FA(I)
90    CONTINUE
C
      RETURN
      END
C
C ****************************************************************************
C
      SUBROUTINE FCMSH(NDIM,NVAR,N,N1,X,F,G,B,XA,FA,GA,GB,GC,XW)
C
C ****************************************************************************
C
C MULTIPLE SHOOTING METHOD FOR PERIODIC SOLUTIONS
C
C SUBPROGRAMS CALLED: INTEG,CHRULE
```

```
C FORTRAN LIBRARY FUNCTIONS: MOD
C
C -----------------------------------------------------------------------
      IMPLICIT REAL*8(A-H,O-Z)
      DIMENSION X(N1),F(N1),G(N,N1),XA(NDIM),FA(NDIM),GA(NDIM,NVAR),
     & GB(NDIM,NDIM),GC(NDIM,NVAR),B(NDIM,NDIM),XW(1)
      COMMON/TYPE/ITYP,ICON,ISHT,NS,IPER
      COMMON/PNTR5/NXINT,NFINT,NXWR2,NFWR2,NYK2,NYK22,NYK32
C -----------------------------------------------------------------------
C
      DO 10 I=1,NS
      DO 10 J=1,NS
       G(I,J)=0.0
10    CONTINUE
      DO 30 I=1,NDIM
        DO 20 J=1,NDIM
         GB(J,I)=0.0
20      CONTINUE
       GB(I,I)=1.0
30    CONTINUE
      DO 80 I=1,ISHT
       L=(ISHT-2+I)*NDIM-1
       DO 40 J=1,NDIM
        JJ=MOD(L+J,NS)+1
        XA(J)=X(JJ)
40     CONTINUE
       IST=MOD(ISHT-2+I,ISHT)+1
       CALL INTEG(NDIM,NVAR,IST,XA,FA,GA,XW(NXINT),XW(NFINT),XW)
       DO 60 J=1,NDIM
        JJ=(I-1)*NDIM+J
        DO 50 K=1,NVAR
          KK=MOD(L+K,NS)+1
          IF(K.GT.NDIM) KK=NS+K-NDIM
          G(JJ,KK)=GA(J,K)
50      CONTINUE
        F(JJ)=FA(J)
60     CONTINUE
       CALL CHRULE(NDIM,NVAR,NDIM,NDIM,GA,GB,GC)
       DO 70 J=1,NDIM
       DO 70 K=1,NDIM
        GB(K,J)=GC(K,J)
70     CONTINUE
80    CONTINUE
      DO 90 J=1,NDIM
      DO 90 K=1,NDIM
       B(K,J)=GB(K,J)
90    CONTINUE
C
      RETURN
      END
```

```
C
C
C  *********************************************************************
C
      SUBROUTINE FCDER(NDIM,NVAR,N,N1,IST,X,G,B,XA,FA,GA,GB,GC,XW)
C
C  *********************************************************************
C
C
C DERIVATIVES FOR ADDITIONAL EQUATIONS WHEN MULTIPLE SHOOTING IS USED
C
C SUBPROGRAMS CALLED: INTEG,CHRULE
C FORTRAN LIBRARY FUNCTIONS: MOD
C
C  -------------------------------------------------------------------
      IMPLICIT REAL*8 (A-H,O-Z)
      DIMENSION X(N1),G(N,N1),XW(1),B(NDIM,NDIM),XA(NDIM),FA(NDIM),
     & GA(NDIM,NVAR),GB(NDIM,NDIM),GC(NDIM,NVAR)
      COMMON/TYPE/ITYP,ICON,ISHT,NS,IPER
      COMMON/PNTR5/NXINT,NFINT,NXWR2,NFWR2,NYK2,NYK22,NYK32
C  -------------------------------------------------------------------
C
      DO 20 I=1,NDIM
        II=(IST-1)*NDIM+I
        XA(I)=X(II)
        DO 10 J=1,NDIM
          GB(J,I)=0.0
10      CONTINUE
        GB(I,I)=1.0
20    CONTINUE
      DO 60 I=1,ISHT
        L=I*NDIM-1
        DO 40 J=1,NDIM
          JJ=(I-1)*NDIM+J
          DO 30 K=1,NDIM
            KK=MOD(L+K,NS)+1
            GA(K,J)=G(KK,JJ)
30        CONTINUE
40      CONTINUE
        IF(I.EQ.IST) CALL INTEG(NDIM,NVAR,IST,XA,FA,GA,
     &     XW(NXINT),XW(NFINT),XW)
        CALL CHRULE(NDIM,NVAR,NDIM,NDIM,GA,GB,GC)
        DO 50 J=1,NDIM
        DO 50 K=1,NDIM
          GB(K,J)=GC(K,J)
50      CONTINUE
60    CONTINUE
      DO 70 J=1,NDIM
      DO 70 K=1,NDIM
        B(K,J)=GC(K,J)
70    CONTINUE
C
      RETURN
      END
```

```
C
C ****************************************************************
C
      SUBROUTINE FCAODE(NDIM,NVAR,N,N1,X,F,G,XA,FA,GA)
C
C ****************************************************************
C
C ADDITIONAL EQUATION FOR PERIODIC SOLUTIONS OF AODES WHICH FIXES A
C POINT ON THE ORBIT (THE ANCHOR EQUATION) + ASSOCIATED DERIVATIVES
C
C SUBPROGRAMS CALLED: PRSTR
C FORTRAN LIBRARY FUNCTIONS: DFLOAT
C
C ----------------------------------------------------------------
      IMPLICIT REAL*8(A-H,O-Z)
      DIMENSION X(N1),F(N1),G(N,N1),XA(NDIM),FA(NDIM),GA(NDIM,NVAR)
      COMMON/TYPE/ITYP,ICON,ISHT,NS,IPER
C ----------------------------------------------------------------
C
      DO 10 I=1,N1
         G(N,I)=0.0
10    CONTINUE
      DO 20 I=1,NDIM
         XA(I)=X(I)
20    CONTINUE
      CALL PRSTR(NDIM,NVAR,N,0.0D0,XA,FA,GA)
      S1=0.0
      DO 40 I=1,NDIM
         S1=S1+XA(I)*FA(I)
         S2=0.0
         DO 30 J=1,NDIM
            S2=S2+GA(J,I)*XA(J)
30       CONTINUE
         G(N,I)=S2+FA(I)
40    CONTINUE
      F(N)=S1
      NDIM1=NDIM+1
      DO 60 I=NDIM1,NVAR
         S1=0.0
         DO 50 J=1,NDIM
            S1=S1+XA(J)*GA(J,I)
50       CONTINUE
         NSI=NS+I-NDIM
         G(N,NSI)=S1
60    CONTINUE
      IF(ICON.GT.2)GOTO 70
         IF(ICON.EQ.2)G(N,NS+1)=G(N,NS+2)
         G(N,NS+2)=0.0
70    CONTINUE
      DO 100 I=1,ISHT
         DO 80 J=1,NDIM
            II=NDIM*(I-1)+J
```

```
           XA(J)=F(II)
80         CONTINUE
           CALL PRSTR(NDIM,NVAR,NDIM,0.0D0,XA,FA,GA)
           DO 90 J=1,NDIM
             II=NDIM*(I-1)+J
             G(II,N1)=FA(J)/DFLOAT(ISHT)
90         CONTINUE
100     CONTINUE
C
        RETURN
        END
C
C ********************************************************************
C
        SUBROUTINE CHRULE(NDIM,NVAR,N,M,GA,GB,GC)
C
C ********************************************************************
C
C CHAIN RULE FOR DERIVATIVES OF VECTOR VALUED FUNCTIONS
C
C SUBPROGRAMS CALLED: NONE
C FORTRAN LIBRARY FUNCTIONS: NONE
C
C -------------------------------------------------------------------
        IMPLICIT REAL*8 (A-H,O-Z)
        DIMENSION GA(NDIM,NVAR),GB(N,M),GC(NDIM,NVAR)
C -------------------------------------------------------------------
C
        DO 30 J=1,M
          DO 20 I=1,NDIM
            SUM=0.0
            DO 10 K=1,N
              SUM=SUM+GA(I,K)*GB(K,J)
10          CONTINUE
            GC(I,J)=SUM
20        CONTINUE
30      CONTINUE
C
        RETURN
        END
C
C ********************************************************************
C
        SUBROUTINE FCAD(NDIM,N1,F,B,A,XW,IW,IERR)
C
C ********************************************************************
C
C ADDITIONAL EQUATIONS FOR FINDING BIFURCATION SOLUTIONS
C
C SUBPROGRAMS CALLED: COEF
C FORTRAN LIBRARY FUNCTIONS: DFLOAT,DCOS,DSIN,MOD
C
```

```
C  ------------------------------------------------------------------
      IMPLICIT REAL*8(A-H,O-Z)
      DIMENSION F(N1),B(NDIM,NDIM),A(NDIM),XW(1),IW(1)
      COMMON/TYPE/ITYP,ICON,ISHT,NS,IPER
      COMMON/VARP/ALPHA,BETA,ARG,PER
      COMMON/PNTR4/NA,NX0,NXA,NFA,NGA,NGB,NGC,NRES4
      DATA DP /6.2831853071795864D0/
C  ------------------------------------------------------------------
C
      CALL COEF(NDIM,B,A,IERR,XW(NXA),XW(NGA),XW(NX0),XW,IW)
      IF(IERR.LT.0) RETURN
      S1=0.0
      S2=0.0
      IF(ITYP.EQ.1)GOTO 90
        GOTO(140,140,10,50,70),ICON
10      IF(ITYP.EQ.2)GOTO 30
          DO 20 I=1,NDIM
            II=NDIM+1-I
            S1=S1+A(II)
20        CONTINUE
          F(NS+1)=S1+1.0
          GOTO 140
30      DO 40 I=2,NDIM
          II=NDIM+1-I
          S1=S1+DFLOAT(I-1)*A(II)
40      CONTINUE
        F(NS+1)=S1+DFLOAT(NDIM)
        GOTO 140
50      J=1
        DO 60 I=1,NDIM
          II=NDIM+1-I
          S1=S1+A(II)*DFLOAT(J)
          J=-J
60      CONTINUE
        F(NS+1)=S1+DFLOAT(J)
        GOTO 140
70      DO 80 I=1,NDIM
          II=NDIM+1-I
          S1=S1+A(II)*DCOS(DP*ARG*DFLOAT(I-1))
          S2=S2+A(II)*DSIN(DP*ARG*DFLOAT(I-1))
80      CONTINUE
        F(NS+1)=S1+DCOS(DP*ARG*DFLOAT(NDIM))
        F(NS+2)=S2+DSIN(DP*ARG*DFLOAT(NDIM))
        GOTO 140
90    CONTINUE
        GOTO (140,140,100,140,110),ICON
100     F(NS+1)=A(NDIM)
        GOTO 140
110     NI=NDIM-MOD(NDIM,2)
        J=1
        DO 120 I=1,NI,2
          II=NDIM-I
```

```
             S1=S1+DFLOAT(J)*A(II)*ARG**I
             J=-J
120     CONTINUE
        F(NS+1)=S1+DFLOAT(MOD(NDIM,2)*J)*ARG**NDIM
        NI=NDIM+MOD(NDIM,2)
        J=1
        DO 130 I=2,NI,2
          II=NDIM+2-I
          S2=S2+DFLOAT(J)*A(II)*ARG**(I-2)
          J=-J
130     CONTINUE
        F(NS+2)=S2+DFLOAT(MOD(NDIM+1,2)*J)*ARG**NDIM
140     CONTINUE
C
        RETURN
        END
C
C **************************************************************************
C
        SUBROUTINE COEF(N,B,A,IERR,DET,C,X0,XW,IW)
C
C **************************************************************************
C
C COEFFICIENTS OF CHARACTERISTIC POLYNOMIAL OBTAINED BY THE METHOD OF
C INTERPOLATION POLYNOMIAL
C
C SUBPROGRAMS CALLED: DETER,GAUSD
C FORTRAN LIBRARY FUNCTIONS: NONE
C
C --------------------------------------------------------------------------
        IMPLICIT REAL*8 (A-H,O-Z)
        DIMENSION DET(N),A(N),B(N,N),C(N,N),X0(N),XW(1),IW(1)
        COMMON/PNTR4/NA,NX0,NXA,NFA,NGA,NGB,NGC,NRES4
        COMMON/PNTR6/NNDIR1,NMARK,NIRR,NIRK,NIVR
C --------------------------------------------------------------------------
C
        DO 30 I=1,N
          DO 20 J=1,N
            DO 10 K=1,N
              C(J,K)=B(J,K)
10          CONTINUE
            C(J,J)=B(J,J)-X0(I)
20        CONTINUE
          CALL DETER(N,C,DET(I),IW(NIVR))
30      CONTINUE
        DO 50 I=1,N
          DO 40 J=1,N
            K=N-I
            C(J,I)=X0(J)**K
40        CONTINUE
          A(I)=(-1)**N*DET(I)-X0(I)**N
50      CONTINUE
```

```
      CALL GAUSD(N,C,A,IERR,XW(NFA),IW(NIVR))
C
      RETURN
      END
C
C     *****************************************************************
C
      SUBROUTINE DETER(N,A,DET,IRS)
C
C     *****************************************************************
C
C DETERMINANT OF MATRIX OBTAINED BY GAUSS-JORDAN ELIMINATION
C
C SUBPROGRAMS CALLED: NONE
C FORTRAN LIBRARY FUNCTIONS: DABS,DSIGN,DLOG,DEXP
C
C     -----------------------------------------------------------------
      IMPLICIT REAL*8(A-H,O-Z)
      DIMENSION A(N,N),IRS(N)
C     -----------------------------------------------------------------
C
      DET=1.0
      DETLOG=0.0
      ID=1
      DO 10 I=1,N
        IRS(I)=0
10    CONTINUE
20    IR=1
      IS=1
      AMAX=0.0
      DO 60 I=1,N
        IF (IRS(I)) 60,30,60
30      DO 50 J=1,N
          P=DABS(A(I,J))
          IF (P-AMAX) 50,50,40
40        IR=I
          IS=J
          AMAX=P
50      CONTINUE
60    CONTINUE
      IF (AMAX.NE.0.0) GO TO 70
      DET=0.0
      RETURN
C
70    IRS(IR)=IS
      DO 90 I=1,N
        IF (I.EQ.IR.OR.A(I,IS).EQ.0.0) GO TO 90
        P=A(I,IS)/A(IR,IS)
        DO 80 J=1,N
          A(I,J)=A(I,J)-P*A(IR,J)
80      CONTINUE
        A(I,IS)=0.0
```

```
90      CONTINUE
        DET=DET*A(IR,IS)
        DETLOG=DETLOG+DLOG(DABS(A(IR,IS)))
        ID=ID+1
        IF (ID.LE.N) GO TO 20
        DET=DSIGN(DEXP(DETLOG),DET)
        DO 110 I=1,N
          IF(IRS(I).EQ.I)GOTO 110
          DO 100 J=I,N
            IF(IRS(J).EQ.I)IRS(J)=IRS(I)
100       CONTINUE
          DET=-DET
110     CONTINUE
C
        RETURN
        END
C
C ********************************************************************
C
        SUBROUTINE GAUSD(N,A,B,IERR,X,IRS)
C
C ********************************************************************
C
C SOLUTION OF N LINEAR EQUATIONS FOR N UNKNOWNS BY GAUSS-JORDAN
C ELIMINATION
C
C A(,)   : N BY N MATRIX OF THE SYSTEM
C B()    : RIGHT HAND SIDES,AFTER RETURN RESULTING SOLUTION
C IERR   : IF (IERR.NE.0) AFTER RETURN, THEN MATRIX A WAS NUMERICALLY
C          SINGULAR
C
C SUBPROGRAMS CALLED: NONE
C FORTRAN LIBRARY FUNCTIONS: DABS
C
C -------------------------------------------------------------------
        IMPLICIT REAL*8(A-H,O-Z)
        DIMENSION A(N,N),B(N),X(N),IRS(N)
        COMMON/TOLS/EMACH,EMIN,EACC
C -------------------------------------------------------------------
C
        IERR=0
        ID=1
        DO 10 I=1,N
          IRS(I)=0
10      CONTINUE
20      IR=1
        IS=1
        AMAX=0.0
        DO 60 I=1,N
          IF (IRS(I)) 60,30,60
30        DO 50 J=1,N
            P=DABS(A(I,J))
```

```
              IF (P-AMAX) 50,50,40
40              IR=I
                IS=J
                AMAX=P
50          CONTINUE
60      CONTINUE
        IF (AMAX.GT.EMACH) GO TO 70
        IERR=-1
        RETURN
C
70      IRS(IR)=IS
        DO 90 I=1,N
          IF (I.EQ.IR.OR.A(I,IS).EQ.0.0) GO TO 90
          P=A(I,IS)/A(IR,IS)
          DO 80 J=1,N
            IF (A(IR,J).NE.0.0) A(I,J)=A(I,J)-P*A(IR,J)
80        CONTINUE
          A(I,IS)=0.0
          B(I)=B(I)-P*B(IR)
90      CONTINUE
        ID=ID+1
        IF (ID.LE.N) GO TO 20
        DO 100 I=1,N
          IR=IRS(I)
          X(IR)=B(I)/A(I,IR)
100     CONTINUE
        DO 110 I=1,N
          B(I)=X(I)
110     CONTINUE
C
        RETURN
        END
C
C ********************************************************************
C
        SUBROUTINE INTEG(NDIM,NVAR,IST,XA,FA,GA,XINT,FINT,XW)
C
C ********************************************************************
C
C SPECIFICATION OF BOUNDARY VALUE PROBLEM FOR PERIODIC ORBITS FOLLOWED
C BY NUMERICAL INTEGRATION OF AODES OR FODES
C
C SUBPROGRAMS CALLED: PRSTR,ODEDRV
C FORTRAN LIBRARY FUNCTIONS: DFLOAT
C
C -------------------------------------------------------------------
        IMPLICIT REAL*8(A-H,O-Z)
        DIMENSION XA(NDIM),FA(NDIM),GA(NDIM,NVAR),XINT(1),FINT(1),XW(1)
        COMMON/VARP/ALPHA,BETA,ARG,PER
        COMMON/TYPE/ITYP,ICON,ISHT,NS,IPER
        COMMON/MER2/H2,HMIN2,HMAX2,ACCUR2
C -------------------------------------------------------------------
```

```
C
      NMERS=NDIM*(NVAR+1)
      GOTO (10,40,10,40,20,30),ITYP
10    CALL PRSTR(NDIM,NVAR,NMERS,0.0D0,XA,FA,GA)
      RETURN
20    PER=ALPHA
      IPRM=1
      GO TO 40
30    PER=BETA
      IPRM=2
40    CNPB=DFLOAT(IST-1)/DFLOAT(ISHT)
      CNPE=CNPB+1.0/DFLOAT(ISHT)
      IF(ISHT.EQ.1) GO TO 50
        CNPB=CNPB*DFLOAT(IPER)
        CNPE=CNPE*DFLOAT(IPER)
50    CONTINUE
      BGNT=CNPB*PER
      ENDT=CNPE*PER
      DO 60 I=1,NMERS
        XINT(I)=0.0
60    CONTINUE
      DO 70 I=1,NDIM
        XINT(I)=XA(I)
        K=I*NDIM+I
        XINT(K)=1.0
70    CONTINUE
      CALL ODEDRV(NDIM,NVAR,NMERS,BGNT,ENDT,H2,HMIN2,HMAX2,XINT,FINT,
     &            ACCUR2,NDIM,IMSG,0,XW)
      DO 80 I=1,NDIM
        FA(I)=XINT(I)
80    CONTINUE
      IF(ITYP.EQ.5.OR.ITYP.EQ.6) CALL PRSTR(NDIM,NVAR,NDIM,BGNT,XA,XINT,
     &                                      GA)
      DO 90 I=1,NVAR
      DO 90 J=1,NDIM
        K=I*NDIM+J
        GA(J,I)=XINT(K)
90    CONTINUE
      IF(ITYP.NE.5.OR.ITYP.NE.6) GO TO 110
        DO 100 I=1,NDIM
          K=NDIM+IPRM
          GA(I,K)=GA(I,K)+CNPE*FINT(I)-CNPB*XINT(I)
100     CONTINUE
C
110   RETURN
      END
C
```

```
C  **********************************************************************
C
       SUBROUTINE ODEDRV(NDIM,NVAR,N,BGNT,ENDT,H,HMIN,HMAX,X,F,ACCUR,
      &                  KACC,IMSG,NCROS,XW)
C
C  **********************************************************************
C
C DRIVING SUBPROGRAM FOR RUNGE-KUTTA-MERSON METHOD
C
C SUBPROGRAMS CALLED: PRST2,SURF,MERS2,INSECT
C FORTRAN LIBRARY FUNCTIONS: DABS
C
C  --------------------------------------------------------------------
       IMPLICIT DOUBLE PRECISION (A-H,O-Z)
       DIMENSION X(N),F(N),XW(1)
       COMMON/TOLS/EMACH,EMIN,EACC
       COMMON/PNTR5/NXINT,NFINT,NXWR2,NFWR2,NYK2,NYK22,NYK32
C  --------------------------------------------------------------------
C
       ISTOP = 0
       IEND = 0
       IF(IMSG.NE.-1) IMSG = 0
       T = BGNT
       NN=N
       IF (NCROS .NE. 2) GO TO 10
         NN=N+1
         X(NN) = 0.0
10     CONTINUE
       CALL PRST2(NDIM,NVAR,N,T,X,F,XW)
       IF (NCROS .EQ. 0) GO TO 20
         CALL SURF(NDIM,T,X,F,SOLD)
         IF (DABS(SOLD) .LT. DABS(ACCUR)) SOLD = 0.0
         IF(SOLD.NE.0.0) GO TO 20
         IF(IMSG.NE.-1) GO TO 20
         IMSG = 2
         RETURN
20     CONTINUE
C
C DIRECTION OF INTEGRATION
C
       IF (ENDT .NE. BGNT) GO TO 30
         IMSG = 1
         RETURN
30     CONTINUE
       IF (ENDT .LT. BGNT) DRCTN = -1.0
       IF (ENDT .GT. BGNT) DRCTN = 1.0
C
C CHECK INPUT DATA
C
       IF (DABS(ENDT-BGNT) .GE. EMACH) GO TO 40
         IMSG = 1
         RETURN
```

```
40       CONTINUE
         IF (DABS(H) .LT. DABS(HMIN)) H =  HMIN
         IF (DABS(H) .GT. DABS(HMAX)) H =  HMAX
         IF (DABS(ENDT-BGNT) .GE. DABS(H)) GO TO 50
           ISTOP = 1
           H = ENDT - BGNT
           IF (DABS(H) .LT. DABS(HMIN)) IWRN = 1
50       CONTINUE
         H = DRCTN * DABS(H)
C
60       CALL MERS2(NDIM,NVAR,NN,T,X,F,H,HMIN,HMAX,ACCUR,DRCTN,KACC,IWRN,
     &              NCROS,ISTOP,IEND,XW(NXWR2),XW(NFWR2),XW(NYK2),
     &              XW(NYK22),XW(NYK32),XW)
C
         IF (DRCTN*T .LE. DRCTN*ENDT) GO TO 70
         IEND = 2
         DRCTN = -DRCTN
         HOLD = H
         GO TO 80
C
70       IF (NCROS .NE. 1) GO TO 60
         CALL INSECT(NDIM,NVAR,NN,T,X,F,ENDT,SOLD,SNEW,H,HOLD,DRCTN,
     &               IMSG,IEND,XW)
           IF (IEND .EQ. 1) GO TO 80
         GO TO 60
C
80       H = DRCTN * DABS(T - ENDT)
         IF (DABS(H) .GT. DABS(HMAX)) GO TO 90
           ISTOP = 1
           IF (DABS(H) .LE. DABS(HMIN)) IWRN = 1
         GO TO 100
90       H = DRCTN * DABS(HMAX)
100      CALL MERS2(NDIM,NVAR,NN,T,X,F,H,HMIN,HMAX,0.1D0*ACCUR,DRCTN,KACC,
     &              IWRN,NCROS,ISTOP,IEND,XW(NXWR2),XW(NFWR2),XW(NYK2),
     &              XW(NYK22),XW(NYK32),XW)
         IF (ISTOP .EQ. 0) GO TO 80
C
         IF (NCROS .NE. 1 )GO TO 120
           IF (IEND .EQ. 1) GO TO 110
           CALL INSECT(NDIM,NVAR,NN,T,X,F,ENDT,SOLD,SNEW,H,HOLD,DRCTN,
     &                 IMSG,IEND,XW)
           IF (IEND .EQ. 1) GO TO 80
           IMSG = 3
           GO TO 120
C
110      ENDT = X(NN)
C
120      H = HOLD
         CALL PRST2(NDIM,NVAR,N,ENDT,X,F,XW)
C
         RETURN
         END
```

```
C
C
C **********************************************************************
C
      SUBROUTINE INSECT(NDIM,NVAR,N,T,X,F,ENDT,SOLD,SNEW,H,HOLD,DRCTN
     &                 ,IMSG,IEND,XW)
C
C **********************************************************************
C
C FIND THE INTERSECTION OF THE ORBIT WITH THE POINCARE SURFACE
C
C SUBPROGRAMS CALLED: PRST2,SURF
C FORTRAN LIBRARY FUNCTIONS: NONE
C
C ---------------------------------------------------------------------
      IMPLICIT DOUBLE PRECISION (A-H,O-Z)
      DIMENSION X(N),F(N),XW(1)
C ---------------------------------------------------------------------
C
      CALL PRST2(NDIM,NVAR,N,T,X,F,XW)
      CALL SURF(NDIM,T,X,F,SNEW)
C
      IF (.NOT.(SOLD .LT. 0.0 .AND. SNEW .GE. 0.0)) GO TO 10
         DRCTN = -1.0
         IEND  = 1
         IMSG  = 4
         GO TO 20
C
10    IF (.NOT.(SOLD .GT. 0.0 .AND. SNEW .LE. 0.0)) GO TO 30
         DRCTN = 1.0
         IEND  = 1
         IMSG  = 5
C
20       HOLD = H
         N = N+1
         X(N) = T
         T = SNEW + ENDT
         RETURN
C
30    SOLD = SNEW
C
      RETURN
      END
C
C **********************************************************************
C
      SUBROUTINE SURF(NDIM,T,X,F,S)
C
C **********************************************************************
C
C EVALUATION OF THE FUNCTION S(X) = X * F(X) WHICH DEFINES THE POINCARE
C SURFACE BY S = 0
C
```

```
C ·SUBPROGRAMS CALLED: NONE
C FORTRAN LIBRARY FUNCTIONS: NONE
C
C --------------------------------------------------------------------
      IMPLICIT DOUBLE PRECISION(A-H,O-Z)
      DIMENSION X(NDIM),F(NDIM)
C --------------------------------------------------------------------
C
      S = 0.0
      DO 10 I=1,NDIM
        S = S + X(I) * F(I)
10    CONTINUE
C
      RETURN
      END
C
C ********************************************************************
C
      SUBROUTINE GSURF(NDIM,NVAR,T,X,F,G,GRADS)
C
C ********************************************************************
C
C GRADIENT OF S(X)
C
C SUBPROGRAMS CALLED: NONE
C FORTRAN LIBRARY FUNCTIONS: NONE
C
C --------------------------------------------------------------------
      IMPLICIT DOUBLE PRECISION(A-H,O-Z)
      DIMENSION X(NDIM),F(NDIM),G(NDIM,NVAR),GRADS(NDIM)
C --------------------------------------------------------------------
C
      DO 20 I=1,NDIM
        SUM = 0.0
        DO 10 J=1,NDIM
          SUM = SUM + G(J,I) * X(J)
10      CONTINUE
        GRADS(I) = SUM + F(I)
20    CONTINUE
C
      RETURN
      END
C
C ********************************************************************
C
      SUBROUTINE MERS2(NDIM,NVAR,N,T,X,F,H,HMIN,HMAX,ACCUR,DRCTN,KACC,
     &               IWRN,NCROS,ISTOP,IEND,XWORK,FWORK,YK,YK2,YK3,XW)
C
C ********************************************************************
C
C RUNGE-KUTTA-MERSON METHOD FOR INTEGRATION OF N FIRST ORDER ODES
C IN THE FORM DX(T)/DT = F(X(T),T) OR DX(T)/DT = F(X(T))
```

```
C
C SUBPROGRAMS CALLED: RHS
C FORTRAN LIBRARY FUNCTIONS: DABS
C
C ---------------------------------------------------------------------
      IMPLICIT DOUBLE PRECISION (A-H,O-Z)
      DIMENSION X(N),F(N),XWORK(N),FWORK(N),YK(N),YK2(N),YK3(N),XW(1)
      COMMON/PRNT/LR,LW
      COMMON/TOLS/EMACH,EMIN,EACC
C ---------------------------------------------------------------------
C
      CALL RHS(NDIM,NVAR,N,T,X,F,NCROS,IEND,XW)
   10 DO 20 I=1,N
      YK(I)=H*F(I)/3.0
      XWORK(I)=X(I)+YK(I)
   20 CONTINUE
      CALL RHS(NDIM,NVAR,N,T+H/3.0,XWORK,FWORK,NCROS,IEND,XW)
      DO 30 I=1,N
      YK2(I)=H*FWORK(I)/3.0
      XWORK(I)=X(I)+0.5D0*(YK(I)+YK2(I))
   30 CONTINUE
      CALL RHS(NDIM,NVAR,N,T+H/3.0,XWORK,FWORK,NCROS,IEND,XW)
      DO 40 I=1,N
      YK2(I)=H*FWORK(I)/3.0
      XWORK(I)=X(I)+0.375D0*YK(I)+1.125D0*YK2(I)
   40 CONTINUE
      CALL RHS(NDIM,NVAR,N,T+H/2.0,XWORK,FWORK,NCROS,IEND,XW)
      DO 50 I=1,N
      YK3(I)=H*FWORK(I)/3.0
      XWORK(I)=X(I)+1.5D0*YK(I)+6.0*YK3(I)-4.5D0*YK2(I)
   50 CONTINUE
      CALL RHS(NDIM,NVAR,N,T+H,XWORK,FWORK,NCROS,IEND,XW)
C
      E=EMACH
      DO 60 I=1,N
      IF (I.GT.KACC) GO TO 60
      P=DABS(0.2D0*(YK(I)+4.0*YK3(I)-4.5D0*YK2(I)-
     &   0.5D0*FWORK(I)*H/3.0))
      IF(ACCUR.GT.0.0) P=P/(DABS(XWORK(I))+EMIN)
      IF (P.GT.E) E=P
   60 CONTINUE
      EREL = E / DABS(ACCUR)
      IF (E-DABS(ACCUR)) 80,70,70
   70 IF (IWRN.EQ.1) GO TO 80
C
      ISTOP=0
      H=0.8*H*EREL**(-0.2)
      IF (DABS(H).GE.DABS(HMIN)) GO TO 10
      IWRN=1
      WRITE (LW,1000) H,HMIN,T
      H=DABS(HMIN)*DRCTN
      GO TO 10
```

```
C
   80 DO   90 I=1,N
         X(I)=X(I)+0.5D0*(YK3(I)*4.0+YK(I)+H*FWORK(I)/3.0)
   90 CONTINUE
      T=T+H
      H=0.8*H*EREL**(-0.2)
      IF (DABS(H).GT.DABS(HMAX)) H=DABS(HMAX)*DRCTN
      IF (DABS(H).GE.DABS(HMIN)) GO TO 100
         H=DABS(HMIN)*DRCTN
         RETURN
  100 IF (IWRN.EQ.0) RETURN
         IF (ISTOP.EQ.1) RETURN
         IWRN=0
         WRITE (LW,1100) H,T
C
 1000 FORMAT (' MERS2 H=',E12.3,' IS TOO SMALL, REPLACED BY H=',E12.3/
     &  ' AT T=',E14.6)
 1100 FORMAT (' MERS2 H=',E12.3,' ALREADY SATISFIES DESIRED ACCURACY'/
     &  ' AT T=',E14.6)
C
      RETURN
      END
C
C ********************************************************************
C
      SUBROUTINE RHS(NDIM,NVAR,N,T,X,F,NCROS,IEND,XW)
C
C ********************************************************************
C
C DRIVING SUBPROGRAM FOR RIGHT HAND SIDES OF THE ODES
C
C SUBPROGRAMS CALLED: PRST2,GSURF
C FORTRAN LIBRARY FUNCTIONS: DSQRT
C
C --------------------------------------------------------------------
      IMPLICIT DOUBLE PRECISION(A-H,O-Z)
      DIMENSION X(N),F(N),XW(1)
      COMMON/PNTR4/NA,NX0,NXA,NFA,NGA,NGB,NGC,NRES4
C --------------------------------------------------------------------
C
      IF (.NOT.(IEND .EQ. 1 .AND. NCROS .EQ. 1)) GO TO 30
         NN=N-1
         CALL PRST2 (NDIM,NVAR,NN,X(N),X,F,XW)
         CALL GSURF(NDIM,NVAR,X(N),X,F,XW(NGA),XW(NRES4))
         SUM = 0.0
         DO 10 I = 1,NDIM
            II = NRES4+I-1
            SUM = SUM + F(I)*XW(II)
   10    CONTINUE
         SUM = 1.0 / SUM
         DO 20 I = 1,NN
            F(I) = F(I) * SUM
```

```
20      CONTINUE
        F(N) = SUM
        RETURN
30    CONTINUE
C
        IF (NCROS .EQ. 2) GO TO 40
          CALL PRST2 (NDIM,NVAR,N,T,X,F,XW)
          RETURN
40    CONTINUE
C
        NN=N-1
        CALL PRST2(NDIM,NVAR,NN,T,X,F,XW)
        SUM=0.0
        DO 50 I=1,NN
          SUM=SUM+F(I)**2
50      CONTINUE
        F(N)=DSQRT(SUM)+1.0
C
        RETURN
        END
C
C **********************************************************************
C
        SUBROUTINE PRST2(NDIM,NVAR,N,T,X,F,XW)
C
C **********************************************************************
C
C RIGHT HAND SIDES OF MODEL + VARIATIONAL EQUATIONS
C
C SUBPROGRAMS CALLED: PRSTR,CHRULE
C FORTRAN LIBRARY FUNCTIONS: NONE
C
C ----------------------------------------------------------------------
        IMPLICIT REAL*8 (A-H,O-Z)
        DIMENSION X(N),F(N),XW(1)
        COMMON/PNTR4/NA,NX0,NXA,NFA,NGA,NGB,NGC,NRES4
C ----------------------------------------------------------------------
C
        CALL PRSTR(NDIM,NVAR,N,T,X,F,XW(NGA))
        IF(N.EQ.NDIM) RETURN
        CALL CHRULE(NDIM,NVAR,NDIM,NVAR,XW(NGA),X(NDIM+1),F(NDIM+1))
        NPRM=(NVAR-NDIM)*NDIM
        DO 10 I=1,NPRM
          I1=NDIM*(NDIM+1)+I
          I2=NGA+NDIM*NDIM+I-1
          F(I1)=F(I1)+XW(I2)
10      CONTINUE
C
        RETURN
        END
```

Example 357

```
C
C ***********************************************************************
C
      SUBROUTINE PRSTR(NDIM,NVAR,N,T,X,F,G)
C
C ***********************************************************************
C
C SPECIFICATION OF THE USER'S PROBLEM
C RIGHT HAND SIDES AND JACOBI MATRIX OF THE MODEL EQUATIONS ARE
C EVALUATED HERE
C
C T    :  TIME (EXPLICITLY OCCURS ONLY FOR FODES)
C X()  :  ARRAY OF NDIM STATE SPACE VARIABLES
C F()  :  ARRAY OF NDIM RIGHT HAND SIDES DEPENDING ON X;ALPHA,BETA,PAR()
C            (IN ADDITION F DEPENDS EXPLICITLY ON T FOR FODES)
C G(,): NDIM BY NDIM+2 MATRIX OF FIRST DERIVATIVES,
C         G = [DF/DX,DF/DALPHA,DF/DBETA]
C
C ----------------------------------------------------------------------
      IMPLICIT REAL*8(A-H,O-Z)
      DIMENSION X(NDIM),F(NDIM),G(NDIM,NVAR)
      COMMON/FIXP/A,Q,DUMMY(18)
      COMMON/VARP/D1,B,ARG,PER
C ----------------------------------------------------------------------
C
C   EXAMPLE - AN AUTONOMOUS ODES PROBLEM
C   TWO REACTION CELLS WITH BRUSSELATOR KINETICS COUPLED BY DIFFUSION
C   VARIABLE PARAMETERS - D1(FORMAL NAME ALPHA) , B(FORMAL NAME BETA)
C   FIXED PARAMETERS -  A,Q
C
      TMP1=X(1)**2*X(2)
      TMP2=X(3)**2*X(4)
      TMP3=X(3)-X(1)
      TMP4=(X(4)-X(2))/Q
      TMP5=B*X(1)
      TMP6=B*X(3)
      F(1)=A-TMP5-X(1)+TMP1+D1*TMP3
      F(2)=TMP5-TMP1+D1*TMP4
      F(3)=A-TMP6-X(3)+TMP2-D1*TMP3
      F(4)=TMP6-TMP2-D1*TMP4
      IF(N.EQ.NDIM) RETURN
C
      TMP7=2.0*X(1)*X(2)
      TMP8=X(1)**2
      TMP9=2.0*X(3)*X(4)
      TMP10=X(3)**2
      TMP11=D1/Q
      G(1,1)=-B-1.0+TMP7-D1
      G(2,1)= B-TMP7
      G(3,1)= D1
      G(4,1)= 0.0
      G(1,2)= TMP8
```

```
      G(2,2)=-TMP8-TMP11
      G(3,2)= 0.0
      G(4,2)= TMP11
      G(1,3)= D1
      G(2,3)= 0.0
      G(3,3)=-B-1.0+TMP9-D1
      G(4,3)= B-TMP9
      G(1,4)= 0.0
      G(2,4)= TMP11
      G(3,4)= TMP10
      G(4,4)=-TMP10-TMP11
      G(1,5)= TMP3
      G(2,5)= TMP4
      G(3,5)=-TMP3
      G(4,5)=-TMP4
      G(1,6)=-X(1)
      G(2,6)= X(1)
      G(3,6)=-X(3)
      G(4,6)= X(3)
C
      RETURN
      END
C
C ************************************************************************
C
C END OF THE PROGRAM CONT
C
C ************************************************************************
```

Example 359

INPUT DATA WITH STARTING POINTS FOR COMPUTATION OF FAMILIES OF
STATIONARY AND PERIODIC ORBITS SHOWN IN FIGS 6.14 AND 6.17

```
    140
TWO REACTION CELLS WITH BRUSSELATOR KINETICS COUPLED BY DIFFUSION
SH FAMILY (FIG. 6.14)
     1    1    1    0    4    2    0
2.0             0.1
2.0             3.0             2.0             3.0
0.0             0.0             0.0             0.0
10.0            10.0            10.0            10.0
0.1             5.9             0.0             0.0
0.0             0.0             0.0             0.0
1.5             10.0            10.0            50.0
     1    5   -2    2  400    1    0

    140
TWO REACTION CELLS WITH BRUSSELATOR KINETICS COUPLED BY DIFFUSION
SN FAMILY (FIG. 6.14)
     1    1    1    0    4    2    0
2.0             0.1
0.97653824E+00 0.26074117E+01 0.30234618E+01 0.23096773E+01
0.0             0.0             0.0             0.0
10.0            10.0            10.0            10.0
1.1             5.9             0.0             0.0
0.0             0.0             0.0             0.0
1.5             10.0            10.0            50.0
     1    5   -2    2  400    1    0

    140
TWO REACTION CELLS WITH BRUSSELATOR KINETICS COUPLED BY DIFFUSION
PH FAMILY (FIG. 6.17)
     1    2    1    1    4    2    0
2.0             0.1
2.116           2.418           2.116           2.418
0.0             0.0             0.0             0.0
10.0            10.0            10.0            10.0
0.1             5.9             0.0             4.98
0.0             0.0             0.0             0.0
1.5             10.0            10.0            50.0
     1    5   -2    1  400    1    1

    140
TWO REACTION CELLS WITH BRUSSELATOR KINETICS COUPLED BY DIFFUSION
PNI1 FAMILY (FIG. 6.17)
     1    2    1    1    4    2    0
2.0             0.1
0.16033807E+01 0.35282800E+01 0.23368458E+01 0.26586846E+01
0.0             0.0             0.0             0.0
10.0            10.0            10.0            10.0
0.44797795E-01 5.9             0.0             0.34433899E+01
0.0             0.0             0.0             0.0
```

```
1.5              10.0              10.0              50.0
        1     5     2     1   400     1     0

   140
TWO REACTION CELLS WITH BRUSSELATOR KINETICS COUPLED BY DIFFUSION
PNI2 FAMILY (FIG. 6.17)
        1     2     1     1     4     2     0
2.0                   0.1
0.101705E+01      0.337089E+01      0.344603E+01      0.277412E+01
0.0               0.0               0.0               0.0
10.0              10.0              10.0              10.0
0.11E+01          5.9               0.0               0.354977E+01
0.0               0.0               0.0               0.0
1.5               10.0              10.0              50.0
        1     5    -2     1   400     1     1

   140
TWO REACTION CELLS WITH BRUSSELATOR KINETICS COUPLED BY DIFFUSION
PNI3 FAMILY (FIG. 6.17)
        1     2     1     1     4     2     0
2.0                   0.1
0.75552515E+00 0.54580318E+01 0.32465432E+01 0.19445033E+01
0.0               0.0               0.0               0.0
10.0              10.0              10.0              10.0
0.38176493E-01 5.9                  0.0            0.21456645E+02
0.0               0.0               0.0               0.0
1.5               10.0              10.0              50.0
        1     5    -2     1   400     1     0

   140
TWO REACTION CELLS WITH BRUSSELATOR KINETICS COUPLED BY DIFFUSION
PNA1 FAMILY (FIG. 6.17)
        1     2     1     2     4     2     0
2.0                   0.1
0.20400411D+01 0.29022147D+01 0.19600442D+01 0.29986525D+01
0.0               0.0               0.0               0.0
10.0              10.0              10.0              10.0
0.40937169D-01 5.9                  0.0            0.10156529D+02
0.0               0.0               0.0               0.0
1.5               10.0              10.0              50.0
        1     9    -2     1   400     1     0

   140
TWO REACTION CELLS WITH BRUSSELATOR KINETICS COUPLED BY DIFFUSION
PNA2 FAMILY (FIG. 6.17)
        1     2     1     2     4     2     0
2.0                   0.1
0.12521212E+01 0.40796261E+01 0.28630193E+01 0.19663115E+01
0.0               0.0               0.0               0.0
10.0              10.0              10.0              10.0
0.12161163E-01 5.9                  0.0            0.43981591E+01
0.0               0.0               0.0               0.0
```

Example   361

```
1.5            10.0           10.0           50.0
     1    9   -2    1   400    1    0

   140
TWO REACTION CELLS WITH BRUSSELATOR KINETICS COUPLED BY DIFFUSION
PNO FAMILY (FIG. 6.17)
     2    2    1    2    4    2    0
2.0            0.1
2.0            2.95           2.116          2.418
0.0            0.0            0.0            0.0
10.0           10.0           10.0           10.0
0.0            5.9            0.0            4.98
0.0            0.0            0.0            0.0
1.5            10.0           10.0           50.0
     1    9    1    1   400    1    1
0.001          1.0E-06        1.0            1.0E-04
0.01           1.0E-10        1.0            1.0E-07
-1.0E-6        1.0E-10        1.0E-15
    20    5    1
```

SUBJECT INDEX

Printed in the United States
By Bookmasters